T0192803

Undergraduate Lecture Notes in Physics (ULNP) publishes authoritative texts covering topics throughout pure and applied physics. Each title in the series is suitable as a basis for undergraduate instruction, typically containing practice problems, worked examples, chapter summaries, and suggestions for further reading.

ULNP titles must provide at least one of the following:

- An exceptionally clear and concise treatment of a standard undergraduate subject.
- A solid undergraduate-level introduction to a graduate, advanced, or non-standard subject.
- A novel perspective or an unusual approach to teaching a subject.

ULNP especially encourages new, original, and idiosyncratic approaches to physics teaching at the undergraduate level.

The purpose of ULNP is to provide intriguing, absorbing books that will continue to be the reader's preferred reference throughout their academic career.

More information about this series at http://www.springer.com/series/8917

Kerry Kuehn

A Student's Guide Through the Great Physics Texts

Volume III: Electricity, Magnetism and Light

 Springer

Kerry Kuehn
Wisconsin Lutheran College
Milwaukee
Wisconsin
USA

ISSN 2192-4791   ISSN 2192-4805 (electronic)
Undergraduate Lecture Notes in Physics
ISBN 978-3-319-79363-4   ISBN 978-3-319-21816-8 (eBook)
DOI 10.1007/978-3-319-21816-8

Springer Cham Heidelberg New York Dordrecht London
© Springer International Publishing Switzerland 2016
Softcover re-print of the Hardcover 1st edition 2016

Printed on acid-free paper

Springer International Publishing AG Switzerland is part of Springer Science+Business Media (www.springer.com)

For Cindy.

Preface

What is the nature of this book?

This four-volume book grew from a four-semester general physics curriculum which I developed and taught for the past decade to undergraduate students at Wisconsin Lutheran College in Milwaukee. The curriculum is designed to encourage a critical and circumspect approach to natural science while at the same time providing a suitable foundation for advanced coursework in physics. This is accomplished by holding before the student some of the best thinking about nature that has been committed to writing. The scientific texts found herein are considered classics precisely because they address timeless questions in a particularly honest and convincing manner. This does not mean that everything they say is true—in fact many classic scientific texts contradict one another—but it is by the careful reading, analysis and discussion of the most reputable observations and opinions that one may begin to discern truth from error.

Who is this book for?

Like fine wine, the classic texts in any discipline can be enjoyed by both the novice and the connoisseur. For example, Sophocles' tragic play *Antigone* can be appreciated by the young student who is drawn to the story of the heroine who braves the righteous wrath of King Creon by choosing to illegally bury the corpse of her slain brother, and also by the seasoned scholar who carefully evaluates the relationship between justice, divine law and the state. Likewise, Galileo's *Dialogues Concerning Two New Sciences* can be enjoyed by the young student who seeks a clear geometrical description of the speed of falling bodies, and also by the seasoned scholar who is amused by Galileo's wit and sarcasm, or who finds in his *Dialogues* the progressive Aristotelianism of certain late medieval scholastics.[1]

Having said this, I believe that this book is particularly suitable for the following audiences. First, it could serve as the primary textbook in an introductory discussion-based physics course at the university level. It was designed to appeal to a broad constituency of students at small liberal arts colleges which often lack the resources to offer the separate and specialized introductory physics courses found at many state-funded universities (*e.g. Physics for poets, Physics for engineers, Physics for health-care-professionals, Physics of sports, etc.*). Indeed, at my institution it is common to have history and fine arts students sitting in the course alongside biology and physics majors. Advanced high-school or home-school students will find in this book a physics curriculum that emphasizes reading comprehension, and which can serve as a bridge into college-level work. It might also be adopted as a supplementary text for an advanced placement course in physics, astronomy or the history and philosophy of science. Many practicing physicists, especially those at the beginning of their scientific careers, may not have taken the opportunity to carefully study some of the foundational texts of physics and astronomy. Perhaps this is because they have (quite understandably) focused their attention on acquiring a strong technical proficiency in a narrow subfield. Such individuals will find herein a structured review of such foundational texts. This book will also likely appeal to humanists, social scientists and motivated lay-readers who seek a thematically-organized anthology of texts which offer insight into the historical development and cultural significance of contemporary scientific theories. Finally, and most importantly, this book is designed for the benefit of the teaching professor. Early in my career as a faculty member, I was afforded considerable freedom to develop a physics curriculum at my institution which would sustain my interest for the foreseeable future—perhaps until retirement. Indeed, reading and re-reading the classic texts assembled herein has provided me countless hours of enjoyment, reflection and inspiration.

How is this book unique?

Here I will offer a mild critique of textbooks typically employed in introductory university physics courses. While what follows is admittedly a bit of a caricature, I believe it to be a quite plausible one. I do this in order to highlight the unique features and emphases of the present book. In many university-level physics textbooks, the chapter format follows a standard recipe. First, accepted scientific laws are presented in the form of one or more mathematical equations. This is followed by a

---

[1] See Wallace, W. A., The Problem of Causality in Galileo's Science, *The Review of Metaphysics*, *36*(3), 607–632, 1983.

few example problems so the student can learn how to plug numbers into the afore-mentioned equations and how to avoid common conceptual or computational errors. Finally, the student is presented with contemporary applications which illustrate the relevance of these equations for various industrial or diagnostic technologies.

While this method often succeeds in preparing students to pass certain stan-dardized tests or to solve fairly straightforward technical problems, it is lacking in important respects. First, it is quite bland. Although memorizing formulas and learning how to perform numerical calculations is certainly crucial for acquiring a working knowledge of physical theories, it is often the more general questions about the assumptions and the methods of science that students find particularly stimulating and enticing. For instance, in his famous *Mathematical Principles of Natural Philosophy*, Newton enumerates four general rules for doing philosophy. Now the reader may certainly choose to reject Newton's rules, but Newton himself suggests that they are necessary for the subsequent development of his universal theory of gravitation. Is he correct? For instance, if one rejects Rules III and IV—which articulate the principle of induction—then in what sense can his theory of gravity be considered universal? Questions like "is Newton's theory of gravity cor-rect?" and "how do you know?" can appeal to the innate sense of inquisitiveness and wonder that attracted many students to the study of natural science in the first place. Moreover, in seeking a solution to these questions, the student must typically acquire a deeper understanding of the technical aspects of the theory. In this way, broadly posed questions can serve as a motivation and a guide to obtaining a detailed understanding of physical theories.

Second, and perhaps more importantly, the method employed by most standard textbooks does not prepare the student to become a practicing scientist precisely because it tends to mask the way science is actually done. The science is presented as an accomplished fact; the prescribed questions revolve largely around techno-logical applications of accepted laws. On the contrary, by carefully studying the foundational texts themselves the student is exposed to the polemical debates, the technical difficulties and the creative inspirations which accompanied the develop-ment of scientific theories. For example, when studying the motion of falling bodies in Galileo's *Dialogues*, the student must consider alternative explanations of the observed phenomena; must understand the strengths and weaknesses of competing theories; and must ultimately accept—or reject—Galileo's proposal on the basis of evidence and reason. Through this process the student gains a deeper understanding of Galileo's ideas, their significance, and their limitations.

Moreover, when studying the foundational texts, the student is obliged to thoughtfully address issues of language and terminology—issues which simply do not arise when learning from standard textbooks. In fact, when scientific the-ories are being developed the scientists themselves are usually struggling to define terms which capture the essential features of their discoveries. For example, Oersted coined a term which is translated as "electric conflict" to describe the effect that an electrical current has on a nearby magnetic compass needle. He was attempting to distinguish between the properties of stationary and moving charges, but he lacked the modern concept of the magnetic field which was later introduced by Faraday.

When students encounter a familiar term such as "magnetic field," they typically accept it as settled terminology, and thereby presume that they understand the phenomenon by virtue of recognizing and memorizing the canonical term. But when they encounter an unfamiliar term such as "electric conflict," as part of the scientific argument from which it derives and wherein it is situated, they are tutored into the original argument and are thus obliged to think scientifically, along with the great scientist. In other words, when reading the foundational texts, the student is led into *doing* science and not merely into memorizing and applying nomenclature.

Generally speaking, this book draws upon two things that we have in common: (i) a shared conversation recorded in the foundational scientific texts, and (ii) an innate faculty of reason. The careful reading and analysis of the foundational texts is extremely valuable in learning how to think clearly and accurately about natural science. It encourages the student to carefully distinguish between observation and speculation, and finally, between truth and falsehood. The ability to do this is essential when considering the practical and even philosophical implications of various scientific theories. Indeed, one of the central aims of this book is to help the student grow not only as a potential scientist, but as an educated person. More specifically, it will help the student develop important intellectual virtues (*i.e.* good habits), which will serve him or her in any vocation, whether in the marketplace, in the family, or in society.

How is this book organized?

This book is divided into four separate volumes; volumes I and II were concurrently published in the autumn of 2014, and volumes III and IV are due to be published approximately a year later. Within each volume, the readings are centered on a particular theme and proceed chronologically. For example, Volume I is entitled *The Heavens and the Earth*. It provides an introduction to astronomy and cosmology beginning with the geocentrism of Aristotle's *On the Heavens* and Ptolemy's *Almagest*, proceeding through heliocentrism advanced in Copernicus' *Revolutions of the Heavenly Spheres* and Kepler's *Epitome of Copernican Astronomy*, and arriving finally at big bang cosmology with Lemaître's *The Primeval Atom*. Volume II, *Space, Time and Motion*, provides a careful look at the science of motion and rest. Here, students engage in a detailed analysis of significant portions of Galileo's *Dialogues Concerning Two New Sciences*, Pascal's *Treatise on the Equilibrium of Fluids and the Weight of the Mass of Air*, Newton's *Mathematical Principles of Natural Philosophy* and Einstein's *Relativity*.

Volume III traces the theoretical and experimental development of the electromagnetic theory of light using texts by William Gilbert, Benjamin Franklin, Charles Coulomb, André Marie Ampère, Christiaan Huygens, James Clerk Maxwell, Heinrich Hertz, Albert Michelson, and others. Volume IV provides an exploration of modern physics, focusing on the mechanical theory of heat, radio-activity and the development of modern quantum theory. Selections are taken from works by

Joseph Fourier, William Thomson, Rudolph Clausius, Joseph Thomson, James Clerk Maxwell, Ernest Rutherford, Max Planck, James Chadwick, Niels Bohr, Erwin Schrödinger and Werner Heisenberg.

While the four volumes of the book are arranged around distinct themes, the readings themselves are not strictly constrained in this way. For example, in his *Treatise on Light*, Huygens is primarily interested in demonstrating that light can be best understood as a wave propagating through an aethereal medium comprised of tiny, hard elastic particles. In so doing, he spends some time discussing the speed of light measurements performed earlier by Ole Rømer. These measurements, in turn, relied upon an understanding of the motion of the moons of Jupiter which had recently been reported by Galileo in his *Sidereal Messenger*. So here, in this *Treatise on Light*, we find references to a variety of inter-related topics. Huygens does not artificially restrict his discussion to a narrow topic—nor does Galileo, or Newton or the other great thinkers. Instead, the reader will find in this book recurring concepts and problems which cut across different themes and which are naturally addressed in a historical context with increasing levels of sophistication and care. Science is a conversation which stretches backwards in time to antiquity.

How might this book be used?

This book is designed for college classrooms, small-group discussions and individual study. Each of the four volumes of the book contains roughly 30 chapters, providing more than enough material for a one-semester undergraduate-level physics course; this is the context in which this book was originally implemented. In such a setting, one or two 50-min classroom sessions should be devoted to analyzing and discussing each chapter. This assumes that the student has read the assigned text before coming to class. When teaching such a course, I typically improvise—leaving out a chapter here or there (in the interest of time) and occasionally adding a reading selection from another source that would be particularly interesting or appropriate.

Each chapter of each volume has five main components. First, at the beginning of each chapter, I include a short introduction to the reading. If this is the first encounter with a particular author, the introduction includes a biographical sketch of the author and some historical context. The introduction will often contain a summary of some important concepts from the previous chapter and will conclude with a few provocative questions to sharpen the reader's attention while reading the upcoming text.

Next comes the reading selection. There are two basic criteria which I used for selecting each text: it must be *significant* in the development of physical theory, and it must be *appropriate* for beginning undergraduate students. Balancing these criteria was very difficult. Over the past decade, I have continually refined the selections so that they might comprise the most critical contribution of each scientist, while at the same time not overwhelming the students by virtue of their length, language or complexity. The readings are not easy, so the student should not feel overwhelmed if he or she does not grasp everything on the first (or second, or third...) reading. Nobody does. Rather—like classic literature—these texts must be "grown into" (so to speak) by returning to them time and again.

I have found that the most effective way to help students successfully engage foundational texts is to carefully prepare questions which help them identify and understand key concepts. So as the third component of each chapter, I have prepared a study guide in the form of a set of questions which can be used to direct either classroom discussion or individual reading. After the source texts themselves, the study guide is perhaps the most important component of each chapter, so I will spend a bit more time here explaining it.

The study guide typically consists of a few general discussion questions about key topics contained in the text. Each of these general questions is followed by several sub-questions which aid the student by focusing his or her attention on the author's definitions, methods, analysis and conclusions. For example, when students are reading a selection from Albert Michelson's book *Light Waves and their Uses*, I will often initiate classroom discussion with a general question such as "Is it possible to measure the absolute speed of the earth?" This question gets students thinking about the issues addressed in the text in a broad and intuitive way. If the students get stuck, or the discussion falters, I will then prompt them with more detailed follow-up questions such as: "What is meant by the term absolute speed?" "How, exactly, did Michelson attempt to measure the absolute speed of the earth?" "What technical difficulties did Michelson encounter while doing his experiments?" "To what conclusion(s) was Michelson led by his results?" and finally "Are Michelson's conclusions then justified?" After answering such simpler questions, the students are usually more confident and better prepared to address the general question which was initially posed.

In the classroom, I always emphasize that it is critical for participants to carefully read the assigned selections before engaging in discussion. This will help them to make relevant comments and to cite textual evidence to support or contradict assertions made during the course of the discussion. In this way, many assertions will be revealed as problematic—in which case they may then be refined or rejected altogether. Incidentally, this is precisely the method used by scientists themselves in order to discover and evaluate competing ideas or theories. During our discussion, students are encouraged to speak with complete freedom; I stipulate only one classroom rule: any comment or question must be stated publicly so that all others can hear and respond. Many students are initially apprehensive about engaging in public discourse, especially about science. If this becomes a problem, I like to emphasize that students do not need to make an elaborate point in order to engage in classroom discussion. Often, a short question will suffice. For example, the student might say "I am unclear what the author means by the term *inertia*. Can someone please clarify?" Starting like this, I have found that students soon join gamely in classroom discussion.

Fourth, I have prepared a set of exercises which test the student's understanding of the text and his or her ability to apply key concepts in unfamiliar situations.

Some of these are accompanied by a brief explanation of related concepts or formulas. Most of them are numerical exercises, but some are provocative essay prompts. In addition, some of the chapters contain suggested laboratory exercises, a few of which are in fact field exercises which require several days (or even months) of observations. For example, in Chap. 3 of Volume I, there is an astronomy field exercise which involves charting the progression of a planet through the zodiac over the course of a few months. So if this book is being used in a semester-long college or university setting, the instructor may wish to skim through the exercises at the end of each chapter so he or she can identify and assign the longer ones as ongoing exercises early in the semester.

Finally, I have included at the end of each chapter a list of vocabulary words which are drawn from the text and with which the student should become acquainted. Expanding his or her vocabulary will aid the student not only in their comprehension of subsequent texts, but also on many standardized college and university admissions exams.

What mathematics preparation is required?

It is sometime said that mathematics is the "language of science." This sentiment appropriately inspires and encourages the serious study of mathematics. Of course if it were taken literally then many seminal works in physics—and much of biology— would have to be considered either unintelligible or unscientific, since they contain little or no mathematics. Moreover, if mathematics is the *only* language of science, then physics instructors should be stunned whenever students are enlightened by verbal explanations which lack mathematical form. To be sure, mathematics offers a refined and sophisticated language for describing observed phenomena, but many of our most significant observations about nature may be expressed using everyday images, terms and concepts: heavy and light, hot and cold, strong and weak, straight and curved, same and different, before and after, cause and effect, form and function, one and many. So it should come as no surprise that, when studying physics *via* the reading and analysis of foundational texts, one enjoys a considerable degree of flexibility in terms of the mathematical rigor required.

For instance, Faraday's *Experimental Researches in Electricity* are almost entirely devoid of mathematics. Rather, they consist of detailed qualitative descriptions of his observations, such as the relationship between the relative motion of magnets and conductors on the one hand, and the direction and intensity of induced electrical currents on the other hand. So when studying Faraday's work, it is quite natural for the student to aim for a conceptual, as opposed to a quantitative, understanding of electromagnetic induction. Alternatively, the student can certainly attempt to connect Faraday's qualitative descriptions with the mathematical methods which are often used today to describe electromagnetic induction (*i.e.* vector calculus and differential equations). The former method has the advantage of demonstrating the conceptual framework in which the science was actually conceived and developed; the latter method has the advantage of allowing the student to make a more seamless transition to upper-level undergraduate or graduate courses which typically employ sophisticated mathematical methods.

In this book, I approach the issue of mathematical proficiency in the following manner. Each reading selection is followed by both study questions and homework exercises. In the study questions, I do not attempt to force anachronistic concepts or methods into the student's understanding of the text. They are designed to encourage the student to approach the text in the same spirit as the author, insofar as this is possible. In the homework exercises, on the other hand, I often ask the student to employ mathematical methods which go beyond those included in the reading selection itself. For example, one homework exercise associated with a selection from Hertz's book *Electric Waves* requires the student to prove that two counter-propagating waves superimpose to form a standing wave. Although Hertz casually mentions that a standing wave is formed in this way, the problem itself requires that the student use trigonometric identities which are not described in Hertz's text. In cases such as this, a note in the text suggests the mathematical methods which are required. I have found this to work quite well, especially in light of the easy access which today's students have to excellent print and online mathematical resources.

Generally speaking, there is an increasing level of mathematical sophistication required as the student progresses through the curriculum. In Volume I students need little more than a basic understanding of geometry. Euclidean geometry is sufficient in understanding Ptolemy's epicyclic theory of planetary motion and Galileo's calculation of the altitude of lunar mountains. The student will be introduced to some basic ideas of non-Euclidean geometry toward the end of Volume I when studying modern cosmology through the works of Einstein, Hubble and Lemaître, but this is not pushed too hard. In Volume II students will make extensive use of geometrical methods and proofs, especially when analyzing Galileo's work on projectile motion and the application of Newton's laws of motion. Although Newton develops his theory of gravity in the *Principia* using geometrical proofs, the homework problems often require the student to make connections with the methods of calculus. The selections on Einstein's special theory of relativity demand only the use of algebra and geometry. In Volume III, mathematical methods will, for the most part, be limited to geometry and algebra. More sophisticated mathematical methods will be required, however, in solving some of the problems dealing with Maxwell's electromagnetic theory of light. This is because Maxwell's equations are most succinctly presented using vector calculus and differential equations. Finally, in Volume IV, the student will be aided by a working knowledge of calculus, as well as some familiarity with the use of differential equations.

It is my feeling that in a general physics course, such as the one being presented in this book, the extensive use of advanced mathematical methods (beyond geometry, algebra and elementary calculus) is not absolutely necessary. Students who plan to major in physics or engineering will presumably learn more advanced mathematical methods (*e.g.* vector calculus and differential equations) in their collateral mathematics courses, and they will learn to apply these methods in upper-division (junior and senior-level) physics courses. Students who do not plan to major in physics will typically not appreciate the extensive use of such advanced mathematical methods. And it will tend to obscure, rather than clarify, important physical concepts. In any case, I have attempted to provide guidance for the instructor, or for the self-directed student, so that he or she can incorporate an appropriate level of mathematical rigor.

Figures, formulas, and footnotes

One of the difficulties in assembling readings from different sources and publishers into an anthology such as this is how to deal with footnotes, references, formulas and other issues of annotation. For example, for any given text selection, there may be footnotes supplied by the author, the translator and the anthologist. So I have appended a [K.K.] marking to indicate when the footnote is my own; I have not included this marking when there is no danger of confusion, for example in my footnotes appearing in the introduction, study questions and homework exercises of each chapter.

For the sake of clarity and consistency, I have added (or sometimes changed the) numbering for figures appearing in the texts. For example, Fig. 16.3 is the third figure in Chap. 16 of this volume I; this is not necessarily how Kepler or his translator numbered this figure when it appeared in an earlier publication of his *Epitome Astronomae Copernicanae*. For ease of reference, I have also added (or sometimes changed the) numbering of equations appearing in the texts. For example, Eqs. 31.1 and 31.2 are the equations of the Lorentz and Galilei transformations appearing in the reading in Chap. 31 of Volume II, extracted from Einstein's book *Relativity*. This is not necessarily how Einstein numbered them.

In several cases, the translator or editor has included references to page numbers in a previous publication. For example, the translators of Galileo's *Dialogues* have indicated, within their 1914 English translation, the locations of page breaks in the Italian text published in 1638. A similar situation occurs with Faith Wallis's 1999 translation of Bede's *The Reckoning of Time*. For consistency, I have rendered such page numbering in bold type surrounded by slashes. So **/50/** refers to page 50 in some earlier "canonical" publication.

Acknowledgements

I suppose that it is common for a teacher to eventually mull over the idea of compiling his or her thoughts on teaching into a coherent and transmittable form. Committing this curriculum to writing was particularly difficult because I am keenly aware how my own thinking about teaching physics has changed significantly since my first days in front of the classroom—and how it is quite likely to continue to evolve. So this book should be understood as a snapshot, so to speak, of how I am teaching my courses at the time of writing. I would like to add, however, that I believe the evolution of my teaching has reflected a maturing in thought, rather than a mere drifting in opinion. After all, the classic texts themselves are formative: how can a person, whether student or teacher, not become better informed when learning from the best thinkers?

This being said, I would like to offer my apologies to those students who suffered through the birth pains, as it were, of the curriculum presented in this book. The countless corrections and suggestions that they offered are greatly appreciated; any and all remaining errors in the text are my own fault. Many of the reading selections included herein were carefully scanned, edited and typeset by undergraduate students who served as research and editorial assistants on this project: Jaymee Martin-Schnell, Dylan Applin, Samuel Wiepking, Timothy Kriewall, Stephanie Kriewall, Cody Morse, Michaela Otterstatter, and Ethan Jahns deserve special thanks. My home institution, Wisconsin Lutheran College, provided me with considerable time and freedom to develop this book, including a year-long sabbatical leave, for which I am very grateful. During this sabbatical, I received support and encouragement from my trusty colleagues in the Department of Mathematical and Physical Sciences. Also, the Higher Education Initiatives Program of the Wisconsin Space Grant Consortium provided generous funding for this project, as did the Faculty Development Committee of Wisconsin Lutheran College. Greg Schulz has been an invaluable intellectual resource throughout this project. Aaron Jensen conscientiously translated selections of the *Almagest*, included in Volume I of this book, from Heiberg's edition of Ptolemy's Greek manuscript. And Glen Thompson was instrumental in getting this translation project initiated. Starla Siegmann and Jenny

Baker, librarians at the Marvin M. Schwan Library of Wisconsin Lutheran College, were always up to the challenge of speedily procuring obscure resources from remote libraries. I would also like to thank the following individuals who facilitated the complex task of acquiring permissions to reprint the texts included in this book: Jenny Howard at the Liverpool University Press, Elizabeth Sandler, Emilie David and Norma Rosado-Blake at the American Association for the Advancement of Science, Chris Erdmann at the Harvard College Observatory's Wolbach Library, Carmen Pagán at Encyclopædia Britannica, and Michael Fisher, McKenzie Carnahan and and Scarlett Huffman at the Harvard University Press. Cornelia Mutel and Kathryn Hodson very kindly provided digital images for inclusion with the Galileo and Pascal selections from the History of Hydraulics Rare Book Collection at the University of Iowa's IIHR-Hydroscience and Engineering. Also, I would like to thank Jeanine Burke, the acquisition editor at Springer who originally agreed to take on this project with me, and Robert Korec, Tom Spicer and Catherine Rice, who all patiently saw it through to publication. Shortly after submitting my book proposal to Springer, I received very encouraging and helpful comments from several anonymous reviewers, for whom I am thankful. I received similar suggestions from the editors of Springer's Undergraduate Lecture Notes in Physics series for which I am likewise grateful. Finally, I would especially like to thank my wife, Cindy, who has provided unwavering encouragement and support for my work from the very start.

Milwaukee, 2015   Kerry Kuehn

Contents

Chapter 1
Iron, Loadstones and Terrestrial Magnetism

Men of acute intelligence, without actual knowledge of facts,
and in the absence of experiment, easily slip and err.
—William Gilbert

1.1 Introduction

William Gilbert (1544–1603) was born in Colchester, County Essex, England. A
physician in the service of Queen Elizabeth, Gilbert carried out extensive studies on
the properties of loadstones and other magnetic substances. In 1600, he published
De Magnete, wherein he offered a comprehensive review—and a sharp critique—
of his contemporaries' understanding of magnetism. Of particular note in Gilbert's
De Magnete is the careful distinction that he draws between electrical and magnetic phenomena. Previously, this distinction had been either greatly muddled or
entirely unrecognized. According to Gilbert, electrical attractions, such as those
exhibited by precious stones or amber when rubbed, were attributed to a *mate-*
rial cause—a corporeal *effluvium* emitted by the substance which is "awakened
by friction." Magnetic attractions and repulsions, on the other hand, such as those
exhibited by loadstones and magnetized iron, Gilbert attributed to a *formal cause*—
a native strength or *vigor* which ferruginous substances share with the earth itself.[1]
The language of *material* and *formal* causes which Gilbert employs, though uncom-
mon today, was typical of the intellectual heirs of the ancient greek philosopher
Aristotle.[2] But despite this similarity, Gilbert rejected Aristotelian thinking in at
least two very important ways. First, Gilbert explicitly rejected Aristotle's geocentrism in favor of a Copernican worldview.[3] Gilbert's heliocentrism was later shared
by the famous German astronomer Johannes Kepler, who—explicitly invoking

---

[1] Gilbert explains the distinction between electrical and magnetic forces in Chaps. 2–4 of Book II
of his *De Magnete*.

[2] Aristotle's *four causes*, or we might say his "four types of explanations for changes occurring in
nature," are described in Chap. 3 of Book II of his *Physics*. They are the *material*, the *formal*, the
efficient and the *final* causes.

[3] Copernicus' heliocentrism is laid out in his *On the Revolutions of the Heavenly Spheres*, which
was published in 1543; see Chaps. 11–13 of Vol. I of the present series.

© Springer International Publishing Switzerland 2016
K. Kuehn, *A Student's Guide Through the Great Physics Texts*,
Undergraduate Lecture Notes in Physics, DOI 10.1007/978-3-319-21816-8_1

Gilbert's *De Magnete*—would attribute Earth's motion about the sun to some type of magnetic force.[4] After all, Kepler reasoned, the earth (and perhaps the sun also) is a huge magnet—and magnets are known to attract one another. Second, Gilbert rejected the theory of the magnetic compass needle, or *versorium*, which was propounded by Aristotle's intellectual heirs, the so-called "elementarian philosophers."[5] It is here that we begin our reading of Gilbert's *De Magnete*. The text selections that follow are extracted from Book III of the English translation by P. Fleury Mottelay. As you consider this text—and are thereby transported back in time, as it were, to the birth of the modern science of magnetism—you might permit yourself to naively consider the following simple questions: how, exactly, is a magnetic stone different than other stones? And what might be the hidden cause of such strange and wonderful behavior?

1.2   Reading: Gilbert, *On the Loadstone and Magnetic Bodies and on the Great Magnet the Earth*

Gilbert, W., *On the Loadstone and Magnetic Bodies and on the Great Magnet the Earth: a New Physiology, Demonstrated with Many Arguments and Experiments*, Bernard Quaritch, London, 1893. Book III.

1.2.1   Chapter I: Of Direction

In the foregoing books it has been shown that a loadstone has its poles, iron also poles, and rotation and fixed verticity, and finally that loadstone and iron direct their poles toward the poles of the earth. But now we have to set forth the causes of these things and their wonderful efficiencies known aforetime but not demonstrated. Of these rotations all the writers who went before us have given their opinions with such brevity and indefiniteness that, as it would seem, no one could be persuaded thereby, while the authors themselves could hardly be contented with them. By men of intelligence, all their petty reasonings—as being useless, questionable, and absurd, and based on no proofs or premises—are rejected with the result that magnetic science, neglected more and more and understood by none, has been exiled. The true south pole, and not the north (as before our time all believed), of a loadstone placed on

---

[4] Kepler's theory of planetary motion is presented in Part III of Book IV of his *Epitome of Copernican Astronomy*, which was published around 1620; see especially Chap. 17 of volume I of the present series. Kepler's explanation based on magnetism would later be superseded by Newton's universal law of gravity, as described in Book III of his *Principia*, which was published in 1687; see Chap. 27 of volume II of the present series.

[5] The "elementarian philosophers" were those who, like Aristotle, believed that all earthly substances were comprised of just four elements: earth, fire, air and water. Aristotle's classification of substances is described in Book I of his *On the Heavens*; see Chap. 1 of volume I of the present series.

its float in water turns to the north; the south end of a piece of magnetized and of unmagnetized iron also moves to the north. An oblong piece of iron of three or four finger-breadths, properly stroked with a loadstone, quickly turn to north and south. Therefore artificers place such a bar, balanced on a point, in a compass-box or in a sun-dial; or they construct a versorium out of two curved pieces of iron that touch at their extremities so that the movement may be more constant; thus is constructed the mariner's compass, an instrument beneficial, salutary, and fortunate for seamen, showing the way to safety and to port. But it is to be understood at the threshold of their argument, before we proceed farther, that these directions of loadstone or of iron are not ever and always toward the world's true poles, that they do not always seek those fixed and definite points, nor rest on the line of the true meridian, but that at places, more or less far apart, they commonly vary either to the east or to the west; sometimes, too, in certain regions of land or sea, they point to the true poles. This discrepance is known as the variation of the needle and of the loadstone; and as it is produced by other causes and is, as it were, a sort of perturbation and depravation of the true direction, we propose to treat here only of the true direction of the compass and the magnetic needle, which would all over the earth be the same, toward the true poles and in the true meridian, were not hindrances and disturbing causes present to prevent: in the book next following we will treat of its variation and of the cause of perturbation.

They who aforetime wrote of the world and of natural philosophy, in particular those great elementarian philosophers and all their progeny and pupils down to our day; those, I mean, who taught that the earth is ever at rest, and is, as it were, a dead-weight planted in the centre of the universe at equal distance everywhere from the heavens, of simple uncomplex matter possessing only the qualities of dryness and cold—these philosophers were ever seeking the causes of things in the heavens, in the stars, the planets; in fire, air, water, and in the bodies of compounds; but never did they recognize that the terrestrial globe, besides dryness and cold, hath some principal, efficient, predominant potencies that give to it firmness, direction, and movement throughout its entire mass and down to its inmost depths; neither did they make inquiry whether such things were, and, for this reason, the common herd of philosophizers, in search of the causes of magnetic movements, called in causes remote and far away. Martinus Cortesius, who would be content with no cause whatever in the universal world, dreamt of an attractive magnetic point beyond the heavens, acting on iron. Petrus Peregrinus holds that direction has its rise at the celestial poles. Cardan was of the opinion that the rotation of iron is caused by the star in the tail of Ursa Major. The Frenchman Bessard thinks that the magnetic needle turns to the pole of the zodiac. Marsilius Ficinus will have it that the loadstone follows its Arctic pole, and that iron follows the loadstone, and chaff follows amber: as for amber, why, that, mayhaps, follows the Antarctic pole: emptiest of dreams! Others have come down to rocks and I know not what "magnetic mountains"! So has ever been the wont of mankind: homely things are vile; things from abroad and things afar are dear to them and the object of longing. As for us, we are habitants of this very earth, and study it as cause of this mighty effect. Earth, the mother of all, hath these causes shut up in her recesses: all magnetic movements are to be considered with respect to her law, position, constitution, verticity, poles, equator, horizon, meridians, centre, periphery, diameter, and to the form of her whole inward substance. So hath the earth been ordered by the Supreme Artificer and by nature, that it shall have parts unlike in position, terminal points of an entire and absolute body, and such points dignified by distinct functions, whereby it shall itself take a fixed direction. For like as a loadstone, when in a suitable vessel it is floated on water, or when it is suspended in air by a slender thread, does by its native verticity, according to the magnetic laws, conform its poles to the poles of the common mother,—so, were the earth to vary from her natural direction and from her position in the universe, or were her poles to be pulled toward the rising or the setting sun, or other points whatsoever in the visible firmament (were that possible), they would recur again by a magnetic movement to north and south, and halt at the same points where now they stand. But why the terrestrial globe should seem constantly to turn one of its poles toward those points and toward Cynosura [constellation of the Lesser Bear], or why her poles should vary from the poles of the ecliptic by 23° 29′, with some variation not yet sufficiently studied by astronomers,—that depends on the magnetic energy. The causes of the precession of the equinoxes and of the progression of the fixed stars, as well as of change in the declinations of the sun and the tropics, are traceable to magnetic forces: hence we have no further need of Thebit Bencora's "movement of trepidation," which is at wide variance with observations. A rotating needle turns to conformity with the situation of the earth, and, though it be shaken oft, returns still to the same points. For in far northern climes, in latitude 70 to 80° (whither in the milder season our seamen are wont to penetrate without injury from the cold), and in the middle regions, in the torrid zone under the equinoctial line, as also in all maritime regions and lands of the southern hemisphere, at the highest latitudes yet known, the magnetic needle ever finds its direction and ever tends in the same way (barring difference of variation) on this side of the equator where we dwell and in the other, the southern part, which, though less known, has been to some extent explored by our sailors: and the lily of the mariner's compass ever points north. Of this, we are assured by the most illustrious navigators and by many intelligent seamen. The same was pointed out to me and confirmed by our most illustrious Neptune, Francis Drake, and by Thomas Candish (Cavendish), that other world-explorer.

Our terrella teaches the same lesson. The proposition is demonstrated on a spherical loadstone (Fig. 1.1). Let A, B be the poles; $C D$, an iron wire placed on the stone, always tends direct in the meridian to the poles A, B, whether the centre of the wire be in the middle line or equator of the stone, or whether it be in any other region between equator and poles, as H, G, F, E. So the point of a magnetized needle looks north on this side of the equator: on the other side the crotch is directed to the south; but the point or lily does not turn to the south below the equator, as somebody has thought. Some inexperienced persons, however, who, in distant regions below the equator, have at times seen the needle grow sluggish and less prompt, have deemed the distance from the Arctic pole or from the magnetic rocks to be the cause. But they are very much mistaken, for it has the same power and adjusts itself as quickly to the meridian as the point of variation in southern regions as in northern. Yet at times the movement appears to be slower, the point on which the compass needle is poised becoming in time, during a long voyage, rather blunt, or the magnetized

Fig. 1.1 The orientation of a magnetized iron wire near a spherical loadstone.—[*K.K.*]

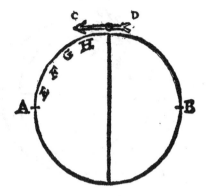

needle itself having lost somewhat of its acquired force through age or from rusting. This, too, may be tested experimentally by poising the versorium of a sun-dial on a rather short-pointed needle rising perpendicularly out of the surface of the terrella. The magnetized needle turns to the poles of the terrella, and quits the earth's poles; for a general cause that is remote is overcome by a particular cause that is present and strong. Magnetized bodies incline of their own accord to the earth's position, and they conform to the terrella. Two loadstones of equal weight and force conform to the terrella in accordance with magnetic laws. Iron gets force from the loadstone and is made to conform to the magnetic movements. Therefore true direction is the movement of a magnetized body in the line of the earth's verticity toward the natural position and unition of both, their forms being in accord and supplying the forces. For we have, after many experiments in various ways, found that the disposing and ranging of the magnetized bodies depends on the differences of position, while the force that gives the motion is the one form common to both; also that in all magnetic bodies there is attraction and repulsion. For both the loadstone and the magnetized iron conform themselves, by rotation and by dip, to the common position of nature and the earth. And the earth's energy, with the force inhering in it as a whole, by pulling toward its poles and by repelling, arranges in order all magnetic bodies that are unattached and lying loose. For in all things do all magnetic bodies conform to the globe of earth in accordance with the same laws and in the same ways in which another loadstone or any magnetic body whatsoever conforms to the terrella.

1.2.2   Chapter II: Directive (or Versorial) Force, Which We Call Verticity: What It Is; How It Resides in the Loadstone; and How It Is Acquired When Not Naturally Produced

The directive force, which by us is also called verticity, is a force distributed by the innate energy from the equator in both directions to the poles. That energy, proceeding north and south to the poles, produces the movement of direction, and produces also constant and permanent station in the system of nature, and that not in the earth

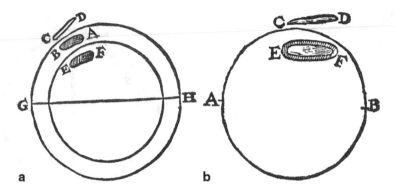

Fig. 1.2 Loadstones **a** mined from the topmost and the inward parts of the earth, and **b** separated from a terrella.—[*K.K.*]

alone but in all magnetic bodies also. Loadstone occurs either in a special vein or in iron mines, for, being a homogenic earth-substance possessing and conceiving a primary form, it becomes converted into or concreted with a stony body which, in addition to the prime virtues of the form, derives from different beds and mines, as from different matrices, various dissimilitudes and differences, and very many secondary qualities and varieties of its substance. A loadstone mined in this debris of the earth's surface and of its projections, whether it be (as sometimes found in China) entire in itself, or whether it be part of a considerable vein, gets from the earth its form and imitates the nature of the whole. All the inner parts of the earth are in union and act in harmony, and produce direction to north and south. Yet the magnetic bodies that in the topmost parts of the earth attract one another are not true united parts of the whole, but are appendages and agnate parts that copy the nature of the whole; hence, when floating free on water, they take the direction they have in the terrestrial order of nature. We once had chiselled and dug out of its vein a loadstone 20 pounds in weight, having first noted and marked its extremities; then, after it had been taken out of the earth, we placed it on a float in water so it could freely turn about; straightway that extremity of it which in the mine looked north turned to the north in water and after a while there abode; for the extremity that in the mine looks north is austral and is attracted by the north parts of earth, just as in the case of iron, which takes verticity from the earth. Of these points we will treat later under the head of "Change of Verticity."

But different is the verticity of the inward parts of the earth that are perfectly united to it and that are not separated from the true substance of the earth by interposition of bodies, as are separated loadstones situated in the outer portion of the globe, where all is defective, spoilt, and irregular. Let AB be a loadstone mine (Fig. 1.2a), and between it and the uniform earthen globe suppose there are various earths and mixtures that in a manner separate the mine from the true globe of the earth.

It is therefore informed by the earth's forces just as CD, a mass of iron, is in air; hence the extremity B of the mine or of any part thereof moves toward the

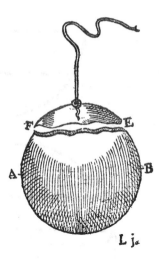

north pole G, just as does C, the extremity of the mass of iron, but not A nor D.
But with the part EF, which comes into existence continuous with the whole and
which is not separated from it by any mixed earthy matter, the case is different. For
if the part EF, being taken out, were to be floated, it is not E that would turn to
the north pole, but F. Thus, in those bodies which acquire verticity in the air, C is
the south extremity and is attracted by the north pole G. In those which come into
existence in the detrital outermost part of the earth, B is south, and so goes to the
north pole. But these parts which, deep below, are of even birth with the earth, have
their verticity regulated differently. For here F turns to the north parts of the earth,
being a south part; and E to the south parts of the earth, being a north part. So the
end C of the magnetic body CD, situate near the earth, turns to the north pole; the
end B of the agnate body BA to the north; the end E of the inborn body EF to the
south pole—as is proved by the following demonstration and as is required by all
magnetic laws.

Describe a terrella with poles A, B (Fig. 1.2b); from its mass separate the small
part EF, and suspend that by a fine thread in a cavity or pit in the terrella. E then
does not seek the pole A but the pole B, and F turns to A, behaving quite differently
from the iron bar CD; for, there, C, touching a north part of the terrella, becomes
magnetized and turns to A, not to B. But here it is to be remarked that if pole A of the
terrella were to be turned toward the southern part of the earth, still the end E of the
solitary part cut out of the terrella and not brought near the rest of the stone would
turn to the south; but the end C of the iron bar would, if placed outside the magnetic
field, turn to the north. Suppose that in the unbroken terrella the part EF gave the
same direction as the whole; now break it off and suspend it by a thread, and E will
turn to B and F to A (Fig. 1.3). Thus parts that when joined with the whole have the
same verticity with it, on being separated take the opposite; for opposite parts attract
opposite parts, yet this is not a true opposition, but a supreme concordance and a
true and genuine conformance of magnetic bodies in nature, if they be but divided

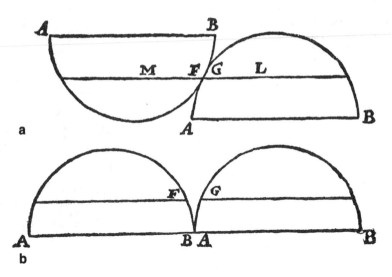

Fig. 1.4 The behavior of a loadstone divided into two equal parts.—[*K.K.*]

and separated; for the parts thus divided must needs be carried away some distance above the whole, as later will appear. Magnetic bodies seek formal unity, and do not so much regard their own mass. Hence the part *FE* is not attracted into its pit, but the moment it wanders abroad and is away from it, is attracted by the opposite pole. But if the part *FE* be again placed in its pit or be brought near without any media interposed, it acquires the original combination, and, being again a united portion of the whole, co-operates with the whole and readily clings in its pristine position, while *E* remains looking toward *A* and *F* toward *B*, and there they rest unchanging.

The case is the same when we divide a loadstone into two equal parts from pole to pole. In the figure (Fig. 1.4), a spherical stone is divided into two equal parts along the axis *AB*; hence, whether the surface *AB* be in one of the two parts supine (as in the first diagram), or prone in both (as in the second), the end *A* tends to *B*. But it is also to be understood that the point *B* does not always tend sure to *A*, for, after the division, the verticity goes to other points, for example to *F*, *G*, as is shown in Chap. 14 of this Third Book. *LM*, too, is now the axis of the two halves, and *AB* is no longer the axis; for, once a magnetic body is divided, the several parts are integral and magnetic, and have vertices proportional to their mass, new poles arising at each end on division. But the axis and poles ever follow the track of a meridian, because the force proceeds along the stone's meridian circles from the equinoctial to the poles invariably, in virtue of an innate energy that belongs to matter, owing to the long and secular position, and bearings toward the earth's poles, of a body possessing the fit properties; and such body is endowed with force from the earth for ages and ages continuously, and has from its first beginning stood firmly and constantly turned toward fixed and determinate points of the same.

1.2.3   Chapter III: How Iron Acquires Verticity from the Loadstone, and How This Verticity Is Lost or Altered

An oblong piece of iron, on being stroked with a loadstone, receives forces magnetic, not corporeal, nor inhering in or consisting with any body, as has been shown in the chapters on coition. Plainly, a body briskly rubbed on one end with a loadstone, and left for a long time in contact with the stone, receives no property of stone, gains nothing in weight; for if you weigh in the smallest and most accurate scales of a goldsmith a piece of iron before it is touched by the loadstone you will find that after the rubbing it has the same precise weight, neither less nor more. And if you wipe the magnetized iron with cloths, or if you rub it with sand or with a whetstone, it loses naught at all of its acquired properties. For the force is diffused through the entire body and through its inmost parts, and can in no wise be washed or wiped away. Test it, therefore, in fire, that fiercest tyrant of nature. Take a piece of iron the length of your hand and as thick as a goose-quill; pass it through a suitable round piece of cork and lay it on the surface of water, and note the end of the bar that looks north. Rub that end with the true smooth end of a loadstone; thus the magnetized iron is made to turn to the north. Take off the cork and put that magnetized end of the iron in the fire till it just begins to glow; on becoming cool again it will retain the virtues of the loadstone and will show verticity, though not so promptly as before, either because the action of the fire was not kept up long enough to do away all its force, or because the whole of the iron was not made hot, for the property is diffused throughout the whole. Take off the cork again, drop the whole of the iron into the fire, and quicken the fire with bellows so that it becomes all alive, and let the glowing iron remain for a little while. After it has grown cool again (but in cooling it must not remain in one position) put iron and cork once more in water, and you shall see that it has lost its acquired verticity. All this shows how difficult it is to do away with the polar property conferred by the loadstone. And were a small loadstone to remain for as long in the same fire, it too would lose its force. Iron, because it is not so easily destroyed or burnt as very many loadstones, retains its powers better, and after they are lost may get them back again from a loadstone; but a burnt loadstone cannot be restored.

Now this iron, stripped of its magnetic form, moves in a way different from any other iron, for it has lost the polar property; and though before contact with the loadstone it may have had a movement to the north, and after contact toward the south, now it turns to no fixed and determinate point; but afterward, very slowly, after a long time, it turns unsteadily toward the poles, having received some measure of force from the earth. There is, I have said, a twofold cause of direction,—one native in the loadstone and in iron, and the other in the earth, derived from the energy that disposes things. For this reason it is that after iron has lost the faculty of distinguishing the poles and verticity, a tardy and feeble power of direction is acquired anew from the earth's verticity. From this we see how difficultly, and how only by the action of intense heat and by protracted firing of the iron till it becomes soft, the magnetic force impressed in it is done away. When this firing has suppressed the acquired polar power, and the same is now quite conquered and as yet has not been called to life again, the iron is left a wanderer, and quite incapable of direction.

But we have to inquire further how it is that iron remains possessed of verticity. It is clear that the presence of a loadstone strongly affects and alters the nature of the iron, also that it draws the iron to itself with wonderful promptness. Nor is it the part rubbed only, but the whole of the iron, that is affected by the friction (applied at one end only), and therefrom the iron acquires a permanent though unequal power, as is thus proved.

Rub with a loadstone a piece of iron wire on one end so as to magnetize it and to make it turn to the north; then cut off part of it, and you shall see it move to the north as before, though weakly. For it is to be understood that the loadstone awakens in the whole mass of the iron a strong verticity (provided the iron rod be not too long), a pretty strong verticity in the shorter piece throughout its entire length, and, as long as the iron remains in contact with the loadstone, one somewhat stronger still. But when the iron is removed from contact it becomes much weaker, especially in the end not touched by the loadstone. And as a long rod, one end of which is thrust into a fire and made red, is very hot at that end, less hot in the parts adjoining and midway, and at the farther end may be held in the hand, that end being only warm,— so the magnetic force grows less from the excited end to the other; but it is there in an instant, and is not introduced in any interval of time nor successively, as when heat enters iron, for the moment the iron is touched by the loadstone it is excited throughout. For example, take an unmagnetized iron rod, 4 or 5 in. long: the instant you simply touch with a loadstone either end, the opposite end straightway, in the twinkling of the eye, repels or attracts a needle, however quickly brought to it.

1.2.4   Chapter IV: Why Magnetized Iron Takes Opposite Verticity; and Why Iron Touched by the True North Side of the Stone Moves to the Earth's North, and When Touched by the True South Side to the Earth's South: Iron Rubbed with the North Point of the Stone Does Not Turn to the South, Nor Vice Versa, as All Writers on the Loadstone Have Erroneously Thought

It has already been shown that the north part of a loadstone does not attract the north part of another stone, but the south part, and that it repels the north end of another stone applied to its north end. That general loadstone, the terrestrial globe, does with its inborn force dispose magnetized iron, and the magnetic iron too does the same with its inborn force, producing movement and determining the direction. For whether we compare together and experiment on two loadstones, or a loadstone and piece of iron, or iron and iron, or earth and loadstone, or earth and iron conformated by the earth or deriving force from the energy of a loadstone, of necessity the forces and movements of each and all agree and harmonize in the same way.

Fig. 1.5 The behavior of
the tip of an iron pointer
placed near a spherical
loadstone.—[*K.K.*]

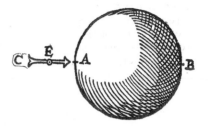

But the question arises, Why does iron touched with loadstone take a direction
of movement toward the earth's opposite pole and not toward that pole of earth
toward which looked the pole of the loadstone with which it was magnetized? Iron
and loadstone, we have said, are of the same primary nature: iron when joined to
a loadstone becomes as it were one body with it, and not only is one extremity of
the iron altered, but the rest of its parts are affected. Let *A* be the north pole of a
loadstone to which is attached the tip of an iron pointer: the tip is now the south
part of the iron, because it is contiguous to the north part of the stone (Fig. 1.5);
the crotch of the pointer becomes north. For were this contiguous magnetic body
separated from the pole of the terrella or the parts nigh the pole, the other extremity
(or the end which when there was conjunction was in contact with the north part
of the stone) is south, while the other end is north. So, too, if a magnetized needle
be divided into any number of parts however minute, those separated parts will take
the same direction which they had before division. Hence, as long as the point of
the needle remains at *A*, the north pole, it is not austral, but is, as it were, part of a
whole; but when it is taken away from the stone it is south, because on being rubbed
it tended toward the north parts of the stone, and the crotch (the other end of the
pointer) is north. The loadstone and the pointer constitute one body: B is the south
pole of the whole mass; *C* (the crotch) is the north extremity of the whole. Even
divide the needle in two at *E*, and *E* will be south as regards the crotch, *E* will also
be north with reference to *B*. *A* is the true north pole of the stone, and is attracted
by the south pole of the earth. The end of a piece of iron touched with the true north
part of the stone is south, and turns to the north pole of the stone *A* if it be near; if
it be at a distance from the stone, it turns to the earth's north. So whenever iron is
magnetized it tends (if free and unrestrained) to the portion of the earth opposite the
part toward which inclines the loadstone at which it was rubbed. For verticity always
enters the iron if only it be magnetized at either end. Hence all the needle points at
B acquire the same verticity after being separated, but it is the opposite verticity to
that of the pole *B* of the stone; and all the crotches in the present figure (Fig. 1.6)
have a verticity opposite to that of the pole *E*, and are made to move and are seized
by *E* when they are in suitable position. The case is as in the oblong stone *F H*, cut
in two at *G*, where *F* and *H*, whether the stone be whole or be broken, move to
opposite poles of the earth, and *O* and *P* mutually attract, one being north, the other
south. For if in the whole stone *H* was south and *F* north, then in the divided stone
P will be north with respect to *H* and *O*, south with respect to *F*; so, too, *F* and
H tend toward connection if they be turned round a little, and at length they come

Fig. 1.6 Iron needles near
one pole of a spherical loadstone (*above*); an oblong
loadstone broken equatorially
(*below*).—[*K.K.*]

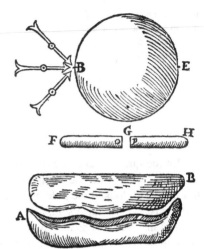

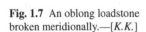

Fig. 1.7 An oblong loadstone
broken meridionally.—[*K.K.*]

together. But if the division be made meridionally, *i.e.*, along the line of the meridian
and not on any parallel circle, then the two parts turn about and *A* pulls *B*, and the
end *B* is attracted to *A*, until, being turned round, they form connection and are held
together (Fig. 1.7). For this reason, iron bars placed on parallels near the equator of
a terrella whose poles are *AB*, do not combine and do not cohere firmly; but when
placed alongside on a meridian line, at once they become firmly joined, not only on
the stone and near it, but at any distance within the magnetic field of the controlling
loadstone. Thus they are held fast together at *E*, but not at *C* of the other figure
(Fig. 1.8). For the opposite ends *C* and *F* of the bars, come together and cohere,
as the ends *A* and *B* of the stone did. But the ends are opposite, because the bars
proceed from opposite poles and parts of the terrella; and *C* is south as regards the
north pole *A*, and *F* is north as regards the south pole *B*. Similarly, too, they cohere
if the rod *C* (not too long) be moved further toward *A*, and the rod *F* toward *B*, and
they will be joined on the terrella just as *A* and *B* of the divided stone were joined.
But now if the magnetized needle point *A* be north, and if with this you touch and
rub the point *B* of another needle that rotates freely but is not magnetized, *B* will be
north and will turn to the south. But if with the north point *B* you touch still another
new rotatory needle on its point, that point again will be south, and will turn to the
north: a piece of iron not only takes from the loadstone, if it be a good loadstone,
the forces needful for itself, but also, after receiving them, infuses them into another
piece, and that into a third, always with due regard to magnetic law.

In all these our demonstrations it is ever to be borne in mind that the poles of
the stone as of the iron, whether magnetized or not, are always in fact and in their
nature opposite to the pole toward which they tend, and that they are thus named
by us, as has been already said. For, everywhere, that is north which tends to the
south of the earth or of a terrella, and that is south which turns to the north of the
stone. Points that are north are attracted by the south part of the earth, and hence
when floated they tend to the south. A piece of iron rubbed with the north end of a

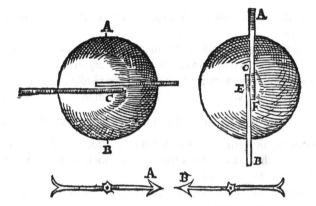

Fig. 1.8 The behavior of iron bars when placed parallel to one another along the equator (*left*) and along a meridian (*right*) of a spherical loadstone.—[*K.K.*]

Fig. 1.9 The interaction of iron needles and load-stones with the poles of the earth.—[*K.K.*]

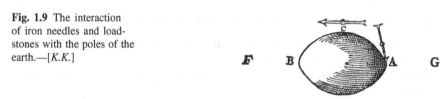

loadstone becomes south at the other end and tends always (if it be within the field of a loadstone and near) to the north part of the loadstone, and to the north part of the earth if it be free to move and stand alone at a distance from the loadstone. The north pole *A* of a loadstone turns to the south of the earth, *G* (Fig. 1.9); a needle magnetized on its point by the part *A* follows *A*, because the point has been made south. But the needle *C*, placed at a distance from the loadstone, turns its point to the earth's north, *F*, for that point was made south by contact with the north part of the loadstone. Thus the ends magnetized by the north part of the stone become south, or are magnetized southerly, and tend to the earth's north; the ends rubbed with the south pole become north, or are magnetized northerly, and tend to the earth's south.

1.3 Study Questions

QUES. 1.1. What is the law to which magnetic bodies conform?

a) How does Gilbert describe the state in which he found magnetic science? What then is his aim?
b) In what direction does a loadstones turn when floating in water? Is this rule followed strictly, or is there any variation? Does iron behave in the same way?
c) Who are the "great elementarian philosophers"? What did they teach regarding the earth and its motion(s)? How did Gilbert's philosophy differ?

d) Who (or what) is responsible for the order of the earth? To what end (if any) does this order exist?

e) Does a rotating compass needle point to a fixed point among the stars? How do you know? And why is this relevant for Gilbert's argument?

f) What is a terrella? How does a rotating compass needle react in the presence of a terrella? What is the significance of this observation?

QUES. 1.2. Is the verticity of all naturally occurring loadstones identical?

a) What is meant by the term verticity? What is its effect? And how is it measured?

b) Where in the earth are loadstones found? Does their verticity depend upon the location at which they are found?

c) How do loadstones orient themselves when freed? Does this differ from objects which acquire verticity in the air?

d) How does a section of a terrella behave when broken away? In which way does the directive force act? Is it attracted to the pit from which it was taken?

QUES. 1.3. Can iron acquire verticity? Can it lose verticity?

a.) What happens when iron is stroked with loadstone? What (significantly) does *not* happen? Why is this important?

b.) Does heat affect the verticity of iron? Of loadstone? If so, under what circumstances?

c.) Can verticity, once lost, be restored? If so, how?

d.) What, according to Gilbert, are the two cause of verticity? Which is predominant?

e.) Is the effect of loadstone on iron instantaneous? Or does it take a significant amount of time, like the conduction of heat through the same bar of iron?

QUES. 1.4. Is it true that a single point within a magnetized needle of iron can be considered both a north and a south pole simultaneously?

a) In what sense are iron and loadstone of the same "primary nature"? And in what sense do an iron needle and a nearby loadstone constitute "one body"?

b) How does an oblong magnetic body behave when cut in half crosswise? Lengthwise?

c) How do iron bars behave when placed on parallels near the equator of a terrella? Do they attract one another? What about if placed side-by-side on a meridian line?

d) Is the pole of a stone (whether magnetized or not) the same as, or different than, the pole toward which it tends? How does Gilbert demonstrate this using Fig. 1.9?

1.4 Exercises

EX. 1.1 (IRON, LOADSTONES AND THE EARTH). Suppose that a chunk of loadstone is extracted from a mine shaft deep within the earth, as shown in Fig. 1.2a. Prior to extraction, the northernmost tip of this oblong black rock is marked with bright white chalk. Later, in the laboratory, the white end of the loadstone is rubbed against the sharp tip of an unmagnetized iron needle. The midpoint of this needle is then painted red, and it is snipped in half with a pair of scissors.

a) If the loadstone is floated on a small boat in the laboratory, in which direction will its white end point?
b) If the two needle segments are likewise floated on separated small boats, in which direction will their red ends point?
c) Are the red ends of both needles of the same polarity as the white end of the loadstone?

EX. 1.2 (LOADSTONE LOGIC). Suppose that pole I of loadstone A is repelled from pole II of loadstone B and attracted to pole I of loadstone C.

a.) What can you conclude regarding the polarity of each of the poles of the three loadstones? For clarity, make a table indicating the possible polarities (north or south) of the poles of each loadstone. How many distinct possibilities are there?
b.) Suppose that you lived in a strange world in which there were three possible magnetic poles: red, green and blue. Like in our world, the two poles of any loadstone must still be different; opposite poles still attract and like poles still repel. If you know (in addition to the above information) that pole I of loadstone C is green and pole I of loadstone A is red, can you make a table, like your previous one, which enumerates the possible polarities of each loadstone? How many distinct possibilities are there?

EX. 1.3 (MAGNETISM LABORATORY). Using the Gilbert reading in this chapter as a guide, perform experiments which demonstrate the nature—and if possible the source—of the forces which act between magnetic bodies. If you are not sure where to begin, you might consider beginning with the following types of questions. (a) How do loadstones, magnets, and magnetized pieces of iron behave in the presence (and in the absence) of one another? (b) Do magnetic forces depend upon the relative orientation of the respective bodies? (c) How do the attractive and repulsive magnetic forces depend upon the size, and the distance between, magnetic bodies? (d) Can certain materials provide magnetic shielding? That is: can they prevent magnetic forces from penetrating them? (e) Are the forces acting on two attracting magnetic bodies the same? How might you determine this? (f) How can objects be electrified? And how do electrically charged objects behave? (g) Do magnetic bodies act upon electrified objects? (h) How are magnetic and electric forces similar? How are they different? Do they have the same origin?

To answer questions such as these use loadstones, iron bars and filings, small magnets, compasses or other materials which are available in your laboratory. You might consider making small boats to hold magnets out of styrofoam to observe the forces between magnetic (or non-magnetic) objects. Also, a magnet

placed on a simple balance can be used to measure attractive or repulsive forces with considerable accuracy. When comparing electrical and magnetic forces, glass and rubber rods can be electrified using silk, wool or rabbit fur. The presence of an electrical charge can be measured using an electroscope or small styrofoam pith balls suspended from threads.

Be sure to carefully describe all of your experimental procedures; be as precise and comprehensive as possible. Use drawings whenever they would be helpful in explaining your observations. Some of your experiments will probably be qualitative, but try to be as quantitative as possible. That is: measure masses, forces and direction to the best of your ability. Make plots when possible; for example, you might make a plot relating the attractive force as a function of distance between magnets. What conclusions can you draw from your observations, data and plots? Do your conclusions agree with those of Gilbert? Be sure to support any of your assertions you make or conclusions you draw with detailed and unambiguous experimental evidence.

1.5   Vocabulary

1. Verticity
2. Loadstone
3. Artificer
4. Versorium
5. Salutary
6. Meridian
7. Progeny
8. Potency
9. Zodiac
10. Ecliptic
11. Precession
12. Equinox
13. Declination
14. Torrid
15. Equinoctial
16. Illustrious
17. Terrella
18. Homogenic
19. Virtue
20. Dissimilitude
21. Agnate
22. Austral
23. Detrital

24. Concordance
25. Secular
26. Corporeal
27. Coition
28. Contiguous

Chapter 2
The Life and Death of a Magnet

The loadstone and iron present and exhibit to us wonderful subtile properties.

—William Gilbert

2.1 Introduction

In the first four chapters of Book III of his *De Magnete*, William Gilbert described the laws that govern magnetic bodies. For instance, he noted how a magnetized needle, which is ordinarily oriented along Earth's magnetic meridian, "quits the earth's poles" when placed near a small magnetized sphere, or *terrella*, and reorients itself along the meridian of the terrella. He also explained how the fragments of a loadstone orient themselves after having been separated, and how iron may acquire "verticity," the tendency to orient itself between the poles of a nearby loadstone. In the next eight chapters, included below, Gilbert expands on these topics, carefully explaining the motion of iron wires and compass needles placed near, or even inside of, small magnetized spheres. He also explains how magnetism can be born, so to speak, during the industrial process of smelting, and under what circumstances magnetic properties can change and die.

2.2 Reading: Gilbert, *On the Loadstone and Magnetic Bodies and on the Great Magnet the Earth*

Gilbert, W., *On the Loadstone and Magnetic Bodies and on the Great Magnet the Earth: a New Physiology, Demonstrated with Many Arguments and Experiments*, Bernard Quaritch, London, 1893. Book III.

© Springer International Publishing Switzerland 2016
K. Kuehn, *A Student's Guide Through the Great Physics Texts*,
Undergraduate Lecture Notes in Physics, DOI 10.1007/978-3-319-21816-8_2

Fig. 2.1 The behavior of a
magnetized rod when broken
in half.—[*K.K.*]

2.2.1   Chapter V: Of Magnetizing Stones of Different Shapes

Of a magnetized piece of iron one extremity is north, the other south, and midway is the limit of verticity: such limit, in the globe of the terrella or in a globe of iron, is the equinoctial circle. But if an iron ring be rubbed at one part with a loadstone, then one of the poles is at the point of friction, and the other pole at the opposite side; the magnetic force divides the ring into two parts by a natural line of demarkation, which, though not in form, is in its power and effect equinoctial. But if a straight rod be bent into the form of a ring without welding and unition of the ends, and it be touched in the middle with a loadstone, the ends will be both of the same verticity. Take a ring, whole and unbroken, rubbed with a loadstone at one point; then cut it across at the opposite point and stretch it out straight: again both ends will be of the same verticity,—just like an iron rod magnetized in the middle, or a ring not cohering at the joint.

2.2.2   Chapter VI: What Seems to be a Contrary Movement of Magnetic Bodies is the Regular Tendence to Union

In magnetic bodies nature ever tends to union—not merely to confluence and agglomeration, but to agreement, so that the force that causes rotation and bearing toward the poles may not be disordered, as is shown in various ways in the following example. Let CD be an unbroken magnetic body, with C looking toward B, the earth's north, and D toward A, the earth's south (Fig. 2.1). Now cut it in two in the middle, in the equator, and then E will tend to A and F to B. For, as in the whole, so in the divided stone, nature seeks to have these bodies united; hence the end E properly and eagerly comes together again with F, and the two combine, but E is never joined to D nor F to C, for, in that case, C would have to turn, in opposition to nature, to A, the south, or D to B, the north—which were abnormal and incongruous. Separate the halves of the stone and turn D toward C: they come together nicely and combine. For D tends to the south, as before, and C to the north; E and F, which in the mine were connate parts, are now greatly at variance, for they do not come together on account of material affinity, but take movement and tendence from the form. Hence the ends, whether they be conjoined or separate, tend in the same way, in accordance with magnetic law, toward the earth's poles in the first figure of the stone, whether unbroken or divided as in the second figure; and FE of the second figure, when the two parts come together and form one body, is as perfect a

Fig. 2.2 Gilbert explains the proper procedure for joining magnets using the grafting of plants.—
[*K.K.*]

magnetic mass as was CD when first produced in the mine; and FE, placed on a
float, turn to the earth's poles, and conform thereto in the same way as the unbroken
stone.

This agreement of the magnetic form is seen in the shapes of plants. Let AB be
a branch of ozier or other tree that sprouts readily; and let A be the upper part of
the branch and be the part rootward (Fig. 2.2). Divide the branch at CD. Now, the
extremity CD, if skillfully grafted again on D, begins to grow, just as B and A,
when united, become consolidated and germinate. But if D be grafted in A, or C
on B, they are at variance and grow not at all, but one of them dies because of the
preposterous and unsuitable apposition, the vegetative force, which tends in a fixed
direction, being now forced into a contrary one.

2.2.3   Chapter VII: A Determinate Verticity and a Directive Power Make Magnetic Bodies Accord, and not an Attractional or a Repulsative Force, Nor Strong Coition Alone or Unition

In the equinoctial circle A there is no coition of the ends of a piece of iron wire with
the terrella; at the poles the coition is very strong. The greater the distance from
the equinoctial the stronger is the coition with the terrella itself, and with any part
thereof, not with the pole only. But the pieces of iron are not made to stand because
of any peculiar attracting force or any strong combined force, but because of the
common energy that gives to them direction, conformity, and rotation. For in the
region B not even the minutest bit of iron that weighs almost nothing can be reared
to the perpendicular by the strongest of loadstones, but adheres obliquely (Fig. 2.3).
And just as the terrella attracts variously, with unlike force, magnetic bodies, so, too,
an iron hump (or protuberance-*nasus*) attached to the stone has a different potency
according to the latitude: thus the hump L, as being strongly adherent, will carry a
greater weight than M; and M a heavier weight than N. But neither does the hump
rear to perpendicular a bit of iron except at the poles, as is shown in the figure. The
hump L will hold and lift from the ground 2 ounces of solid iron, yet it is unable
to make a piece of iron wire weighing two grains stand erect; but that would not be

Fig. 2.3 The verticity of an
iron needle depends on its
location on a magnetized
sphere.—[*K.K.*]

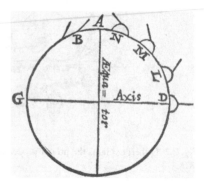

the case if verticity arose from strong attraction, or more properly coition, or from
unition.

2.2.4   Chapter VIII: Of Disagreements Between Pieces of Iron on the Same Pole of a Loadstone; How They May Come Together and be Conjoined

If two pieces of iron wire or two needles above the poles of a terrella adhere, when
about to be raised to the perpendicular they repel each other at their upper ends and
present a furcate appearance; and if one end be forcibly pushed toward the other,
that other retreats and bends back to avoid the association, as shown in the figure
(Fig. 2.4a). *A* and *B*, small iron rods, adhere to the pole obliquely because of their
nearness to each other: either one alone would stand erect and perpendicular. The
reason of the obliquity is that *A* and *B*, having the same verticity, retreat from each
other and fly apart. For if *C* be the north pole of a terrella, then the ends *A* and *B* of
the rods are also north, while the ends in contact with and held fast by the pole *C* are
both south. But let the rods be rather long (say two finger-breadths), and let them be
held together by force: then they cohere and stand together like friends, nor can they
be separated save by force, for they are held fast to each other magnetically, and are
no longer two distinct terminals but one only and one body, like a piece of wire bent
double and made to stand erect.

But here we notice another curious fact, *viz.*, that if the rods be rather short, not
quite a finger's breadth in length, or as long as a barley-corn, they will not unite
on any terms, nor will they stand up together at all, for in short pieces of wire the
verticity at the ends farthest from the terrella is stronger and the magnetic strife
more intense than in longer pieces. Therefore they do not permit any association,
any fellowship. Again, if two light pieces of wire, *A* and *B*, be suspended by a very
slender thread of silk filaments not twisted but laid together, and held at the distance
of one barley-corn's length from the loadstone (Fig. 2.4b), then the opposite ends, *A*
and *E*, situate within the sphere of influence above the pole, go a little apart for the

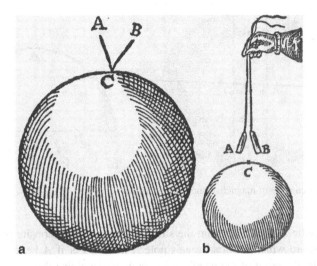

Fig. 2.4 a The upper ends of two short iron wires at the pole of a magnetized sphere stand apart; a single wire would stand erect. **b** The ends of two wires suspended by silk thread above the pole of a magnetized sphere are repelled.—[*K.K.*]

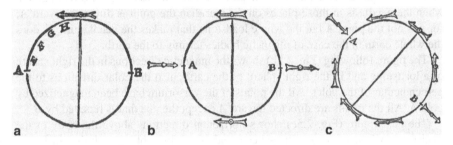

Fig. 2.5 The orientation of a magnetized needle near a spherical loadstone.—[*K.K.*]

same reason, except when they are very near the pole *C* of the stone: in that position the stone attracts them to the one point.

2.2.5   Chapter IX: Directional Figures Showing the Varieties of Rotation

Having now sufficiently shown, according to magnetic laws and principles, the demonstrable cause of the motion toward determinate points, we have next to show the movements. On a spherical loadstone having the poles *A*, *B*, place a rotating needle whose point has been magnetized by the pole *A* (Fig. 2.5a): that point will be directed steadily toward *A* and attracted by *A*, because, having been magnetized by *A*, it accords truly and combines with *A*; and yet it is said to be opposite because

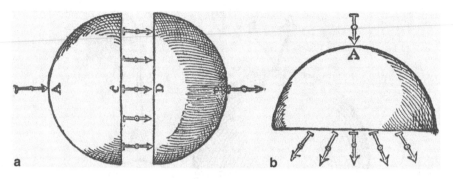

Fig. 2.6 The orientation of magnetized needles placed near a split terrella.—[*K.K.*]

when the needle is separated from the stone it moves to the opposite part of earth from that toward which the loadstone's pole *A* moves. For if *A* be the north pole of the terrella, the point of the needle is its south end, and its other end, the crotch, points to *B*: thus *B* is the loadstone's south pole, while the crotch of the needle is the needle's north end. So, too, the point is attracted by *EFGH* and by every part of a meridian from the equator to the pole, because of the power of directing; and when the needle is in those places on the meridian the point is directed toward *A*; for it is not the point *A* but the whole loadstone that makes the needle turn, as does the whole earth in the case of magnetic bodies turning to the earth.

The figure following (Fig. 2.5b) shows the magnetic directions in the right sphere of a loadstone and in the right sphere of the earth, also the polar directions to the perpendicular of the poles. All the points of the versorium have been magnetized by pole *A*. All the points are directed toward *A* except the one that is repelled by *B*.

The next figure (Fig. 2.5c) shows horizontal directions above the body of the loadstone. All the points that have been made south by rubbing with the north pole or some point around the north pole *A*, turn to the pole *A* and turn away from the south pole *B*, toward which all the crotches are directed.

I call the direction horizontal because it coincides with the plane of the horizon; for nautical and horological instruments are so constructed that the needle shall be suspended or supported in equilibrium on a sharp point, which prevents the dip of the needle, as we shall explain later. And in this way it best serves man's use, noting and distinguishing all the points of the horizon and all the winds. Otherwise in every oblique sphere (whether terrella or earth) the needle and all magnetized bodies would dip below the horizon, and, at the poles, the directions would be perpendicular, as appears from our account of the dip.

The next figure (Fig. 2.6a) shows a spherical loadstone cut in two at the equator; all the points of the needles have been magnetized by pole *A*. The points are directed in the centre of the earth and between the two halves of the terrella, divided in the plane of the equator as shown in the diagram. The case would be the same if the division were made through the plane of a tropic and the separation and distance of the two parts were as above, with the division and separation of the loadstone through the plane of the equinoctial. For the points are repelled by *C*, attracted by *D*, and the needles are parallel, the poles or the verticity at both ends controlling them.

The next figure (Fig. 2.6b) shows half of a terrella by itself, and its directions differing from the directions given by the two parts in the preceding figure, which were placed alongside.

All the points have been magnetized by *A*; all the crotches below, except the middle one, tend not in a right line but obliquely, to the loadstone, for the pole is in the middle of the plane that before was the plane of the equinoctial. All points magnetized by parts of the loadstone away from the pole move to the pole (just as though they had been magnetized by the pole itself) and not to the place of friction, wherever that may be in the whole stone at any latitude betwixt pole and equator. And for this reason there are only two differences of regions—they are north and south as well in the terrella as in the great globe of earth; and there is no east, no west place, no regions truly eastern or western, but, with respect to each other, east and west are simply terms signifying toward the east or west part of the heavens. Hence Ptolemy seems in the *Quadripartitum* to err in laying out eastern and western division, to which he improperly annexes the planets; he is followed by the rabble of philosophasters and astrologers.

2.2.6   Chapter X: Of the Mutation of Verticity and Magnetic Properties, or of the Alteration of the Force Awakened by the Loadstone

Iron excited by the magnetic influx has a verticity that is pretty strong, yet not so stable but that the opposite parts may be altered by the friction not only of a stronger but of the same loadstone, and may lose all their first verticity and take on the opposite. Procure a piece of iron wire and with the self-same poles of a loadstone rub each end equally; pass the wire through a suitable cork float and put it in the water. Then one end of the wire will look toward a pole of the earth whereto that end of the loadstone does not look. But which end of the wire? It will be just the one that was rubbed last. Now rub with the same pole the other end again, and straightway that end will turn in the opposite direction. Again rub the end that first pointed to the pole of the loadstone, and at once that, having, as it were, obtained its orders (*imperium nactus*), will go in the direction opposite to the one it took last. Thus you will be able to alter again and again the property of the iron, and the extremity of it that is last rubbed is master. And now merely hold for a while the north end of the stone near the north end of the wire that was last rubbed, not bringing the two into contact, but at the distance of one, two, or even three finger-breadths, if the stone be a powerful one; again the iron will change its property and will turn to the opposite direction: so it will too, though rather more feebly, if the loadstone be four finger-breadths away. The same results are had in all these experiments whether you employ the south or the north part of the stone. Verticity can also be acquired or altered with plates of gold, silver, and glass between the loadstone and the end of the piece of iron or wire, provided the stone be rather powerful, though the plates of metal be touched neither by the stone nor by the iron. And these changes of verticity occur in cast-iron. But what is imparted or excited by one pole of the loadstone is expelled and annulled by the other, which confers new force. Nor is a stronger loadstone needed to make the iron put off the weaker and sluggish force and to put on a new. Neither is the iron "made drunken" (*inebriatur*) by equal forces of loadstone, so that it becomes "undecided and neutral," as Baptista Porta maintains. But by one same loadstone, and by loadstones endowed with equal power and strength, the force is altered, changed, incited, renewed, driven out. The loadstone itself, however, is not robbed, by friction with another bigger or stronger stone, of its property and verticity, nor is it turned, when on a float, to the opposite direction or to another pole different from that toward which, by its own nature and verticity, it tends. For forces that are innate and long implanted inhere more closely, nor do they easily retire from their ancient seats; and what is the growth of a long period of time is not in an instant reduced to nothing unless that in which it inheres perishes. Nevertheless change comes about in a considerable interval of time, *e.g.*, a year or two, sometimes in a few months—to wit, when a weaker loadstone remains applied, in a way contrary to the order of nature, to a stronger, *i.e.*, with the north pole of one touching the north pole of the other, or the south of one touching the other's south. Under such conditions, in the lapse of time the weaker force declines.

2.2.7 Chapter XI: Of Friction of Iron with the Mid Parts of a Loadstone Between the Poles, and at the Equinoctial Circle of a Terrella

Take a piece of iron wire not magnetized, three finger-widths long (twill be better if its acquired verticity be rather weak or deformed by some process); touch and rub it with the equator of the terrella exactly on the equinoctial line along its whole tract and length, only one end, or both ends, or the whole of the iron, being brought into contact. The wire thus rubbed, run through a cork and float it in water. It will go wandering about without any acquired verticity, and the verticity it had before will be disordered. But if by chance it should be borne in its wavering toward the poles, it will be feebly held still by the earth's poles, and finally will be endowed with verticity by the energy of the earth.

2.2.8 Chapter XII: How Verticity Exists in All Smelted Iron not Excited by the Loadstone

Hitherto we have declared the natural and innate causes and the powers acquired through the loadstone; but now we are to investigate the causes of the magnetic virtue existing in manufactured iron not magnetized by the loadstone. The loadstone and iron present and exhibit to us wonderful subtle properties. It has already oft been shown that iron not excited by the loadstone turns to north and south; further, that it possesses verticity, *i.e.*, distinct poles proper and peculiar to itself, even as the loadstone or iron rubbed with the loadstone. This seemed to us at first strange and incredible: the metal, iron, is smelted out of the ore in the furnace, flows out of the furnace, and hardens in a great mass; the mass is cut up in great workshops and drawn out into iron bars, and from these again the smith fashions all sorts of necessary implements and objects of iron. Thus the same mass is variously worked and transformed into many shapes. What, then, is it that preserves the verticity, or whence is it derived? First take a mass of iron as produced in the first iron-works. Get a smith to shape a mass weighing 2 or 3 ounces, on the anvil, into an iron bar one palm or 9 in. long. Let the smith stand facing the north, with back to the south, so that as he hammers the red-hot iron it may have a motion of extension northward; and so let him complete the task at one or two heatings of the iron (if needed); but ever while he hammers and lengthens it, have him keep the same point of the iron looking north, and lay the finished bar aside in the same direction. In this way fashion 2, 3, or more, yea 100 or 400 bars: it is plain that all the bars so hammered out toward the north and so laid down while cooling will rotate round their centres and when afloat (being passed through suitable pieces of cork) will move about in water, and, when the end is duly reached, will point north. And as an iron bar takes verticity from the direction in which it lies while being stretched, or hammered, or pulled, so too will iron wire when drawn out toward any point of the horizon between east and south or between south and west, or conversely. Nevertheless, when the iron is directed and stretched rather to a point east or west, it takes almost no verticity, or a very faint verticity. This verticity is acquired chiefly through the lengthening. But when inferior iron ore, in which no magnetic properties are apparent, is put in the fire (its position with reference to the world's poles being noted) and there heated for 8 or 10 h, then cooled away from the fire and in the same position with regard to the poles, it acquires verticity according to its position during heating and cooling (Fig. 2.7).

Let a bar of iron be brought to a white heat in a strong fire, in which it lies meridionally, *i.e.*, along the track of a meridian circle; then take it out of the fire and let it cool and return to the original temperature, lying the while in the same position as before: it will come about that, through the like extremities having been directed toward the same poles of the earth, it will acquire verticity; and that the extremity that looked north when the bar, before the firing, was floated in water by means of a cork, if now the same end during the firing and the cooling looked southward, will point to the south. If perchance the turning to the pole should at any time be weak

Fig. 2.7 An iron smelting shop.—[*K.K.*]

and uncertain, put the bar in the fire again, take it out when it has reached white heat,
cool it perfectly as it lies pointing in the direction of the pole from which you wish
it to take verticity, and the verticity will be acquired. Let it be heated again, lying
in the contrary direction, and while yet white-hot lay it down till it cools; for, from
the position in cooling (the earth's verticity acting on it), verticity is infused into
the iron and it turns toward points opposite to the former verticity. So the extremity
that before looked north now turns to the south. For these reasons and in these ways
does the north pole of the earth give to that extremity of the iron which is turned
toward it south verticity; hence, too, that extremity is attracted by the north pole.
And here it is to be observed that this happens with iron not only when it cools lying
in the plane of the horizon, but also at any inclination thereto, even almost up to
perpendicular to the centre of the earth. Thus heated iron more quickly gets energy
(strength) and verticity from the earth in the very process of returning to soundness
in its renascence, so to speak (wherein it is transformated), than when it simply rests
in position. This experiment is best made in winter and in a cold atmosphere, when
the metal returns more surely to the natural temperature than in summer and in warm
climates.

   Let us see also what position alone, without fire and heat, and what mere giving
to the iron a direction toward the earth's poles may do. Iron bars that for a long

time—20 years or more—have lain fixed in the north and south position, as bars are often fixed in buildings and in glass windows-such bars, in the lapse of time, acquire verticity, and whether suspended in air or floated by corks on water turn to the pole toward which they used to be directed, and magnetically attract and repel iron in equilibrium; for great is the effect of long-continued direction of a body toward the poles. This, though made clear by plain experiment, gets confirmation for what we find in a letter written in Italian and appended to a work by Master Philip Costa, of Mantua, also in Italian, *Of the Compounding of Antidotes*, which, translated, is as follows: "At Mantua, an apothecary showed to me a piece of iron completely turned to loadstone, so attracting other iron that it might be compared to a loadstone. But this piece of iron, after it had for a long time supported a terra-cotta ornament on the tower of the Church of San Agostino at Rimini, was at last bent by the force of the winds and so remained for 10 years. The friars, wishing to have it restored to its original shape, gave it to a blacksmith, and in the smithy Master Giulio Cesare, prominent surgeon, discovered that it resembled loadstone and attracted iron." The effect was produced by long-continued lying in the direction of the poles. It is well, therefore, to recall what has already been laid down with regard to alteration of verticity, *viz.*, how that the poles of iron bars are changed when a loadstone simply presents its pole to them and faces them even from some distance. Surely in a like way does that great loadstone the earth affect iron and change verticity. For albeit the iron does not touch the earth's pole nor any magnetic portion of the earth, still the verticity is acquired and altered—not that the earth's pole, that identical point lying 39° of latitude, so great a number of miles, away from this City of London, changes the verticity, but that the entire deeper magnetic mass of the earth which rises between us and the pole, and over which stands the iron—that this, with the energy residing within the field of the magnetic force, the matter of the entire orb conspiring, produces verticity in bodies. For everywhere within the sphere of the magnetic force does the earth's magnetic effluence reign, everywhere does it alter bodies. But those bodies that are most like to it and most closely allied, it rules and controls, as loadstone and iron. For this reason it is not altogether superstitious and silly in many of our affairs and businesses to note the positions and configurations of countries, the points of the horizon and the locations of the stars. For as when the babe is given forth to the light from the mother's womb and gains the power of respiration and certain animal functions, and as the planets and other heavenly bodies, according to their positions in the universe and according to their configuration with the horizon and the earth, do then impart to the new-comer special and peculiar qualities; so a piece of iron, while it is being wrought and lengthened, is affected by the general cause, the earth, to wit; and while it is coming back from the fiery state to its original temperature it becomes imbued with a special verticity according to its position. Long bars have sometimes the same verticity at both ends, and hence they have a wavering and ill-regulated motion on account of their length and of the aforesaid manipulations, just as when an iron wire 4 ft long is rubbed at both ends with one same pole of a loadstone.

2.3   Study Questions

QUES. 2.1. Do the poles of a magnetic body necessarily lie on the extremities of the body?

QUES. 2.2. How does a magnetic body behave when broken in half? Can it be put back together? And in what way is the "magnetic form" of a magnetized body similar to the shape of plants?

QUES. 2.3. Do magnetic bodies experience primarily attractive or repulsive forces?

a) How are iron wires oriented when placed near the surface of a loadstone? Is the force which acts upon them attractional? repulsive? Does the direction of the force depend upon the strength of the loadstone?
b) Do short adjacent iron wires placed at the pole of a magnetized sphere attract one another? Why? Does this demonstrate that the force between two magnetic bodies is primarily a directive, rather than an attractive or repulsive force?

QUES. 2.4. How does a compass needle behave when placed inside a spherical magnet?

a) In Fig. 2.5a, where is the north pole of the terrella? Of the compass needle? Label these on the diagram. What about in Figs. 2.5b and 2.5c?
b) What is the "dip" of a compass needle? Where on the surface of the earth is the dip most pronounced?
c) When a terrella is cut in half, and a compass needle is placed between the halves, in which direction does the needle point?
d) Are both poles of a magnetized hemisphere of equal size and strength?

QUES. 2.5. Can the verticity of previously magnetized iron be altered?

a) Can a very strong loadstone change the verticity of magnetized iron? Can a very weak one? Does the passage of time play a significant role? Can gold or silver plates shield, or block, magnetic effects?
b) What is the process of smelting? Does the verticity of smelted iron depend upon the conditions under which it was smelted? How, specifically?
c) How does Gilbert liken the acquisition of verticity in iron to the configuration of heavenly bodies at the time of an infant's birth? Is this an instructive analogy?

2.4   Exercises

EX. 2.1 (COMPASS AND TERRELLA).   Carefully sketch the orientation of a magnetized compass needle when placed (at several locations) in the vicinity of a spherical terrella (a) which is unbroken, (b) which has been split in half along its equator, and (c) which has been split in half along its meridian. Be sure to clearly mark on your sketches both the point on the terrella and the end of the needle which would seek the earth's north pole.

EX. 2.2 (CURIE TEMPERATURE). What is the nature and the significance of the *Curie temperature* of iron? In particular, what happens to iron at this temperature (from both a macroscopic and a microscopic perspective)? Can you relate the modern definition of the Curie temperature to Chap. 12 of Gilbert's *De Magnete*. Which substance has the highest Curie temperature? Why might this be? And how does a relatively high Curie temperature affect the substance's general behavior?

2.5   Vocabulary

1. Incongruous
2. Confluence
3. Agglomeration
4. Connate
5. Ozier
6. Protuberance
7. Adherent
8. Furcate
9. Nautical
10. Horological
11. Equilibrium
12. Oblique
13. Innate
14. Incite
15. Renascence

Chapter 3
Conservation of Electrical Charge

The electrical fire was not created by friction, but collected,
being really an element diffused among, and attracted by other
matter, particularly by water and metals.

—Benjamin Franklin

3.1   Introduction

Magnetism, though mysterious, was far simpler to systematically study than electricity. This was due largely to the more permanent character of magnetism. For example, a piece of iron, once magnetized, was fairly immune to demagnetization. Electricity, on the other hand was a far more ephemeral phenomenon. For example, a piece of amber electrified by friction retains its electrification for but a short time, especially in humid weather conditions or if not properly insulated. This rendered electrical effects very difficult to reproduce. Perhaps it is no surprise, then, that the science of magnetism proceeded more rapidly than that of electricity. It was not until the eighteenth century—and the development of the frictional machine by Otto von Guericke—that the science of electricity began to proceed more rapidly. So we turn now to one of the most famous of researchers of electricity.

Benjamin Franklin (1706–1790) was born in Boston, Massachusetts. He was one of ten children. He became a printer's apprentice at the age of 12 and then moved to Philadelphia as a young man, where he acquired considerable wealth as the printer and editor of the widely read *Pennsylvania Gazette* and *Poor Richard's Almanac*.[1] An avid reader since youth, Franklin conceived of and founded the first subscription library. He also founded Philadelphia's first fire department, the colonies' first hospital and postal service, the American Philosophical Society, and the Academy and College of Philadelphia, which would later become the University of Pennsylvania. Through his activities as a writer and entrepreneur, Franklin had influenced

---

[1] Much of the biographical information in this brief introduction is from Lemay, J. A. L., *The Life of Benjamin Franklin*, vol. 3, University of Pennsylvania Press, 2009. For a more "bawdy, scurrilous" depiction of his character, see Franklin's satirical essays included in Franklin, B., *Fart Proudly*, Enthea Press, Columbus, Ohio, 1990.

© Springer International Publishing Switzerland 2016
K. Kuehn, *A Student's Guide Through the Great Physics Texts*,
Undergraduate Lecture Notes in Physics, DOI 10.1007/978-3-319-21816-8_3

civic life for many years, but his formal political career began when he was elected
to the Pennsylvania House in 1751. An outspoken critic of British policies—such
as exporting convicted felons to the American colonies—Franklin went on to play
a pivotal role in the American Revolution as a propagandist, political strategist and
foreign diplomat. In 1776, he and four others were appointed by the Second Continental Congress to draft the Declaration of Independence which would sever the
American States from the British Empire.

As indicated in the first letter to his friend Mr. Peter Collinson, included
below, Franklin devoted much of his leisure time to the careful study of natural
philosophy—especially the new science of electricity. It was this work, begun when
he was 40 years old, that brought him fame in continental Europe. It had been known
since ancient times that a small mass of amber, when stroked, would weakly attract
small bits of paper or chaff. Indeed, the greek word for amber is ἤλεκτρον, or
electron, from which the word *electricity* is derived.[2] William Gilbert recognized in
1600 that electrical and magnetic forces were fundamentally different, since magnets did not need to be rubbed in order to produce their forces.[3] Subsequent research
had revealed that different substances, when rubbed, could sometimes even exhibit
subtle *repulsive* forces. Moreover, such frictionally-induced electrical forces could
be amplified: a few decades before Franklin's birth Otto von Guericke had invented
a frictional machine, consisting of a sulphur globe the size of a child's head that
would crackle, spark and attract little fragments of all sorts of metal or chaff when
set spinning and rubbed with dry hands. Immediately before Franklin began his
own electrical researches, von Guericke's frictional machine had been used at the
University of Leyden to electrify "Muschenbroek's wonderful bottle." This also-
called "Leyden jar," consisting of a round glass water-filled flask with a thin metal
wire sticking through a cork at the top, could store electricity and deliver a terrible
shock—like a stroke of lightning—to unsuspecting experimenters who approached
its protruding metal wire.

All of these wonderful electrical phenomena inspired much speculation regarding
their nature and cause. By 1734, the French scientist Charles François de Cisternay
du Fay had supposed that electrical phenomena may be attributed to the behavior
of two distinct and invisible fluids: a *vitreous electricity*, produced by rubbing substances such as glass and gemstones; and *resinous electricity*, produced by rubbing
substances such as amber and hard resin rods. According to Du Fay, each of these
two fluids repelled the same, and attracted the opposite, kind of fluid. It is in this
historical context that Franklin begins his famous *Experiments and Observations on
Electricity*. As mentioned at the outset of his first letter to Mr. Collinson, Franklin
has just received an "electrical tube" in the mail with which to carry out his very
own experimental researches. By rubbing this glass tube with leather or silk and
then drawing a knuckle along the tube, one could readily collect the "electrical fire"
from the tube. When exploring the text below, try to put yourself in Franklin's shoes,

---

[2] Thus, referring to *electricity* is like referring to *ambricity*, as was noted by Franklin himself.

[3] See Chap. 1 of the present volume.

so to speak. That is: try to imagine that you yourself are approaching electricity for the first time. Is it conceivable that electricity is some kind of fluid? What experiments does Franklin perform? What conclusions does he draw? Does he adopt du Fay's two-fluid theory of electricity, or something different? Is he correct? How do you know?

3.2   Reading: Franklin, *Experiments and Observations on Electricity*

Franklin, B., *Experiments and Observations on Electricity*, fourth ed., Henry, David, London, 1769.

3.2.1   Extract of Letter I

From Benj. Franklin, Esq; at Philadelphia, To Peter Collinson, Esq; F.R.S. London. Philadelphia, March 28, 1747.

Sir,

Your kind present of an electric tube, with directions for using it, has put several of us on making electrical experiments, in which we have observed some particular phænomena that we look upon to be new. I shall, therefore communicate them to you in my next, though possibly they may not be new to you, as among the numbers daily employed in those experiments on your side the water, 'tis probable some one or other has hit on the same observations. For my own part, I never was before engaged in any study that so totally engrossed my attention and my time as this has lately done; for what with making experiments when I can be alone, and repeating them to my Friends and Acquaintance, who, from the novelty of the thing, come continually in crouds to see them, I have, during some months past, had little leisure for any thing else.

I am, &c.

B. FRANKLIN.

3.2.2   Letter II

From Mr. Benj. Franklin, Esq; at Philadelphia, To Peter Collinson, Esq; F.R.S. London. July 11, 1747.

Sir,

In my last I informed you that, in pursuing our electrical enquiries, we had observed some particular phænomena, which we looked upon to be new, and of which I promised to give you some account, though I apprehended they might not possibly be new to you, as so many hands are daily employed in electrical experiments on your side the water, some or other of which would probably hit on the same observations.

The first is the wonderful effect of pointed bodies, both in *drawing off* and *throwing off* the electrical fire. For example,

Place an iron shot of 3 or 4 in. diameter on the mouth of a clean dry glass bottle. By a fine silken thread from the cieling, right over the mouth of the bottle, suspend a small cork-ball, about the bigness of a marble; the thread of such a length, as that the cork-ball may rest against the side of the shot. Electrify the shot, and the ball will be repelled to the distance of 4 or 5 in., more or less, according to the quantity of Electricity.—When in this state, if you present to the shot the point of a long, slender, sharp bodkin, at 6 or 8 in. distance, the repellency is instantly destroyed, and the cork flies to the shot. A blunt body must be brought within an inch, and draw a spark to produce the same effect. To prove that the electrical fire is *drawn off* by the point, if you take the blade of the bodkin out of the wooden handle, and fix it in a stick of sealing-wax, and then present it at the distance aforesaid, or if you bring it very near, no such effect follows; but sliding one finger along the wax till you touch the blade, and the ball flies to the shot immediately.—If you present the point in the dark, you will see, sometimes at a foot distance and more, a light gather upon it, like that of a fire-fly, or glow-worm; the less sharp the point, the nearer you must bring it to observe the light; and at whatever distance you see the light, you may draw off the electrical fire, and destroy the repellency.—If a cork-ball so suspended be repelled by the tube, and a point be presented quick to it, though at a considerable distance, 'tis surprizing to see how suddenly it flies back to the tube. Points of wood will do near as well as those of iron, provided the wood is not dry; for perfectly dry wood will no more conduct electricity than sealing-wax.

To shew that points will *throw off*[4] as well as *draw off* the electrical fire; lay a long sharp needle upon the shot, and you cannot electrise the shot so as to make it repel the cork-ball.[5]—Or fix a needle to the end of a suspended gun-barrel, or iron-rod, so as to point beyond it like a little bayonet; and while it remains there, the gun-barrel, or rod, cannot by applying the tube to the other end be electrised so as to give a spark, the fire continually running out silently at the point. In the dark you may see it make the same appearance as it does in the case before-mentioned.

---

[4] This power of points to *throw off* the electrical fire, was first communicated to me by my ingenious friend Mr *Thomas Hopkinson*; since deceased, whose virtue and integrity, in every station of life, public and private, will ever make his Memory dear to those who knew him, and knew how to value him.

[5] This was Mr *Hopkinson's* Experiment, made with an expectation of drawing a more sharp and powerful spark from the point, as from a kind of focus, and he was surprised to find little or none.

The repellency between the cork-ball and the shot is likewise destroyed. (1) By sifting fine sand on it; this does it gradually. (2) By breathing on it. (3) By making a smoke about it from burning wood.[6] (4) By candle-light, even though the candle is at a foot distance: these do it suddenly.—The light of a bright coal from a wood fire; and the light of a red-hot iron do it likewise; but not at so great a distance. Smoke from dry rosin dropt on hot iron, does not destroy the repellency; but is attracted by both shot and cork-ball, forming proportionable atmospheres round them, making them look beautifully, somewhat like some of the figures in *Burnet's* or *Whiston's* Theory of the Earth.

N. B. This experiment should be made in a closet, where the air is very still, or it will be apt to fail.

The light of the sun thrown strongly on both cork and shot by a looking-glass for a long time together, does not impair the repellency in the least. This difference between fire-light and sun-light is another thing that seems new and extraordinary to us.[7]

We had for some time been of opinion, that the electrical fire was not created by friction, but collected, being really an element diffused among, and attracted by other matter, particularly by water and metals. We had even discovered and demonstrated its afflux to the electrical sphere, as well as its efflux, by means of little light windmill wheels made of stiff paper vanes, fixed obliquely, and turning freely on fine wire axes. Also by little wheels of the same matter, but formed like water-wheels. Of the disposition and application of which wheels, and the various phænomena resulting, I could, if I had time, fill you a sheet.[8] The impossibility of electrising one's self (though standing on wax) by rubbing the tube, and drawing the fire from it; and the manner of doing it, by passing the tube near a person or thing standing on the floor, &c. had also occurred to us some months before Mr. *Watson's* ingenious *Sequel* came to hand, and these were some of the new things I intended to have communicated to you.—But now I need only mention some particulars not hinted in that piece, with our reasonings thereupon: though perhaps the latter might well enough be spared.

---

[6] We suppose every particle of sand, moisture, or smoke, being first attracted and then repelled, carries off with it a portion of the electrical fire; but that the same still subsists in those particles, till they communicate it to something else, and that it is never really destroyed.—So when water is thrown on common fire, we do not imagine the element is thereby destroyed or annihilated, but only dispersed, each particle of water carrying off in vapour its portion of the fire, which it had attracted and attached to itself.

[7] This different Effect probably did not arise from any difference in the light, but rather from the particles separated from the candle, being first attracted and then repelled, carrying off the electric matter with them; and from the rarefying the air, between the glowing coal or red-hot iron, and the electrified shot, through which rarified air the electric fluid could more readily pass.

[8] These experiments with the wheels were made and communicated to me by my worthy and ingenious friend Mr *Philip Syng*; but we afterwards discovered that the motion of those wheels was not owing to any afflux or efflux of the electric fluid, but to various circumstances of attraction and repulsion. 1750.

1. A person standing on wax, and rubbing the tube, and another person on wax drawing the fire, they will both of them, (provided they do not stand so as to touch one another) appear to be electrised, to a person standing on the floor; that is, he will perceive a spark on approaching each of them with his knuckle.
2. But if the persons on wax touch one another during the exciting of the tube, neither of them will appear to be electrised.
3. If they touch one another after exciting the tube, and drawing the fire as aforesaid, there will be a stronger spark between them than was between either of them and the person on the floor.
4. After such strong spark, neither of them discover any electricity.

These appearances we attempt to account for thus: We suppose, as aforesaid, that electrical fire is a common element, of which everyone of the three persons above mentioned has his equal share, before any operation is begun with the tube. A, who stands on wax and rubs the tube, collects the electrical fire from himself into the glass; and his communication with the common stock being cut off by the wax, his body is not again immediately supply'd. B, (who stands on wax likewise) passing his knuckle along near the tube, receives the fire which was collected by the glass from A; and his communication with the common stock being likewise cut off, he retains the additional quantity received.—To C, standing on the floor, both appear to be electrised: for he having only the middle quantity of electrical fire, receives a spark upon approaching B, who has an over quantity; but gives one to A, who has an under quantity. If A and B approach to touch each other, the spark is stronger, because the difference between them is greater: After such touch there is no spark between either of them and C, because the electrical fire in all is reduced to the original equality. If they touch while electrising, the equality is never destroy'd, the fire only circulating. Hence have arisen some new terms among us: we say B, (and bodies like circumstanced) is electrised *positively*; A, *negatively*. Or rather, B is electrised *plus*; A, *minus*. And we daily in our experiments electrise bodies *plus* or *minus*, as we think proper.—To electrise *plus* or *minus*, no more needs to be known than this, that the parts of the tube or sphere that are rubbed, do, in the instant of the friction, attract the electrical fire, and therefore take it from the thing rubbing: the same parts immediately, as the friction upon them ceases, are disposed to give the fire they have received, to any body that has less. Thus you may circulate it, as Mr. *Watson* has shewn; you may also accumulate or subtract it, upon, or from any body, as you connect that body with the rubber or with the receiver, the communication with the common stock being cut off. We think that ingenious gentleman was deceived when he imagined (in his *Sequel*) that the electrical fire came down the wire from the cieling to the gun barrel, thence to the sphere, and so electrised the machine and the man turning the wheel, &c. We suppose it was driven off, and not brought on through that wire; and that the machine and man, &c; were electrised *minus*; *i.e.* had less electrical fire in them than things in common.

As the vessel is just upon failing, I cannot give you so large an account of *American* Electricity as I intended: I shall only mention a few particulars more.—We find granulated lead better to fill the phial with, than water, being easily warmed, and keeping warm and dry in damp air.—We fire spirits with the wire of the phial.—We light candles, just blown out, by drawing a spark among the smoke between the wire and snuffers.—We represent lightning, by passing the wire in the dark, over a china plate that has gilt flowers, or applying it to gilt frames of looking-glasses, &c.—We electrise a person 20 or more times running, with a touch of the finger on the wire, thus: He stands on wax. Give him the electrised bottle in his hand. Touch the wire with your finger, and then touch his hand or face; there are sparks every time.[9]—We increase the force of the electrical kiss vastly, thus: Let *A* and *B* stand on wax; or *A* on wax, and *B* on the floor; give one of them the electrised phial in hand; let the other take hold of the wire; there will be a small spark; but when their lips approach, they will be struck and shock'd. The same if another gentleman and lady, *C* and *D*, standing also on wax, and joining hands with *A* and *B*, salute or shake hands. We suspend by fine silk thread a counterfeit spider, made of a small piece of burnt cork, with legs of linnen thread, and a grain or two of lead stuck in him, to give him more weight. Upon the table, over which he hangs, we stick a wire upright, as high as the phial and wire, 4 or 5 in. from the spider: then we animate him, by setting the elec-trified phial at the same distance on the other side of him; he will immediately fly to the wire of the phial, bend his legs in touching it; then spring off, and fly to the wire in the table: thence again to the wire of the phial, playing with his legs against both, in a very entertaining manner, appearing perfectly alive to persons unacquainted. He will continue this motion an hour or more in dry weather.—We electrify, upon wax in the dark, a book that has a double line of gold round upon the covers, and then apply a knuckle to the gilding; the fire appears every where upon the gold like a flash of lightning: not upon the leather, nor, if you touch the leather instead of the gold. We rub our tubes with buckskin, and observe always to keep the same side to the tube, and never to sully the tube by handling; thus they work readily and easily, without the least fatigue, especially if kept in tight pasteboard cases, lined with flan-nel, and fitting close to the tube.[10] This I mention, because the European papers on Electricity frequently speak of rubbing the tube as a fatiguing exercise. Our spheres are fixed on iron axes, which pass through them. At one end of the axis there is a small handle, with which you turn the sphere like a common grindstone. This we find very commodious, as the machine takes up but little room, is portable, and may be enclosed in a tight box, when not in use. Tis true, the sphere does not turn so swift as when the great wheel is used: but swiftness we think of little importance, since a few turns will charge the phial, &c. sufficiently.[11]

I am, &c.

B. FRANKLIN.

---

[9] By taking a spark from the wire, the electricity within the bottle is diminished; the outside of the bottle then draws some from the person holding it, and leaves him in the negative state. Then when his hand or face is touch'd, an equal quantity is restored to him from the person touching.

[10] Our tubes are made here of green glass, 27 or 30 in. long, as big as can be grasped.

[11] This simple easily-made machine was a contrivance of Mr. *Syng's.*

3.3 Study Questions

QUES. 3.1. How can the so-called "electrical fire" be drawn off, or perhaps thrown off, of an electrified object?

a) Describe Franklin's apparatus. How does he know that the lead ball is, in fact, electrified? Why does the bottle need to be clean and dry?
b) What methods are able to remove the electrical fire from the lead ball? How does he know when this has occurred? Which is more effective, sharp or blunt objects? Do they need to touch the ball?

QUES. 3.2. Can electrical fire be created or destroyed? Or does its overall quantity remain unalterable (*i.e.* conserved)?

a) Is electrical fire created when a person standing on a wax block rubs a glass tube with a strip of leather or silk? Is he or she electrified? Is the tube?
b) Now if another person, also standing on the block of wax, "draws the fire" from the rubbed glass tube with his knuckle, is he or she electrified?
c) Under what circumstances can each of the three individuals involved in Franklin's experiment receive a shock? How does Franklin interpret all of these observations?
d) What new terms does he introduce in so doing? In what sense does Franklin establish a "conservation law" for electrical fire? Is Franklin's theory sensible? Is it true?

3.4 Exercises

EX. 3.1 (CHARGED BALLS).   Suppose that a lead ball the size of your fist is placed atop the mouth of a dry glass bottle. A glass rod is rubbed briefly with a sheet of dry buckskin.

a) What is the charge of the glass rod? The buckskin?
b) If the glass rod is briefly touched to the lead ball, what happens? Is the lead ball electrified? Plus or minus? Is the glass rod still electrified?
c) If the buckskin is briefly touched to a second, identical, lead ball (sitting atop a second bottle), what happens? Is the lead ball electrified? Plus or minus? Is the buckskin still electrified?
d) What can be said about the sum of all the charge on the two balls, the rod and the buckskin?
e) Now suppose that the two bottles are situated so that the charged balls are separated by just 3 in. A tiny neutral aluminum ball is suspended from a silk thread in the region between them. What happens? Does it move? Which way? If so, when does it stop?

3.5   Vocabulary

1. Engross
2. Apprehend
3. Leisure
4. Bodkin
5. Buckskin
6. Sully
7. Fatigue
8. Commodious

Chapter 4
Müschenbroek's Wonderful Bottle

So wonderfully are these two states of Electricity, the plus and
the minus, combined and balanced in this miraculous bottle!
—Benjamin Franklin

4.1 Introduction

In his second letter to Mr. Collinson, Franklin presented his one-fluid theory of electricity. Contrary to Du Fay's two-fluid theory, Franklin suggested that all objects contain within them a certain quantity of the so-called "electrical fire". While the overall quantity of this electrical fire is unchanged—it is a conserved quantity—its distribution need not be uniform everywhere. Objects having a (perhaps temporary) surplus are said to be "plus"; objects having a deficit are said to be "minus". Such charge separation commonly occurs with friction between different objects, as when glass is rubbed with silk (the glass becomes "plus" and the silk becomes "minus"), or amber with fur (the amber becomes "minus" and the fur becomes "plus"). In this way, Franklin was able to account for many electrical phenomena by introducing the concept of *positive and negative electricity*. At this point, you might pause to consider whether (or to what extent) Franklin's theory of electricity is different than the modern view.

Now, in his third letter to Mr. Collinson, Franklin deploys his one-fluid theory so as to understand the working of "Mueschenbroek's wonderful bottle". Where might the bottle store its electricity when electrified? In the water? In the glass? In the central wire? When reading through these letters, it helps to recognize that materials may be divided into two classes: *conductors*, such as iron and saltwater, which readily transport electricity through them; and *insulators*, such as wax and glass, which do not. In a subsequent letter Franklin employs these terms, but in the letter below Franklin uses the older terminology, referring to insulators as *electrics per se* and to conductors as *non-electrics*.

© Springer International Publishing Switzerland 2016
K. Kuehn, *A Student's Guide Through the Great Physics Texts,*
Undergraduate Lecture Notes in Physics, DOI 10.1007/978-3-319-21816-8_4

4.2   Reading: Franklin, *Experiments and Observations on Electricity*

Franklin, B., *Experiments and Observations on Electricity*, fourth ed., Henry, David, London, 1769. Letter III, and selections from Letters IV and XXXVIII.

4.2.1   Letter III

From Benj. Franklin, Esq; at Philadelphia, To Peter Collinson, Esq; F.R.S. London. Sept. 1, 1747.

Sir,

The necessary trouble of copying long letters, which, perhaps, when they come to your hands may contain nothing new, or worth your reading, (so quick is the progress made with you in Electricity) half discourages me from writing any more on that subject. Yet I cannot forbear adding a few observations on M. *Muschenbroek*'s wonderful bottle.

1. The non-electric contain'd in the bottle differs when electrised from a non-electric electrised out of the bottle, in this: that the electrical fire of the latter is accumulated *on its surface*, and forms an electrical atmosphere round it of considerable extent; but the electrical fire is crowded *into the substance* of the former, the glass confining it.[1]

2. At the same time that the wire and top of the bottle, &c. is electrised *positively* or *plus*, the bottom of the bottle is electrised *negatively* or *minus*, in exact proportion: *i.e.* whatever quantity of electrical fire is thrown in at the top, an equal quantity goes out of the bottom.[2] To understand this, suppose the common quantity of electricity in each part of the bottle, before the operation begins, is equal to 20; and at every stroke of the tube, suppose a quantity equal to 1 is thrown in; then, after the first stroke, the quantity contain'd in the wire and upper part of the bottle will be 21, in the bottom 19. After the second, the upper part will have 22, the lower 18, and so on, till, after 20 strokes the upper part will have a quantity of electrical fire equal to 40, the lower part none: and then the operation ends: for no more can be thrown into the upper part, when no more can be driven out of the lower part. If you attempt to throw more in, it is spued back through the wire, or flies out in loud cracks through the sides of the bottle.

---

[1] See this opinion rectified in Letter IV §16 and 17. The fire in the bottle was found by subsequent experiments not to be contained in the non-electric, but *in the glass*. 1748.

[2] What is said here, and after, of the *top* and *bottom* of the bottle, is true of the *inside* and *outside* surfaces, and should have been so expressed.

3. The equilibrium cannot be restored in the bottle by *inward* communication or contact of the parts; but it must be done by a communication formed *without* the bottle between the top and bottom, by some non-electric, touching or approaching both at the same time; in which case it is restored with a violence and quickness inexpressible; or, touching each alternately, in which case the equilibrium is restored by degrees.

4. As no more electrical fire can be thrown into the top of the bottle, when all is driven out of the bottom, so in a bottle not yet electrised, none can be thrown into the top, when none *can* get out at the bottom; which happens either when the bottom is too thick, or when the bottle is placed on an electric *per se*. Again, when the bottle is electrised, but little of the electrical fire can be *drawn out* from the top, by touching the wire, unless an equal quantity can at the same time *get in* at the bottom.[3] Thus, place an electrised bottle on clean glass or dry wax, and you will not, by touching the wire, get out the fire from the top. Place it on a non-electric, and touch the wire, you will get it out in a short time; but soonest when you form a direct communication as above.

   So wonderfully are these two states of Electricity, the *plus* and *minus*, combined and balanced in this miraculous bottle! situated and related to each other in a manner that I can by no means comprehend! If it were possible that a bottle should in one part contain a quantity of air strongly comprest, and in another part a perfect vacuum, we know the equilibrium would be instantly restored *within*. But here we have a bottle containing at the same time a *plenum* of electrical fire, and a *vacuum* of the same fire; and yet the equilibrium cannot be restored between them but by a communication *without*! though the *plenum* presses violently to expand, and the hungry vacuum seems to attract as violently in order to be filled.

5. The shock to the nerves (or convulsion rather) is occasioned by the sudden passing of the fire through the body in its way from the top to the bottom of the bottle. The fire takes the shortest course, as Mr. *Watson* justly observes: But it does not appear from experiment that in order for a person to be shocked, a communication with the floor is necessary: for he that holds the bottle with one hand, and touches the wire with the other, will be shock'd as much, though his shoes be dry, or even standing on wax, as otherwise. And on the touch of the wire, (or of the gunbarrel, which is the same thing) the fire does not proceed from the touching finger to the wire, as is supposed, but from the wire to the finger, and passes through the body to the other hand, and so into the bottom of the bottle.

---

[3] See the preceding note, relating to *top* and *bottom*.

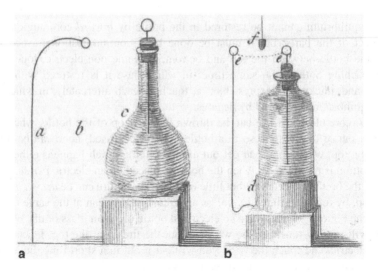

Fig. 4.1 Apparatus for Franklin's Experiments II (**a**) and III (**b**).—[*K.K.*]

4.2.2   Experiments Confirming the Above

Experiment I.   Place an electrised phial on wax; a small cork-ball suspended by a dry silk thread held in your hand, and brought near to the wire, will first be attracted, and then repelled: when in this state of repellency, sink your hand, that the ball may be brought towards the bottom of the bottle; it will be there instantly and strongly attracted, till it has parted with its fire.

If the bottle had a positive electrical atmosphere, as well as the wire, an electrified cork would be repelled from one as well as from the other.

Experiment II.   (Fig. 4.1a) From a bent wire (*a*) sticking in the table, let a small linen thread (*b*) hang down within half an inch of the electrised phial (*c*). Touch the wire of the phial repeatedly with your finger, and at every touch you will see the thread instantly attracted by the bottle. (This is best done by a vinegar cruet, or some such belly'd bottle). As soon as you draw any fire out from the upper part, by touching the wire, the lower part of the bottle draws an equal quantity in by the thread.

Experiment III.   (Fig. 4.1b) Fix a wire in the lead, with which the bottom of the bottle is armed (*d*) so as that bending upwards, its ring end may be level with the top or ring-end of the wire in the cork (*e*) and at 3 or 4 in. distance. Then electricise the bottle, and place it on wax. If a cork suspended by a silk thread (*f*) hang between these two wires, it will play incessantly from one to the other, till the bottle is no longer electrised; that is, it fetches

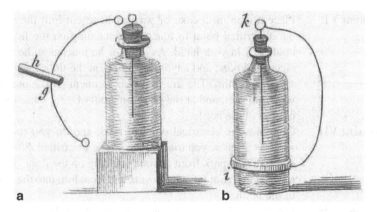

Fig. 4.2 Apparatus for Franklin's Experiments IV (**a**) and V (**b**).—[*K.K.*]

and carries fire from the top to the bottom[4] of the bottle, till the equilibrium is restored.

Experiment IV. (Fig. 4.2a) Place an electrised phial on wax; take a wire (*g*) in form of a *C*, the ends at such a distance when bent, as that the upper may touch the wire of the bottle, when the lower touches the bottom: stick the outer part on a stick of sealing-wax (*h*), which will serve as a handle; then apply the lower end to the bottom of the bottle, and gradually bring the upper end near the wire in the cork. The consequence is, spark follows spark till the equilibrium is restored. Touch the top first, and on approaching the bottom with the other end, you have a constant stream of fire from the wire entering the bottle. Touch the top and bottom together, and the equilibrium will instantly be restored; the crooked wire forming the communication.

Experiment V. (Fig. 4.2b) Let a ring of thin lead, or paper, surround a bottle (*i*) even at some distance from or above the bottom. From that ring let a wire proceed up, till it touch the wire of the cork (*k*). A bottle so fixt cannot by any means be electrised: the equilibrium is never destroyed: for while the communication between the upper and lower parts of the bottle is continued by the outside wire, the fire only circulates: what is driven out at bottom, is constantly supplied from the top.[5] Hence a bottle cannot be electrised that is foul or moist on the outside, if such moisture continue up to the cork or wire.

---

[4] *i.e.* from the inside to the outside.

[5] See the preceding note.

Experiment VI.   Place a man on a cake of wax, and present him the wire of the electrified phial to touch, you standing on the floor, and holding it in your hand. As often as he touches it, he will be electrified *plus*; and anyone standing on the floor may draw a spark from him. The fire in this experiment passes out of the wire into him; and at the same time out of your hand into the bottom of the bottle.

Experiment VII.   Give him the electrical phial to hold; and do you touch the wire; as often as you touch it he will be electrified *minus*, and may draw a spark from anyone standing on the floor. The fire now passes from the wire to you, and from him into the bottom of the bottle.

Experiment VIII. Lay two books on two glasses, back towards back, 2 or 3 in. distant. Set the electrified phial on one, and then touch the wire; that book will be electrified *minus*; the electrical fire being drawn out of it by the bottom of the bottle. Take off the bottle, and holding it in your hand, touch the other with the wire; that book will be electrified *plus*; the fire passing into it from the wire, and the bottle at the same time supplied from your hand. A suspended small cork-ball will play between these books till the equilibrium is restored.

Experiment IX.   When a body is electrised *plus*, it will repel an electrified feather or small cork-ball. When *minus* (or when in the common state) it will attract them, but stronger when *minus* than when in the common state, the difference being greater.

Experiment X.   Though, as in *Experiment* VI, a man standing on wax may be electrised a number of times by repeatedly touching the wire of an electrised bottle (held in the hand of one standing on the floor) he receiving the fire from the wire each time: yet holding it in his own hand, and touching the wire, though he draws a strong spark, and is violently shocked, no electricity remains in him; the fire only passing through him, from the upper to the lower part of the bottle. Observe, before the shock, to let some one on the floor touch him to restore the equilibrium in his body; for in taking hold of the bottom of the bottle, he sometimes becomes a little electrised *minus*, which will continue after the shock, as would also any *plus* Electricity, which he might have given him before the shock. For restoring the equilibrium in the bottle, does not at all affect the Electricity in the man through whom the fire passes; that Electricity is neither increased nor diminished.

Fig. 4.3 Apparatus for
Franklin's Experiment
XI.—[*K.K.*]

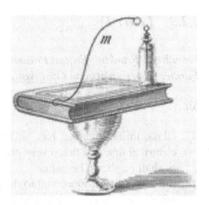

Experiment XI.   The passing of the electrical fire from the upper to the lower
part[6] of the bottle, to restore the equilibrium, is rendered
strongly visible by the following pretty experiment. Take a
book whose covering is filletted with gold; bend a wire of 8 or
10 in. long, in the form of (m) Fig. 4.3. Slip it on the end of
the cover of the book, over the gold line, so as that the shoulder of it may press upon one end of the gold line, the ring up,
but leaning towards the other end of the book. Lay the book
on a glass or wax, and on the other end of the gold lines set
the bottle electrised: then bend the springing wire, by pressing it with a stick of wax till its ring approaches the ring of
the bottle wire, instantly there is a strong spark and stroke,
and the whole line of gold, which completes the communication, between the top and bottom of the bottle, will appear a
vivid flame, like the sharpest lightning. The closer the contact
between the shoulder of the wire, and the gold at one end of
the line, and between the bottom of the bottle and the gold at
the other end, the better the experiment succeeds. The room
should be darkened. If you would have the whole filletting
round the cover appear in fire at once, let the bottle and wire
touch the gold in the diagonally opposite corners.

I am, &c.

B. FRANKLIN.

---

[6] *i.e.* from the *inside* to the *outside*.

4.2.3 Letter IV

From Benj. Franklin, Esq; at Philadelphia, To Peter Collinson, Esq; F.R.S. London.
Further Experiments and Observations in Electricity. 1748

Sir,...[7]

13. Glass, in like manner, has, within its substance, always the same quantity of electrical fire, and that a very great quantity in proportion to the mass of glass, as shall be shewn hereafter.

14. This quantity, proportional to the glass, it strongly and obstinately retains, and will have neither more nor less though it will suffer a change to be made in its parts and situation; *i.e.* we may take away part of it from one of the sides, provided we throw an equal quantity into the other.

15. Yet when the situation of the electrical fire is thus altered in the glass; when some has been taken from one side, and some added to the other, it will not be at rest or in its natural state, till it is restored to its original equality. And this restitution cannot be made through the substance of the glass, but must be done by a non-electric communication formed without, from surface to surface.

16. Thus, the whole force of the bottle, and power of giving a shock, is in the GLASS ITSELF; the non-electrics in contact with the two surfaces, serving only to *give* and *receive* to and from the several parts of the glass; that is, to give on one side, and take away from the other.

17. This was discovered here in the following manner: Purposing to analyze the electrified bottle, in order to find wherein its strength lay, we placed it on glass, and drew out the cork and wire which for that purpose had been loosely put in. Then taking the bottle in one hand, and bringing a finger of the other near its mouth, a strong spark came from the water, and the shock was as violent as if the wire had remained in it, which shewed that the force did not lie in the wire. Then to find if it resided in the water, being crouded into and condensed in it, as confined by the glass, which had been our former opinion, we electrified the bottle again, and placing it on glass, drew out the wire and cork as before; then taking up the bottle, we decanted all its water into an empty bottle, which likewise stood on glass; and taking up that other bottle, we expected, if the force resided in the water, to find a shock from it; but there was none. We judged then that it must either be lost in decanting, or remain in the first bottle. The latter we found to be true; for that bottle on trial gave the shock, though filled up as it stood with fresh unelectrified water from a tea-pot. To find, then, whether glass had this property merely as glass, or whether the form contributed anything to it; we took a pane of sash-glass, and laying it on the hand, placed a plate of lead on its upper surface; then electrified that plate, and bringing a finger to it, there was a spark and shock. we then took two plates of lead of

---

[7] The first twelve sections of this letter have been omitted for the sake of brevity.—[*K.K.*]

equal dimensions, but less than the glass by 2 in. every way, and electrified the glass between them, by electrifying the uppermost lead; then separated the glass from the lead, in doing which, what little fire might be in the lead was taken out, and the glass being touched in the electrified parts with a finger, afforded only very small pricking sparks, but a great number of them might be taken from different places. Then dexterously placing it again between the leaden plates, and completing a circle between the two surfaces, a violent shock ensued. Which demonstrated the power to reside in glass as glass, and that the non-electrics in contact served only, like the armature of a loadstone, to unite the force of the several parts, and bring them at once to any point desired: it being the property of a non-electric, that the whole body instantly receives or gives what electrical fire is given to or taken from any one of its parts.

18. Upon this we made what we called an *electrical-battery*, consisting of eleven panes of large sash-glass, arm'd with thin leaden plates, pasted on each side, placed vertically, and supported at 2 in. distance on silk cords, with thick hooks of leaden wire, one from each side, standing upright, distant from each other, and convenient communications of wire and chain, from the giving side of one pane, to the receiving side of the other; that so the whole might be charged together, and with the same labour as one single pane; and another contrivance to bring the giving sides, after charging, in contact with one long wire, and the receivers with another, which two long wires would give the force of all the plates of glass at once through the body of any animal forming the circle with them.[8] The plates may also be discharged separately, or any number together that is required. But this machine is not much used, as not perfectly answering our intention with regard to the ease of charging, for the reason given, Sect. 10.[9] We made also of large glass panes, magical pictures, and self-moving animated wheels, presently to be described.

4.2.4   Letter XXXVIII

To Mr. Kinnersley, in answer to the foregoing. London, Feb. 20, 1762.

Sir,...[10]

You know I have always look'd upon and mentioned the equal repulsion in cases of positive and of negative electricity, as a phænomenon difficult to be explained.

---

[8] Franklin has built a set of parallel-plate capacitors. Each capacitor consists of a single pane of glass sandwiched between two sheets of lead. In the first configuration, Franklin strings these capacitors together in *series*; in the second he strings them together in *parallel*. The latter configuration provides the higher total capacitance.—[K.K.]

[9] Franklin's Sect. 10 has been omitted for the sake of brevity.—[K.K.]

[10] Only a few of the middle paragraphs from this letter have been included in this volume for the sake of brevity.—[K.K.]

I have sometimes, too, been inclined, with you, to resolve all into attraction; but besides that attraction seems in itself as unintelligible as repulsion, there are some appearances of repulsion that I cannot so easily explain by attraction; this for one instance. When the pair of cork balls are suspended by flaxen threads, from the end of the prime conductor, if you bring a rubbed glass tube near the conductor, but without touching it, you see the balls separate, as being electrified positively;[11] and yet you have communicated no electricity to the conductor, for, if you had, it would have remained there, after withdrawing the tube; but the closing of the balls immediately thereupon, shows that the conductor has no more left in it than its natural quantity. Then again approaching the conductor with the rubbed tube, if, while the balls are separated, you touch with a finger that end of the conductor to which they hang, they will come together again, as being, with that part of the conductor, brought to the same state with your finger, *i.e.* the natural state. But the other end of the conductor, near which the tube is held, is not in that state, but in the negative state, as appears on removing the tube; for then part of the natural quantity left at the end near the balls, leaving that end to supply what is wanting at the other, the whole conductor is found to be equally in the negative state. Does not this indicate that the electricity of the rubbed tube had repelled the electric fluid, which was diffused in the conductor while in its natural state, and forced it to quit the end to which the tube was brought near, accumulating itself on the end to which the balls were suspended? I own I find it difficult to account for its quitting that end, on the approach of the rubbed tube, but on the supposition of repulsion; for, while the conductor was in the same state with the air, *i.e.* the natural state, it does not seem to me easy to suppose, that an attraction should suddenly take place between the air and the natural quantity of the electric fluid in the conductor, so as to draw it to, and accumulate it on the end opposite to that approached by the tube; since bodies, possessing only their natural quantity of that fluid, are not usually seen to attract each other, or to affect mutually the quantities of electricity each contains.

There are likewise appearances of repulsion in other parts of nature. Not to mention the violent force with which the particles of water, heated to a certain degree, separate from each other, or those of gunpowder, when touch'd with the smallest spark of fire, there is the seeming repulsion between the same poles of the magnet, a body containing a subtle moveable fluid, in many respects analogous to the electric fluid. If two magnets are so suspended by strings, as that their poles of the same denomination are opposite to each other, they will separate, and continue so; or if you lay a magnetic steel bar on a smooth table, and approach it with another parallel to it, the poles of both in the same position, the first will recede from the second, so as to avoid the contact, and may thus be push'd (or at least appear to be push'd) off the table. Can this be ascribed to the attraction of any surrounding body or matter drawing them asunder, or drawing the one away from the other? If not,

---

[11] The Leyden jar is here acting as a rudimentary electrometer. By observing whether the two balls suspended from the prime conductor (the wire passing through the top cork), one can detect the presence of electrification.—[*K.K.*]

and repulsion exists in nature, and in magnetism, why may it not exist in electricity? We should not, indeed, multiply causes in philosophy without necessity; and the greater simplicity of your hypothesis would recommend it to me, if I could see that all appearances might be solved by it.[12] But I find, or think I find, the two causes more convenient than one of them alone. Thus I would solve the circular motion of your horizontal stick, supported on a pivot, with two pins at their ends, pointing contrary ways, and moving in the same direction when electrified, whether positively or negatively: When positively, the air opposite the points being electrified positively, repels the points; when negatively, the air opposite the points being also, by their means, electrified negatively, attraction takes place between the electricity in the air behind the heads of the pins, and the negative pins, and so they are, in this case, drawn in the same direction that in the other they were driven. You see I am willing to meet you half way, a complaisance I have not met with in our brother *Nollet*, or any other hypothesis-maker, and therefore may value myself a little upon it, especially as they say I have some ability in defending even the wrong side of a question, when I think fit to take it in hand.

What you give as an established law of the electric fluid, "That quantities of different densities mutually attract each other, in order to restore the equilibrium," is, I think, not well founded, or else not well express'd. Two large cork balls, suspended by silk strings, and both well and equally electrified, separate to a great distance. By bringing into contact with one of them, another ball of the same size, suspended likewise by silk, you will take from it half its electricity. It will then, indeed, hang at a less distance from the other, but the full and the half quantities will not appear to attract each other, that is, the balls will not come together. Indeed, I do not know any proof we have, that one quantity of electric fluid is attracted by another quantity of that fluid, whatever difference there may be in their densities. And, supposing in nature, a mutual attraction between two parcels of any kind of matter, it would be strange if this attraction should subsist strongly while those parcels were unequal, and cease when more matter of the same kind was added to the smallest parcel, so as to make it equal to the biggest. By all the laws of attraction in matter, that we are acquainted with, the attraction is stronger in proportion to the increase of the masses, and never in proportion to the difference of the masses. I should rather think the law would be, "That the electric fluid is attracted strongly by all other matter that we know of, while the parts of that fluid mutually repel each other." Hence its being equally diffused (except in particular circumstances) throughout all other matter. But this you jokingly call "electrical orthodoxy." It is so with some at present, but not with all; and, perhaps, it may not always be orthodoxy with any body. Opinions are continually varying, where we cannot have mathematical evidence of the nature of things; and they must vary. Nor is that variation without its use, since it occasions a more thorough discussion, whereby error is often dissipated, true knowledge is encreased, and its principles become better understood and more firmly established.

---

[12] See Newton's Rules of Reasoning, described at the outset of Book III of his *Principia*; this text is included in Chap. 25 of Vol. II.—[*K.K.*]

4.3 Study Questions

QUES. 4.1. How, exactly, does Mueschenbroek's bottle store electrification?

a) How is the bottle constructed? How can it be electrified?
b) How is the electrification distributed on, or in, the bottle? Is there a maximum amount of electrification that it can hold? What happens if one attempts to add more?
c) In what sense are positive and negative electricity balanced when the bottle is electrified? Why does Franklin find this distribution of electrification wonderful—and a bit puzzling?
d) When a person standing on a block of wax holds the bottle in his hand, can he experience a shock from the bottle? How?

QUES. 4.2. How does Franklin demonstrate that the outside and inside of a Leyden jar have opposite charges?

a) Is a small neutral cork-ball attracted to the wire projecting from an electrified Leyden jar? What happens when the cork-ball touches the wire? What does this suggest?
b) What happens to a cork suspended between the ring-ends of two wires attached to the outside and inside, respectively, of an electrified Leyden jar (as in Fig. 4.1b of Franklin's third experiment)? How is this similar to Franklin's fourth experiment?
c) Can a fouled or wet Leyden jar be electrified? Why do you suppose this is the case, in light of Franklin's fifth experiment?
d) Why does the gold foil coating of an old book momentarily catch fire when, as in Franklin's eleventh experiment, it connects the outside and inside of a Leyden jar?

QUES. 4.3. Where, exactly, is the electrification of a Leyden jar stored? How do you know? And can a flat plate of glass (rather than a glass bottle) store electricity?

QUES. 4.4. Describe Franklin's electrical battery. What are the two different configurations he considers? Does Franklin's battery generate electrification, or merely store electrification?

QUES. 4.5. Can all electrical forces be somehow understood as purely attractive?

a) What happens to two balls suspended from the prime conductor inside a Leyden jar when a charged rod approaches the end of the conductor protruding through the cork? What happens when the charged rod is withdrawn? What does this imply?
b) What happens to the balls if a finger is briefly touched to the prime conductor while the charged rod is held near it? What does this imply?
c) Are there examples of non-electrical repulsive forces in nature? Why does Franklin mention this?

d) Which is simpler, a purely repulsive theory of electricity, or one that involves both attraction and repulsion? Which is more convenient? By what standard should scientific theories be judged?

QUES. 4.6. What is Mr. Kinnersley's established, "orthodox" law of the electric fluid? What problem(s) does Franklin find with this law? And how does Franklin qualify, or restate the law?

4.4 Exercises

EX. 4.1 (MÜSCHENBROEK'S BOTTLE). Suppose that an uncharged Müschenbroek bottle in placed on a wax block. Describe the state of electrification of both the interior and the exterior of the bottle (a) initially, (b) after a plastic rod is rubbed with rabbit fur and brought near the wire protruding through the cork of the jar, (c) after the rod has briefly touched the wire, and (d) after an external copper wire has briefly connected the protruding wire to the exterior of the bottle.

EX. 4.2 (CAPACITANCE, CHARGE AND ELECTRIC POTENTIAL LABORATORY). A Leyden jar is a scientific apparatus specifically designed to store electricity. But as a matter of fact, all bodies—whether Müschenbroek bottles, metallic spheres, or fleshy persons—have the ability to store an excess of positive or negative electrical charge. Nonetheless some have the capacity to store more charge than others. The so-called electrical *capacitance* of a body depends on its size and shape as well as on the type of substance surrounding the body. As a simple example, the capacitance, C, of an isolated metallic (conducting) sphere of radius R is given by

$$C = 4\pi \varepsilon R. \tag{4.1}$$

Here, ε is the electrical *permittivity* of the medium surrounding the sphere. In the international system of units (SI) the permittivity of a region of space devoid of all matter (a vacuum) is $\varepsilon_0 = 8.854 \times 10^{-12}$ F/m; the *farad* is then the SI unit of capacitance. As another example, the capacitance of a parallel-plate capacitor, which consists of two flat parallel conducting plates, is (approximately)

$$C = \frac{\varepsilon A}{d}, \tag{4.2}$$

where A is the area of the plates, d is their separation, and ε is the permittivity of the insulator sandwiched between them.[13] More generally, the capacitance of a body expresses the relationship between the electric potential, V, which is used to

---

[13] Franklin himself built a parallel-plate capacitor consisting of a flat pane of glass between two lead plates. The *relative permittivity* of glass (the ratio of the electrical permittivity of glass to that of a vacuum) is about four.

Fig. 4.4 Electrostatics equipment consisting of two conductive spheres (*top*), a volt power supply (*middle*), parallel-plate capacitor (*right*), charge producers and proof plane (*bottom*), and an electrometer wired to a faraday ice pail (*left*)

electrify the body and the quantity of unbalanced electrical charge, Q on the body. This relation is expressed mathematically as

$$Q = CV. \tag{4.3}$$

To better understand Eq. 4.3, consider an analogous situation from hydrostatics involving two upright cylindrical vessels having different cross-sectional areas. When both vessels are filled to the same height with fluid, the one with the larger cross sectional area will store a larger volume of fluid. Similarly, when two bodies are charged to the same electric potential, the one with the larger capacitance will store a larger quantity of charge.

In the following laboratory experiments, we will explore the relationship between electric potential (measured in volts), electric charge (measured in coulombs) and capacitance (measured in farads). We will make use of an electrometer, an adjustable parallel-plate capacitor, a Faraday ice pail, charge producers and a proof-plane, conductive spheres, and a volt electrostatics voltage source (see Fig. 4.4).[14]

Electrometer and Faraday Ice Pail The electrometer is a device used to measure the presence of electrical charge; it consists of a voltmeter hooked up in parallel

---

[14] I have here used the Basic Electrostatic System, manufactured by PASCO Scientific in Roseville, CA.

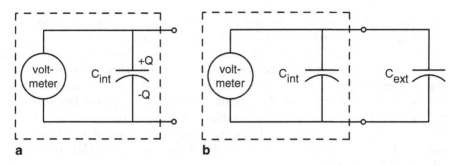

Fig. 4.5 a A schematic diagram of the interior of an electrometer. **b** An external capacitor has been attached across the leads

with an internal capacitor (see Fig. 4.5). When a charge Q is placed on the plates of the interior capacitor, C_{int}, an electric potential, V, is registered by the voltmeter.[15] An external capacitor, C_{ext}, may be attached to the leads of an electrometer so as to allow charge measurements by induction. This is depicted on the right side of Fig. 4.5. When charging by induction, a charged object is brought near the upper plate of the external capacitor, attracting negative charges to that plate. These negative charges must come from somewhere. Since the upper plate of the external capacitor is connected to the upper plate of the internal capacitor by a conducting wire, negative charge will flow from the upper plate of the internal to the upper plate of the external capacitor. Hence, the upper plate of the internal capacitor is left with a positive charge. In addition, the negative charge on the upper plate of the external capacitor will attract positive charge to the lower plate of the external capacitor. This leaves the lower plate of the internal capacitor with a negative charge. The net result of all this is that the voltmeter will register a positive voltage if a positively charged object is brought near the upper plate of the external capacitor and a negative voltage if a negatively charged object is brought near the upper plate of the external capacitor. In our experiments using the electrometer, a Faraday ice pail will act as the external capacitor. It consists of two concentric conducting cylinders: an "outer shield" and an "inner pail". The outer shield and inner pails then serve as the lower and upper capacitor plates respectively. The region between the cylinders is the region between the capacitor plates.

Measuring Charge Let us begin by becoming acquainted with the measurement of charge using the electrometer and ice pail. Connect the electrometer to the ice pail using low-capacitance test-leads. The black (ground) clip should be attached to the outer shield; the red clip should be attached to the inner pail. Briefly touch a finger to the inner and outer pails simultaneously so as to discharge the ice pail; the electrometer should read zero. The sensitivity knob may be adjusted to keep the

---

[15] To convert this to a charge reading (in coulombs), the internal capacitance needs to determined; for some electrometers the internal capacitance is kindly provided.

electrometer readings on scale. Now rub the charge producers together, insert one of them deep into the ice pail (without touching the sides!), and read the electrometer. Remove the object, and again note the reading. Next, note the reading when you insert and remove the other charged object from the ice pail. Then touch one of the objects to the ice pail, remove it and note the reading. Discharge the ice pail and reinsert the same object. Did the object retain any charge after once touching the ice pail? You might also try inserting both (initially uncharged) charge producers into the ice pail and rubbing them together, then noting the electrometer reading when neither, one, or both charge producers are removed from the ice pail. Can you make sense of all these results?

Charge Distribution Is any excess charge on an electrified object distributed uniformly over its surface? Let us explore the charge distribution on a conducting sphere by sampling the charge at various locations with a small hand-held proof-plane. Connect one terminal of the volt power supply to ground and the other terminal to an aluminized sphere perched atop an acrylic (insulating) rod. The other, identical, sphere should be placed at least half a meter from the first (electrified) sphere and then momentarily grounded. Now touch the face of the proof plane to various points on the second (un-electrified) sphere, insert it into the ice pail (without touching the pail), and read the voltage on the electrometer. Are your readings what you'd expect? Now move the first sphere so that it is just 1 cm from the second sphere. Again, sample the charge at various locations on the second sphere using the proof plane, ice pail and electrometer. Is the charge distribution symmetric?[16] Next, with the first and second spheres nearby, momentarily ground the second sphere and then repeat the process of sampling the charge on the second sphere. Is the charge distribution (and polarity) the same as before? Finally, without touching the second sphere, move the first sphere at least half a meter away. What is the charge distribution on the second sphere now? Does any charge remain? Is it symmetric?

Parallel Plates, Constant Spacing Next, we will explore the relationship between the electric potential and charge contained on a variable-width parallel plate capacitor. First, we will keep the plate separation—and hence the capacitance—constant, and measure how the voltage across the capacitor plates (as measured by the electrometer) varies as charge is gradually transferred to the capacitor plates. Begin by grounding the stationary capacitor plate and attaching the the movable plate to the (ungrounded) electrode of the electrometer. Also, electrify one of the aluminum spheres using a volt power supply. Briefly touch your finger simultaneously to both capacitor plates to remove any excess charge from the plates. Now, keeping the plate separation at about 2 mm, use the proof plane to "scoop" charge from the electrified sphere and onto the ungrounded capacitor plate. Note the electrometer reading each time charge is added to the capacitor. Consider: when scooping charge onto

---

[16] In these experiments, you should be sure–not–to ground the proof plane between measurements; this might deplete any charge which might exist on the second sphere.

the ungrounded plate, does the grounded plate become electrified? If so, where does its charge come from? Next, double the plate separation and repeat the previous experiments. Are your results the same?

Parallel Plates, Constant Potential In the previous experiment, we kept the plate spacing constant and measured the potential difference between the plates as we gradually charged the capacitor. Now we will keep the potential difference between the plates constant and measure how the charge on the plates varies with plate separation. Using Eq. 4.3, try to predict what might happen to the amount of charge held by the plates as the plate separation is varied. Now begin your experiments with a plate separation of about 5 cm. Establish a potential difference across the plates using the volt power supply. Then use a proof-plane to sample the charge at various points on the inner and outer surfaces of the capacitor plates: touch the proof plane to various points and insert it into an ice pail attached to an electrometer.[17] Is the charge distribution uniform? Now investigate how the charge at the center point of of one of the plates varies as the plate separation is varied. Do your results match your expectations? What do you think would happen if a pane of glass were inserted between the plates?

Ex. 4.3 (POTENTIAL OF N CHARGED SPHERES). Suppose you have a 1 kV power supply. Its hot (high-voltage) terminal is touched briefly to a 1 cm diameter aluminized sphere which is suspended from an insulating silk thread. This first sphere is then touched briefly to a second identical suspended sphere. Likewise, this second sphere is then touched briefly to a third identical sphere, the third to a fourth, and so on until N spheres are charged. After this series of operations, how much charge (in coulombs) resides on each of the N spheres, and what is the electric potential (in volts) of each sphere? (HINT: See Eq. 4.3 and the surrounding discussion.)

Ex. 4.4 (ELECTRICITY ESSAY). Is Franklin's theory of electricity different than the modern theory of electricity? If so, how? And which is correct? In answering these questions, you might consider the following: How does Franklin define *positive* and *negative* electrification? Does he assume the existence of two distinct electrical fluids? What role does the concept of equilibrium play in Franklin's theory? Can his theory explain both electrical attraction and electrical repulsion? If not, then how might his theory be modified in order to do so?

---

[17] In this case, it is preferable to ground the proof plane after each sample (try just breathing on it), since the capacitor plates are held at a constant potential and thus will not be depleted of charge.

4.5   Vocabulary

1. Equilibrium
2. Vacuum
3. Plenum
4. Phial
5. Fillet
6. Obstinate
7. Restitution
8. Decant
9. Dexterous
10. Armature
11. Contrivance
12. Unintelligible
13. Flax
14. Denomination
15. Analogous
16. Ascribe
17. Asunder
18. Orthodoxy

Chapter 5
Thunder and Lightning

> *This kite is to be raised when a thunder-gust appears to be coming on.*
>
> —Benjamin Franklin

5.1 Introduction

In the previous two chapters, we focused on Franklin's careful laboratory experiments which culminated in his theory of electricity. But he also had a strong interested in the weather, and he wrote many scientific correspondences on meteorological topics such as waterspouts, trade winds and thunderstorms. Franklin's interests in meteorology and electricity converged in his famous kite experiment. This perilous experiment is described in detail in letter XI, transcribed below. On a closely related topic, Franklin describes the use of lightning rods to protects buildings from damage in letter LIX, also included below. Sandwiched between these two scientific letters I have included one of Franklin's more whimsical correspondences: in letter XLVIII, he provides friendly advice to Miss Polly Stevenson, the daughter of a widow with whom Franklin lodged while serving as a diplomat in London before the onset of the Revolution.[1]

5.2 Reading: Franklin, *Experiments and Observations on Electricity*

Franklin, B., *Experiments and Observations on Electricity*, fourth ed., Henry, David, London, 1769. Letters XI, XLVIII and LIX.

---

[1] The Benjamin Franklin House at 36 Craven Street, Franklin's only surviving residence, is now a historical and educational facility in London.

© Springer International Publishing Switzerland 2016
K. Kuehn, *A Student's Guide Through the Great Physics Texts*,
Undergraduate Lecture Notes in Physics, DOI 10.1007/978-3-319-21816-8_5

5.2.1   Letter XI

From Benj. Franklin, Esq; of Philadelphia. Oct. 19, 1752.

As frequent mention is made in public papers from *Europe* of the success of the *Philadelphia* experiment for drawing the electric fire from the clouds by means of point rods of iron erected on high buildings, &c. it may be agreeable to the curious to be informed that the same experiment has succeeded in *Philadelphia*, though made in a different and more easy manner, which is as follows:

Make a small cross of two light strips of cedar, the arms so long as to reach to the four corners of a large thin silk handkerchief when extended; tie the corners of the handkerchief to the extremities of the cross, so you have the body of a kite; which being properly accommodated with a tail, loop and string, will rise in the air, like those made of paper; but this being of silk, is fitter to bear the wet and wind of a thunder-gust without tearing. To the top of the upright stick of the cross is to be fixed a very sharp pointed wire, rising a foot or more above the wood. To the end of the twine, next the hand, is to be tied a silk ribbon, and where the silk and twine join, a key may be fastened. This kite is to be raised when a thunder-gust appears to be coming on, and the person who holds the string must stand within a door or window, or under some cover, so that the silk ribbon may not be wet; and care must be taken that the twine does not touch the frame of the door or window. As soon as any of the thunder clouds come over the kite, the pointed wire will draw the electric fire from them, and the kite, with all the twine, will be electrified, and the loose filaments of the twine will stand out every way, and be attracted by an approaching finger. And when the rain has wet the kite and twine, so that it can conduct the electric fire freely, you will find it stream out plentifully from the key on the approach of your knuckle. At this key the phial may be charged; and from electric fire thus obtained, spirits may be kindled, and all the other electric experiments be performed, which are usually done by the help of a rubbed glass globe or tube, and thereby the sameness of the electric matter with that of lightning be completely demonstrated.

B.F

5.2.2   Letter XLVIII

To Miss S—n, at Wanstead. Craven-Street, May 17, 1760.

I send my dear good girl the books I mentioned to her last night. I beg her to accept them as a small mark of my esteem and friendship. They are written in the familiar easy manner for which the French are so remarkable, and afford a good deal of philosophic and practical knowledge, unembarrassed with the dry mathematics used by more exact reasoners, but which is apt to discourage young beginners.—I would advise you to read with a pen in your hand, and enter in a little book short hints of what you find that is curious, or that may be useful; for this will be the best method of imprinting such particulars in your memory, where they will be ready, either for practice or some future occasion, if they are matters of utility; or at least to adorn and improve your conversation, if they are rather points of curiosity.—And, as many of the terms of science are such as you cannot have met with in your common reading, and may therefore be unacquainted with, I think it would be well for you to have a good dictionary at hand, to consult immediately when you meet a word you do not comprehend the precise meaning of. This may at first seem troublesome and interrupting; but 'tis a trouble that will daily diminish, as you will daily find less and less occasion for your Dictionary as you become more acquainted with the terms; and in the mean time you will read with more satisfaction because with more understanding.—When any point occurs in which you would be glad to have farther information than your book affords you, I beg you would not in the least apprehend that I should think it a trouble to receive and answer your questions. It will be a pleasure, and no trouble. For though I may not be able, out of my own little stock of knowledge to afford you what you require, I can easily direct you to the books where it may most readily be found. Adieu, and believe me ever, my dear friend,

Yours affectionately,

B. FRANKLIN.

5.2.3 Letter LIX

Of Lightning, and the Method (now used in America) of securing Buildings and Persons from its mischievous Effects.

Experiments made in electricity first gave philosophers a suspicion that the matter of lightning was the same with the electric matter. Experiments afterwards made on lightning obtained from the clouds by pointed rods, received into bottles, and subjected to every trial, have since proved this suspicion to be perfectly well founded; and that whatever properties we find in electricity, are also the properties of lightning.

This matter of lightning, or of electricity, is an extreme subtile fluid, penetrating other bodies, and subsisting in them, equally diffused.

When by any operation of art or nature, there happens to be a greater proportion of this fluid in one body than in another, the body which has most, will communicate to that which has least, till the proportion becomes equal; or, if it is too great, till there be proper conductors to convey it from one to the other.

If the communication be through air without any conductor, a bright light is seen between the bodies, and a sound is heard. In our small experiments we call this light and sound the electric spark and snap; but in the great operations of nature, the light is what we call *lightning*, and the sound (produced at the same time, tho' generally arriving later at our ears than the light does to our eyes) is, with its echoes, called *thunder*.

If the communication of this fluid is by a conductor, it may be without either light or sound, the subtile fluid passing in the substance of the conductor.

If the conductor be good and of sufficient bigness, the fluid passes thro' it without hurting it. If otherwise, it is damaged or destroyed.

All metals, and water, are good conductors.—Other bodies may become conductors by having some quantity of water in them, as wood, and other materials used in building; but not having much water in them, they are not good conductors, and are therefore often damaged in the operation.

Glass, wax, silk, wool, hair, feathers, and even wood, perfectly dry, are non conductors: that is, they resist instead of facilitating the passage of this subtile fluid.

When this fluid has an opportunity of passing through two conductors, one good, and sufficient, as of metal, the other not so good, it passes in the best, and will follow it in any direction.

The distance at which a body charged with this fluid will discharge itself suddenly, striking through the air into another body that is not charged, or not so highly charg'd, is different according to the quantity of the fluid, the dimensions and form of the bodies themselves, and the state of the air between them.—This distance, whatever it happens to be between any two bodies, is called their *striking distance*, as till they come within that distance of each other, no stroke will be made.

The clouds have often more of this fluid in proportion than the earth; in which case as soon as they come near enough (that is, within the striking distance) or meet with a conductor, the fluid quits them and strikes into the earth. A cloud fully charged with this fluid, if so high as to be beyond the striking distance from the earth, passes quietly without making noise or giving light; unless it meets with other clouds that have less.

Tall trees, and lofty buildings, as the towers and spires of churches, become sometimes conductors between the clouds and the earth; but not being good ones, that is, not conveying the fluid freely, they are often damaged.

Buildings that have their roofs covered with lead, or other metal, and spouts of metal continued from the roof into the ground to carry off the water, are never hurt by lightning, as whenever it falls on such a building, it passes in the metals and not in the walls.

When other buildings happen to be within the striking distance from such clouds, the fluid passes in the walls whether of wood, brick or stone, quitting the walls only when it can find better conductors near them, as metal rods, bolts, and hinges of windows or doors, gilding on wainscot, or frames of pictures; the silvering on the backs of looking-glasses; the wires of bells; and the bodies of animals, as containing watery fluids. And in passing thro' the house it follows the direction of these conductors, taking as many in it's way as can assist it in its passage, whether in a straight or crooked line, leaping from one to the other, only rending the wall in the spaces where these partial good conductors are distant from each other.

An iron rod being placed on the outside of a building, from the highest part continued down into the moist earth, in any direction straight or crooked, following the form of the roof or other parts of the building, will receive the lightning at its upper end, attracting it so as to prevent its striking any other part; and, affording it a good conveyance into the earth, will prevent its damaging any part of the building.

A small quantity of metal is found able to conduct a great quantity of this fluid. A wire no bigger than a goose quill, has been known to conduct (with safety to the building as far as the wire is continued) a quantity of lightning that did prodigious damage, both above and below it; and probably larger rods are not necessary, tho' it is common in America, to make them half an inch, some of three quarters, or an inch in diameter.

The rod may be fastened to the wall, chimney, &c. with staples of iron.—The lightning will not leave the rod (a good conductor) to pass into the walls (a bad conductor), through those staples.—It would rather, if any were in the wall, pass out of it into the rod to get more readily by that conductor into the earth.

If the building be very large and extensive, two or more rods may be placed at different parts, for greater security.

Small ragged parts of clouds suspended in the air between the great body of clouds and the earth (like leaf gold in electrical experiments), often serve as partial conductors of the lightning, which proceeds from one of them to another, and by their help comes within the striking distance to the earth or a building. It therefore strikes through those conductors a building that would otherwise be out of the striking distance.

Long sharp points communicating with the earth, and presented to such parts of clouds, drawing silently from them the fluid they are charged with, they are then attracted to the cloud, and may leave the distance so great as to be beyond the reach of striking.

It is therefore that we elevate the upper end of the rod six or eight feet above the highest part of the building, tapering it gradually to a fine sharp point, which is gilt to prevent its rusting.

Thus the pointed rod either prevents a stroke from the cloud, or, if a stroke is made, conducts it to the earth with safety to the building.

The lower end of the rod should enter the earth so deep as to come at the moist part, perhaps two or three feet; and if bent when under the surface so as to go in a horizontal line six or eight feet from the wall, and then bent again downwards three or four feet, it will prevent damage to any of the stones of the foundation.

A person apprehensive of danger from lightning, happening during the time of thunder to be in a house not so secured, will do well to avoid sitting near the chimney, near a looking glass or any gilt pictures or wainscot; the safest place is in the middle of the room, (so it be not under a metal lustre suspended by a chain) sitting in one chair and laying the feet up in another. It is still safer to bring two or three mattrasses or beds into the middle of the room, and folding them up double, place the chair upon them; for they not being so good conductors as the wall, the lightning will not chuse an interrupted course through the air of the room and the bedding, when it can go thro' a continued better conductor the wall. But where it can be had, a hammock or swinging bed, suspended by silk cords equally distant from the walls on every side, and from the cieling and floor above and below, affords the safest situation a person can have in any room whatever; and what indeed may be deemed quite free from danger of any stroke by lightning.

Paris, Sept. 1767.   B.F.

5.3   Study Questions

QUES. 5.1. Describe Franklin's famous kite experiment. What is the purpose of the pointed wire, the silk ribbon and the key? Why do you suppose the twine must not touch the door or window frame in which the kite flier is standing? Finally, what critical conclusion did Franklin draw from this experiment?

QUES. 5.2. What advice does Franklin offers to young Miss Stevenson?

QUES. 5.3. How can buildings and people be best protected from lightning?

a) What is the cause of lightning? What does it tend to strike? And what factors affect its striking distance?
b) What is the purpose of a lightning rod? What shape should it be, and to what should it be attached? In case such protection is not provided, what path does lightning follow when passing through a building?
c) What safety advice does Franklin offer when residing in such an unprotected building?

5.4   Vocabulary

1. Apprehend
2. Wainscot
3. Convey
4. Prodigious
5. Gilt
6. Apprehensive
7. Lustre

Chapter 6
Coulomb's Law

> *The repulsive force between two small spheres charged with the
> same sort of electricity is in the inverse ratio of the squares of
> the distances between the centers of the two spheres.*
> —Charles Coulomb

6.1 Introduction

In a 1762 letter to Mr. Kinnersley, Benjamin Franklin considered the possibility
that all electrical phenomena are governed by a single principle: attraction so as to
achieve equilibrium.[1] According to this theory, when two bodies contain different
quantities of the so-called "electrical fire"—that subtle and invisible fluid—they are
drawn toward one another until at last they reach a state of equilibrium. Notice, however, that this theory does not offer a simple explanation of the observed repulsion of
certain bodies—those containing equal amounts of positive (or negative) electricity.
After all, they are already in equilibrium with one another. Perhaps, however, the
repulsion of such bodies is due to the fact that they are being drawn away from one
another and toward other nearby bodies with which they are not yet in equilibrium?
Thus, this apparent repulsion is in fact an attraction from behind.

At the risk of "multiplying causes in philosophy," Franklin finally rejects
this simple theory and claims that not one, but *two* principles govern electrical
phenomena—attraction and repulsion. Why shouldn't there be two? After all, both
attraction and repulsion are quite common themes in nature. For example, gravity
is purely attractive, while parcels of heated gas obviously repel one another, and
magnetism exhibits both attraction and repulsion. According to this dualistic theory
of electricity, (i) parcels of the electric fluid naturally repel one another, while (ii)
parcels of electric fluid attracts other types of matter—particularly matter suffering
from a relative deficit of electric fluid. In the course of entertaining these various theories of electricity, Franklin finally bemoaned the fact that "opinions are continually
varying, where we cannot have mathematical evidence of the nature of things."

---

[1] See Franklin's letter XXXVIII, included in Chap. 4 of the present volume.

© Springer International Publishing Switzerland 2016
K. Kuehn, *A Student's Guide Through the Great Physics Texts,*
Undergraduate Lecture Notes in Physics, DOI 10.1007/978-3-319-21816-8_6

A few years later, Coulomb would provide just such a mathematical treatment of the repulsion of like charges. Charles-Augustin Coulomb (1736–1806) was born in Angoulême, France. He graduated from the Royal Engineering School of Mézières in 1761, and as a military engineer, he was stationed in numerous locations, including Martinique, West Indies. Falling ill, he returned to France where he proceeded to carry out a series of notable experiments in applied mechanics. These included studies of the frictional force acting between slipping surfaces and the torsion force exerted by a twisted metal wire attempting to unwind. Coulomb then adapted his torsion apparatus so as to perform an extremely delicate measurement of the force exerted by electrified objects upon one another. This so-called "Coulomb balance" would later be adapted by Henry Cavendish so as to measure the gravitational force of attraction between heavy lead balls.[2] An aristocrat by birth, Coulomb later retired to his estate at Blois—at the outset of the French Revolution—where he continued his scientific work on mechanics and fluid dynamics until the time of his death. In the following reading selection, Coulomb describes how he used his torsion balance to determine the law of electrical force, now known as *Coulomb's law*. Are there any flaws in his method? Are his conclusions convincing?

6.2 Reading: Coulomb, *Law of Electric Force and the Fundamental Law of Electricity*

Coulomb, C., Law of Electric Force and Fundamental Law of Electricity, in *Source Book in Physics*, edited by W. F. Magie, Source books in the history of science, pp. 408–413, Harvard University Press, Cambridge, Massachusetts, 1963.

6.2.1 Law of Electric Force

Construction and use of an electric balance based on the properties of metallic wires of having a force of reaction of torsion proportional to the angle of torsion.

Experimental determination of the law according to which the elements of bodies electrified with the same kind of electricity repel each other.

In a memoir presented to the Academy in 1784, I determined by experiment the laws of the force of torsion of a metallic wire, and I found that this force was in a ratio compounded of the angle of torsion, of the fourth power of the diameter of the suspended wire, and of the reciprocal of its length, all being multiplied by a constant

---

[2] See Ex. 27.4 and the discussion of Newton's universal law of gravitation in Chap. 27 of volume II.

coefficient which depends on the nature of the metal and which is easy to determine by experiment.

I showed in the same memoir that by using this force of torsion it was possible to measure with precision very small forces, as for example, a ten thousandth of a grain. I gave in the same memoir an application of this theory, by attempting to measure the constant force attributed to adhesion in the formula which expresses the friction of the surface of a solid body in motion in a fluid.

I submit today to the Academy an electric balance constructed on the same principle; it measures very exactly the state and the electric force of a body however slightly it is charged.

Construction of Balance

Although I have learned by experience that to carry out several electric experiments in a convenient way I should correct some defects in the first balance of this sort which I have made; nevertheless as it is so far the only one that I have used I shall give its description, simply remarking that its form and size may be and should be changed according to the nature of the experiments that one is planning to make. The first figure represents this balance in perspective and the details of it are as follows:

On a glass cylinder $ABCD$ (Fig. 6.1) 12 in. in diameter and 12 in. high is placed a glass plate 13 in. in diameter, which entirely covers the glass vessel; this plate is pierced with two holes of about 20 lines in diameter, one of them in the middle, at f, above which is placed a glass tube 24 in. high; this tube is cemented over the hole f with the cement ordinarily used in electrical apparatus: at the upper end of the tube at b is placed a torsion micrometer which is seen in detail in Fig. 2. The upper part, No. 1, carries the milled head b, the index io, and the clamp q; this piece fits into the hole G of the piece No. 2; this piece No. 2 is made up of a circle ab divided on its edge into 360° and of a copper tube Φ which fits into the tube H, No. 3, sealed to the interior of the upper end of the glass tube or column fb of figure 1. The clamp q (Fig. 6.1, 2, No. 1), is shaped much like the end of a solid crayon holder, which is closed by means of the ring q. In this holder is clamped the end of a very fine silver wire; the other end of the silver wire (Fig. 6.1, 3) is held at P in a clamp made of a cylinder Po of copper or iron with a diameter of not more than a line, whose upper end P is split so as to form a clamp which is closed by means of the sliding piece Φ. This small cylinder is enlarged at C and a hole bored through it, in which can be inserted (Fig. 6.1, 1) the needle ag: the weight of this little cylinder should be sufficiently great to keep the silver wire stretched without breaking it. The needle that is shown (Fig. 6.1, 1) at ag suspended horizontally about half way up in the large vessel which encloses it, is formed either of a silk thread soaked in Spanish wax or of a straw likewise soaked in Spanish wax and finished off from q to a for 18 lines of its length by a cylindrical rod of shellac; at the end a of this needle is carried a little pith ball two or three lines in diameter; at g there is a little vertical piece of paper soaked in terebinth, which serves as a counterweight for the ball a and which slows down the oscillations.

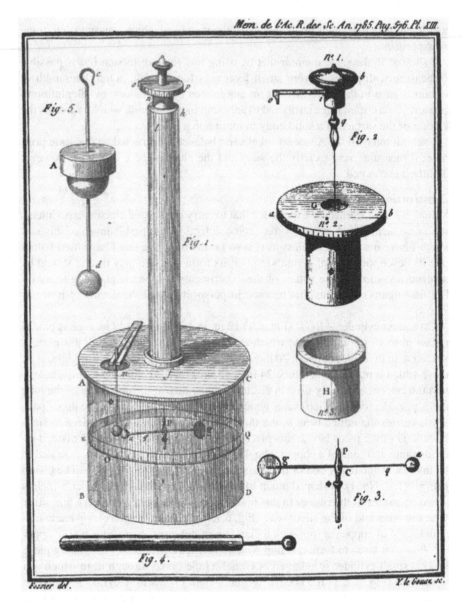

Mem. de l'Ac. R. des Sc. An. 1785. Pag. 576. Pl. XIII.

Fig. 6.1 Coulomb's apparatus

We have said that the cover AC was pierced by a second hole at m. In this second hole there is introduced a small cylinder $m\Phi t$, the lower part of which Φt is made of shellac; at t is another pith ball; about the vessel, at the height of the needle, is described a circle zQ divided into 360° for greater simplicity I use a strip of paper divided into 360° which is pasted around the vessel at the height of the needle.

To arrange this instrument for use I set on the cover so that the hole m practically corresponds to the first division of the circle zoQ traced on the vessel. I place the index oi of the micrometer on the point o or the first division of this micrometer; I then turn the micrometer in the vertical tube fb until, by looking past the vertical wire which suspends the needle and the center of the ball, the needle ag corresponds to the first division of the circle zoQ. I then introduce through the hole m the other ball t suspended by the rod $m\Phi t$, in such a way that it touches the ball a and that by looking past the suspension wire and the ball t we encounter the first division 0 of the circle zoQ. The balance is then in condition to be used for all our operations; as an example we go on to give the method which we have used to determine the fundamental law according to which electrified bodies repel each other.

6.2.2   Fundamental Law of Electricity

The repulsive force between two small spheres charged with the same sort of electricity is in the inverse ratio of the squares of the distances between the centers of the two spheres.

Experiment
We electrify a small conductor, (Fig. 6.1, 4) which is simply a pin with a large head insulated by sinking its point into the end of a rod of Spanish wax; we introduce this pin through the hole m and with it touch the ball t, which is in contact with the ball a; on withdrawing the pin the two balls are electrified with electricity of the same sort and they repel each other to a distance which is measured by looking past the suspension wire and the center of the ball a to the corresponding division of the circle zoQ; then by turning the index of the micrometer in the sense pno we twist the suspension wire lP and exert a force proportional to the angle of torsion, which tends to bring the ball a nearer to the ball t. We observe in this way the distance through which different angles of torsion bring the ball a toward the ball t, and by comparing the forces of torsion with the corresponding distances of the two balls we determine the law of repulsion. I shall here only present some trials which are easy to repeat and which will at once make evident the law of repulsion.

First Trial.   Having electrified the two balls by means of the pin head while the index of the micrometer points to o, the ball a of the needle is separated from the ball t by 36°.

Second Trial.   By twisting the suspension wire through 126° as shown by the pointer o of the micrometer, the two balls approach each other and stand 18° apart.

Third Trial.   By twisting the suspension wire through 567° the two balls approach to a distance of 8° and a half.

Explanation and Result of This Experiment

Before the balls have been electrified they touch, and the center of the ball *a* suspended by the needle is not separated from the point where the torsion of the suspension wire is zero by more than half the diameters of the two balls. It must be mentioned that the silver wire *l P* which formed this suspension was 28 in. long and was so fine that a foot of it weighed only 716 grain. By calculating the force which is needed to twist this wire by acting on the point *a* 4 in. away from the wire *l P* or from the center of suspension, I have found by using the formulas explained in a memoir on the laws of the force of torsion of metallic wires, printed in the Volume of the Academy for 1784, that to twist this wire through 360° the force that was needed when applied at the point *a* so as to act on the lever *an* 4 in. long was only $\frac{1}{340}$ grains: so that since the forces of torsion, as is proved in that memoir, are as the angles of torsion, the least repulsive force between the two balls would separate them sensibly from each other.

We found in our first experiment, in which the index of the micrometer is set on the point *o*, that the balls are separated by 36°, which produces a force of torsion of $360 = \frac{1}{3400}$ of a grain; in the second trial the distance between the balls is 18°, but as the micrometer has been turned through 126° it results that at a distance of 18° the repulsive force was equivalent to 144°; so at half the first distance the repulsion of the balls is quadruple.

In the third trial the suspension wire was twisted through 567° and the two balls are separated by only 8° and a half. The total torsion was consequently 576°, four times that of the second trial, and the distance of the two balls in this third trial lacked only one-half degree of being reduced to half of that at which it stood in the second trial. It results then from these three trials that the repulsive action which the two balls exert on each other when they are electrified similarly is in the inverse ratio of the square of the distances.

6.3 Study Questions

QUES. 6.1. Describe Coulomb's apparatus and measurement procedure.

a) Where are the charged objects, and how did he charge them? Which parts of the apparatus are moveable?
b) What quantities did he measure? Try to reproduce his data table.

What were the results of Coulomb's experiments?

a) How was the distance between the charged objects determined?
b) How did he infer the force acting upon the spheres from his measured quantities?
c) What mathematical relationship did he derive from his data? Did his data "fit" his formula?

QUES. 6.2. Were Coulomb's results reliable? Were they significant?

a) How might his measurement of distances introduce systematic errors into his measurement? In other words, did his distance measurement lead him to overestimate or to underestimate the distance between the spheres?
b) What other systematic errors might have afflicted his experiment? Do you find the conclusion he draws from his experiment convincing? If not, what might have strengthened his conclusion?

6.4 Exercises

EX. 6.1 (TORSION BALANCE). Suppose that the horizontal rod of a torsion balance is 5 cm long, from the torsion wire to the tiny chargeable sphere. A force of 1 μN is required to turn the rod by 1° clockwise. How would the force differ (i) if you were to instead twist the rod by 25°, (ii) if you were to simultaneously twist the rod clockwise, and the torsion wire counterclockwise, each by 1° (iii) if you were to double the tension in the torsion wire, or (iv) if you were to double the diameter of the torsion wire?

EX. 6.2 (FORCE ADDITION). Three charges are fixed to the vertices of an equilateral triangle whose sides are 1 mm long. The top vertex has a charge of 1 nC. The bottom vertices each have a charge of 2 nC. By geometrically (or vectorially) adding up the horizontal and vertical components of each force, find the net (total) force acting on each charge. Hint: Your answer should consist of three magnitudes and three directions (with respect to the horizontal bottom of the triangle). (ANSWER: The net force acting on the 1 nC charge is 0.031 N straight upwards; the force acting on each of the two other charges is 0.047 N directed outwards at an angle 19° below the horizontal.)

EX. 6.3 (EQUILIBRIUM OF CHARGES). Two insulating spheres are fixed 1 cm apart. The left one has a charge of $+2$ C. The right one has a charge of -1 C. Could one place a third sphere, with a charge of $+1$ C, such that there is zero force acting upon it? If so, where (*i.e.* at what distance from each of the other two charges)? If not, why not?

EX. 6.4 (COULOMB'S LAW LABORATORY). Using a torsion balance,[3] repeat Coulomb's experiments. You will need to charge the spheres with a kilovolt power supply[4] and measure the repulsive force on the spheres as a function of distance between the spheres. As Coulomb did, you should also try to do: measure the repulsive force in terms of the number of degrees which the torsion wire must be twisted

---

[3] I have used the Coulomb's Law Apparatus, Model ES-9070, by Pasco Scientiific, Inc., Roseville, CA.

[4] For example, the 6 kV Power Supply, Model ES-9586A, also by Pasco Scientiific, Inc.

so as to maintain a particular inter-sphere separation. To convert this angle into a force, you will need to calibrate your torsion wire by measuring the amount of force required to twist the wire through various angles using milligram weights.

In addition to determining the dependence of the force upon the distance between the spheres, you should also determine the dependence of the force upon the amount of charge on the spheres while keeping the inter-sphere distance constant. The amount of charge on a sphere can be determined from its capacitance and the potential (measured in volts) of the power supplied used to electrify it. After carrying out your experiments, you should be able to produce plots of (i) force versus distance and (ii) force versus charge. What mathematical relationship exists between force and distance, and between force and charge? Are your results consistent with those of Coulomb? Are your experiments better, or worse, than those of Coulomb? How so?

Chapter 7
The Dawn of Electro-Magnetism

It is sufficiently evident from the preceding facts that the electric conflict is not confined to the conductor, but dispersed pretty widely in the circumjacent space.

—Hans Christian Oersted

7.1 Introduction

So far, we have explored magnetism and electricity as separate phenomena. This is the legacy of William Gilbert, who recognized that while magnets and electrically charged bodies can exert forces, they do not exert forces on each other: electrically charged bodies attract or repel other electrically charged bodies; magnetic bodies attract or repel other magnetic bodies.[1] This simple and pleasant demarcation was upset by the work of Hans Christian Oersted (1777–1851), who was born in Rudkøbing, Denmark. After early homeschooling, which included the study of diverse languages, he entered the University of Copenhagen in 1793 to study pharmacy, like his father. In 1799, he presented his doctoral "Dissertation on the Structure of the Elementary Metaphysics of External Nature," receiving highest academic honors in philosophy. Afterwards, he worked for a short time as a lecturer and pharmacist before obtaining funding to travel to Germany. There, his contact with Johann Wilhelm Ritter drew him to the experimental study of physics and chemistry. In 1806, he accepted a position as Professor of Natural Philosophy at the University of Copenhagen. He subsequently published dozens of papers—in Danish, English, French, German and Latin—on various topics in chemistry, electricity, acoustics, thermodynamics, and the magnetic and mechanical properties of substances.[2] He was also an accomplished writer and poet and a close friend of fairy-tale author Hans Christian Andersen.

Today, Oersted is best known for his discovery of the connection between electrical currents and magnetism. This work, described in the reading selection in this

---

[1] See especially Chaps. 1, 4 and 6 of the present volume.

[2] See Oersted, H. C., *Selected Scientific Works of Hans Christian Oersted*, Princeton University Press, 1998.

© Springer International Publishing Switzerland 2016
K. Kuehn, *A Student's Guide Through the Great Physics Texts*,
Undergraduate Lecture Notes in Physics, DOI 10.1007/978-3-319-21816-8_7

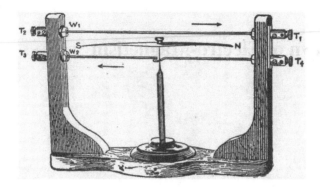

Fig. 7.1 Apparatus for measuring the effect of an electric current on a compass needle. (From Gage, Alfred P., *The principles of physics*, Boston & London: Ginn and Company, 1895, p. 500)

chapter, inspired later experiments on electro-magnetic phenomena by both Faraday and Ampère and culminated in Maxwell's electromagnetic theory of light.[3] Originally written in Latin, Oersted's *Experiments on the Effect of a Current of Electricity on the Magnetic Needle* was translated into English and published in 1820 in Thomson's *Annals of Philosophy*. A few initial comments will help to clarify this famous text (which unfortunately lacks illustrations). First, the straight section of the uniting wire connecting the positive and negative terminals of Oersted's galvanic apparatus is initially aligned parallel to the floor and in the north-south direction. The wire therefore lies in the plane of the magnetic meridian, which is an imaginary plane containing both of the earth's magnetic poles and the point on the earth's surface at which the apparatus is located. Second, the "declination" of the compass needle refers to its eastward (or westward) deviation *out of* the plane of the magnetic meridian; the "inclination" of the compass needle refers to its upward (or downward) deviation *within* the plane of the magnetic meridian. Finally, inside the uniting wire, one might imagine either a current of *positive electricity* moving from the positive to the negative terminal of the battery, or a current of *negative electricity* moving from the negative to the positive terminal. In the next chapter, we will consider Ampère's definition of the direction or "sense of the electrical current" traveling through a conducting wire in more detail (Fig. 7.1).

7.2   Reading: Oersted, *Experiments on the Effect of a Current of Electricity on the Magnetic Needle*

Ørsted, H. C., Experiments on the effect of a current of electricity on the magnetic needle, *Annals of Philosophy*, *16*(4), 273–276, 1820.

---

[3] See Chaps. 8, 25 and 31 of the present volume.

*By John Christian Oersted, Knight of the Order of Danneborg, Professor of Natural
Philosophy, and Secretary to the Royal Society of Copenhagen.*

The first experiments respecting the subject which I mean at present to explain, were
made by me last winter, while lecturing on electricity, galvanism, and magnetism,
in the University. It seemed demonstrated by these experiments that the magnetic
needle was moved from its position by the galvanic apparatus, but that the galvanic
circle must be complete, and not open, which last method was tried in vain some
years ago by very celebrated philosophers. But as these experiments were made
with a feeble apparatus, and were not, therefore, sufficiently conclusive, considering the importance of the subject, I associated myself with my friend Esmarck to
repeat and extend them by means of a very powerful galvanic battery, provided by
us in common. Mr. Wleugel, a Knight of the Order of Danneborg, and at the head
of the Pilots, was present at, and assisted in, the experiments. There were present
likewise Mr. Hauch, a man very well skilled in the Natural Sciences, Mr. Reinhardt,
a Professor of Natural History, Mr. Jacobsen, Professor of Medicine, and that very
skillful chemist, Mr. Zeise, Doctor of Philosophy. I had often made experiments by
myself; but every fact which I had observed was repeated in the presence of these
gentlemen.

The galvanic apparatuswhich we employed consists of 20 copper troughs, the
length and height of each of which was 12 in.; but the breadth scarcely exceeded
2½ in. Every trough is supplied with two plates of copper, so bent that they could
carry a copper rod, which supports the zinc plate in the water of the next trough.
The water of the troughs contained 1/60th of its weight of sulphuric acid, and an equal
quantity of nitric acid. The portion of each zinc plate sunk in the water is a square
whose side is about 10 in. in length. A smaller apparatus will answer provided it be
strong enough to heat a metallic wire red hot.

The opposite ends of the galvanic battery were joined by a metallic wire, which,
for shortness sake, we shall call the *uniting conductor* or the *uniting wire*. To the
effect which takes place in this conductor and in the surrounding space, we shall
give the name of the *conflict of electricity*.

Let the straight part of this wire be placed horizontally above the magnetic needle, properly suspended, and parallel to it. If necessary, the uniting wire is bent so
as to assume a proper position for the experiment. Things being in this state, the
needle will be moved, and the end of it next the negative side of the battery will go
westward.

If the distance of the uniting wire does not exceed three-quarters of an inch from
the needle, the declination of the needle makes an angle of about 45°. If the distance
is increased, the angle diminishes proportionally. The declination likewise varies
with the power of the battery.

The uniting wire may change its place, either towards the east or west, provided it
continue parallel to the needle, without any other change of the effect than in respect
to its quantity. Hence the effect cannot be ascribed to attraction; for the same pole
of the magnetic needle, which approaches the uniting wire, while placed on its east

side, ought to recede from it when on the west side, if these declinations depended on attractions and repulsions. The uniting conductor may consist of several wires, or metallic ribbons, connected together. The nature of the metal does not alter the effect, but merely the quantity. Wires of platinum, gold, silver, brass, iron, ribbons of lead and tin, a mass of mercury were employed with equal success. The conductor does not lose its effect, though interrupted by water, unless the interruption amounts to several inches in length.

The effect of the uniting wire passes to the needle through glass, metals, wood, water, resin, stoneware, stones; for it is not taken away by interposing plates of glass, metal or wood. Even glass, metal, and wood, interposed at once, do not destroy, and indeed scarcely diminish the effect. The disc of the electrophorus, plates of porphyry, a stone-ware vessel, even filled with water, were interposed with the same result. We found the effects unchanged when the needle was included in a brass box filled with water. It is needless to observe that the transmission of effects through all these matters has never before been observed in electricity and galvanism. The effects, therefore, which take place in the conflict of electricity are very different from the effects of either of the electricities.

If the uniting wire be placed in a horizontal plane under the magnetic needle, all the effects are the same as when it is above the needle, only they are in an opposite direction; for the pole of the magnetic needle next to the negative end of the battery declines to the east.

That these facts may be the more easily retained, we may use this formula—the pole *above* which the *negative* electricity enters is turned to the *west*; *under* which, to the *east*.

If the uniting wire is so turned in a horizontal plane as to form a gradually increasing angle with the magnetic meridian, the declination of the needle *increases*, if the motion of the wire is towards the place of the disturbed needle; but it *diminishes* if the wire moves further from that place.

When the uniting wire is situated in the same horizontal plane in which the needle moves by means of the counterpoise, and parallel to it, no declination is produced either to the east or west; but an *inclination* takes place, so that the pole, next which the negative electricity enters the wire, is *depressed* when the wire is situated on the *west* side, and *elevated* when situated on the *east side*.

If the uniting wire be placed perpendicularly to the plane of the magnetic meridian, whether above or below it, the needle remains at rest, unless it be very near the pole; in that case the pole is *elevated* when he entrance is from the *west* side of the wire, and *depressed*, when from the *east* side.

When the uniting wire is placed perpendicularly opposite to the pole of the magnetic needle, and the upper extremity of the wire receives the negative electricity, the pole is moved towards the east; but when the wire is opposite to a point between the pole and the middle of the needle, the pole is most towards the west. When the upper end of the wire receives positive electricity, the phenomena are reversed.

If the uniting wire is bent so as to form two legs parallel to each other, it repels or attracts the magnetic poles according to the different conditions of the case. Suppose the wire placed opposite to either pole of the needle, so that the plane of the parallel legs is is perpendicular to the magnetic meridian, and let the eastern leg be united with the negative end, the western leg with the positive end of the battery: in that case the nearest pole will be repelled either to the east or west, according to the position of the plane of the legs. The eastmost leg being united with the positive, and the westmost with the negative side of the battery, the nearest pole will will be attracted. When the plane of the legs is placed perpendicular to the place between the pole and the middle of the needle, the same effects recur, but reversed.

A brass needle, suspended like a magnetic needle, is not moved by the effect of the uniting wire. Likewise needles of glass and of gum lac remain unacted on.

We may now make a few observations towards explaining these phenomena.

The electric conflict acts only on the magnetic particles of matter. All non-magnetic bodies appear penetrable by the electric conflict, while magnetic bodies, or rather their magnetic particles, resist the passage of this conflict. Hence they can be moved by the impetus of the contending powers.

It is sufficiently evident from the preceding facts that the electric conflict is not confined to the conductor, but dispersed pretty widely in the circumjacent space.

From the preceding facts we may likewise collect that this conflict performs circles; for without this condition, it seems impossible that the one part of the uniting wire, when placed below the magnetic pole, should drive it towards the east, and when placed above it, towards the west; for it is the nature of a circle that the motions in opposite parts should have an opposite direction. Besides, a motion in circles, joined with a progressive motion, according to the length of the conductor, ought to form a conchoidal or spiral line; but this, unless I am mistaken, contributes nothing to explain the phenomena hitherto observed.

All the effects on the north pole above-mentioned are easily understood by supposing that negative electricity moves in a spiral line bent towards the right, and propels the north pole, but does not act on the south pole. The effects on the south pole are explained in a similar manner, if we ascribe to positive electricity a contrary motion and power of acting on the south pole, but not upon the north. The agreement of this law with nature will be better seen by a repetition of the experiments than by a long explanation. The mode of judging of the experiments will be much facilitated if the course of the electricities in the uniting wire be pointed out by marks or figures.

I shall merely add to the above that I have demonstrated in a book published 5 years ago that heat and light consist of the conflict of the electricities. From the observations now stated, we may concluded that a circular motion likewise occurs in these effects. This I think will contribute very much to illustrate the phenomena to which the appellation of polarization of light has been given.

Copenhagen, July 21, 1820

7.3 Study Questions

QUES. 7.1. Can an electrical current passing through a wire affect a magnetic needle?

a) Describe Oersted's apparatus. Who was present during Oersted's experiments? Why is this relevant information?
b) What term does Oersted use to denote the effect which takes place in and around the wire?
c) How did Oersted initially orient the uniting wire and the magnetic needle? What effect does the galvanic apparatus have upon the needle? Is the effect of the uniting wire on the needle one of attraction? How do you know?
d) What factors affect the final orientation of the needle? What factors do *not* affect its orientation? Why is this significant?
e) How does the shape, placement and orientation of the wire affect the motion of the needle?
f) What general law governs the effect of a galvanic current on a magnetic needle? Why had previous philosophers failed to observe effects similar to those here reported by Oersted?

QUES. 7.2. Is Oersted's "electric conflict" produced by stationary (*i.e.* static) electricity?

a) How does Oersted describe electric conflict? Does it act on all objects?
b) Under what conditions does it appear? And what configuration does it adopt?
c) Is Oersted's explanation reasonable? Is it clear? Is it correct?

7.4 Exercises

EX. 7.1 (WIRE AND COMPASS). In what direction will a magnetic needle point if it is placed (a) to the east of a straight wire carrying an electrical current southward, from the positive toward the negative terminal of a battery, (b) at the center of a loop of wire lying on a table and carrying an electrical current clockwise (as viewed from above), and (c) along the axis of a helix which is lying on the table. To be definite, suppose the axis of the helix is oriented such that the electrical current flows clockwise around the loops as it travels from the south to the north end of the helix.

EX. 7.2 (CURRENTS AND COMPASSES LABORATORY). Using a magnetic compass and a magnetometer, observe the direction and strength of the magnetic forces around a current carrying wire.[4] How do these change when you change (i) the current

---

[4] For automated data acquisition, I have used the magnetic field sensor (Model MG-BTA) and a Lab Pro interface, both manufactured by Vernier Software & Technology, Beaverton, OR.

through the wire and (ii) distance between the wire and the sensor, and (iii) the shape of the wire (*i.e.* straight or helix-shaped)? You will want to obtain several data points for each of these so you can make plots. Are your results in agreement with those of Oersted? In what sense? Also, reconsider the magnetic forces near a bar magnet. Can you draw any comparisons between the magnetic forces near a wire and those near the bar magnet? How are they similar or different? Do they have the same cause?[5]

EX. 7.3 (ELECTRIC CONFLICT ESSAY). Can the observed behavior of the magnetic needle near a current-carrying wire be explained using Coulomb's law of electric force?[6] If so, how? If not, what does this imply? In other words, is the phenomenon discovered by Oersted a completely new effect, or can it be understood as a manifestation of the attraction and repulsion of electrical charges?

EX. 7.4 (MAGNETIC FIELD AROUND A STRAIGHT CURRENT-CARRYING WIRE). Shortly after Oersted's discovery, Michael Faraday suggested the concept of the *magnetic field* as a way of understanding magnetic forces, such as the repulsion of bar magnets. According to Faraday, magnetic fields are invisible curved lines of force which physically exist in the region of space surrounding magnets.[7] As suggested by Oersted's experiments, not only magnets but also current-carrying wires can generate magnetic fields. The strength of the magnetic field in the vicinity of curved wires can be quite complicated. But for the special case of a long, straight current-carrying wire the magnetic field strength, B, may be written in a particularly simple mathematical form:[8]

$$B = \frac{\mu}{2\pi} \frac{i}{r}. \tag{7.1}$$

Here, i is the current carried by the wire, r is the perpendicular distance from the wire's axis to the nearby point under consideration, and μ is the so-called *permeability* of the medium surrounding the wire. In the international system of units (SI), the permeability of vacuum is $\mu_0 = 4\pi \times 10^{-7}$ N/A^2.[9]

The magnetic field near a magnet or a current-carrying wire has not only a strength, but also a direction. Its direction at a particular location can be discerned from the orientation of the north-seeking end of an ordinary compass needle when placed at that location; this is precisely what Oersted did. For the special case of a long straight wire, the magnetic field is rather simple: magnetic field lines form

---

[5] See Ampère's theory of magnetism in Chap. 8 of the present volume.

[6] See Chap. 6 of the present volume.

[7] The concept of the magnetic field will be developed in more detail by Michael Faraday in Chaps. 28 and 29 of the present volume. See also the discussion of vector fields in Appendix A.

[8] The magnetic field in the vicinity of arbitrarily shaped current-carrying wires can be computed using the Biot-Savart law; see Ex. 8.4 in the following chapter.

[9] A system of units for measuring magnetic field strength was first developed in the 1830's by the mathematician Carl Friedrich Gauss. Today, magnetic field strength is typically measured in either tesla (mks) or gauss (cgs); one gauss equals 100×10^{-6} T.

closed axisymmetric loops (circles) around the wire. The direction of the magnetic field at a point on the circle is given by the *right-hand rule*: imagine that the fully-extended thumb of your right hand represents the direction of the (positive) current flow in the wire. The fingers of the same hand then wrap around the wire (your thumb) in the direction in which the magnetic field is oriented.

With all this in mind, consider the following questions. Suppose that you orient a very long straight wire in the direction perpendicular to your local magnetic meridian.

a) What must be the size and direction of the electrical current in the wire so as to produce a southward-directed 100 milli-gauss magnetic field at a distance 10 cm above the wire? (ANSWER: 5 A eastward)
b) At what point with respect to your wire (distance from the wire and angle with respect to a vertical) would the magnetic field strength produced by this current be exactly cancelled by Earth's ambient magnetic field? Hint: you will need to look up the strength and inclination (dip) of the ambient magnetic field at your particular location on Earth's surface.

7.5   Vocabulary

1. Galvanism
2. Uniting conductor
3. Declination
4. Meridian
5. Impetus
6. Contend
7. Circumjacent
8. Conchoidal
9. Appellation
10. Polarization

Chapter 8
Electric Currents, Magnetic Forces

> *The attractions and repulsions which occur between two*
> *parallel currents, according as they are directed in the same*
> *sense or in opposite senses, are facts given by an experiment*
> *which is easy to repeat.*
>
> —André Marie Ampère

8.1 Introduction

André Marie Ampère (1775–1836) was born in Lyon, France. He was home-schooled by his father, a merchant who directed his education and was guillotined in 1793 during the Jacobin purges of the French revolution. Ampère taught himself latin and mathematics at an early age in order to understand the scientific and mathematical writings of Euler and Bernoulli. He began his academic career as a mathematics teacher in 1799. He was appointed professor of physics and chemistry in 1802 at the École Centrale at Bourg, professor of analysis in 1809 at the École Polytechnique in Paris, and chair of experimental physics at the Collège de France in 1824. Like many other prominent figures of the French enlightenment, Ampère's academic interests were quite broad, including astronomy, chemistry, mathematics and philosophy. Inspired by Oersted's recent discovery of the effect of an electric current on a magnetized compass needle,[1] Ampère performed a number of careful experiments on the relationship between electricity and magnetism; the mathematical formulation of his results came to be known as Ampère's law. The text that follows is comprised of two extracts. The first is from a paper entitled "Experiments on the New Electrodynamical Phenomena,"[2] In it, Ampère introduces some new terminology. The second is from a memoir presented before the Academy of Sciences on October 2, 1820.[3] It describes his experiments on the force acting between two parallel electrical currents.

---

[1] See Chap. 7 of the present volume.

[2] *Annales de Chimie et de Physique*, Series II, Vol. 20, p. 60, 1822.

[3] *Annales de Chimie et de Physique*, Series II, Vol. 15, p. 59, 1820.

© Springer International Publishing Switzerland 2016
K. Kuehn, *A Student's Guide Through the Great Physics Texts*,
Undergraduate Lecture Notes in Physics, DOI 10.1007/978-3-319-21816-8_8

8.2   Reading: Ampère, *Actions Between Currents*

Ampère, A. M., New Names and Actions Between Currents, in *A Source Book in Physics*, edited by W. F. Magie, Source books in the history of science, pp. 447–460, Harvard University Press, Cambridge, Massachusetts, 1963.

8.2.1   New Names

The word 'electromagnetic' which is used to characterize the phenomena produced by the conducting wires of the voltaic pile, could not suitably describe them except during the period when the only phenomena which were known of this sort were those which M. Oersted discovered, exhibited by an electric current and a magnet. I have determined to use the word electrodynamic in order to unite under a common name all these phenomena, and particularly to designate those which I have observed between two voltaic conductors. It expresses their true character, that of being produced by electricity in motion: while the electric attractions and repulsions, which have been known for a long time, are electrostatic phenomena produced by the unequal distribution of electricity at rest in the bodies in which they are observed.

8.2.2   Actions Between Currents

8.2.2.1   On the Mutual Action of Two Electric Currents

1. Electromotive action is manifested by two sorts of effects which I believe I should first distinguish by precise definitions.

   I shall call the first *electric tension*, the second *electric current*.

   The first is observed when two bodies, between which this action occurs, are separated from each other by non-conducting bodies at all the points of their surfaces except those where it is established; the second occurs when the bodies make a part of a circuit of conducting bodies, which are in contact at points on their surface different from those at which the electromotive action is produced. In the first case, the effect of the electromotive action is to put the two bodies, or the two systems of bodies, between which it exists, in two states of tension, of which the difference is constant when this action is constant, when, for example, it is produced by the contact of two substances of different sorts; this difference may be variable, on the contrary, with the cause which produces it, if it results from friction or from pressure.

   The first case is the only one which can arise when the electromotive action develops between different parts of the same nonconducting body; tourmaline is an example of this when its temperature changes.

In the second case there is no longer any electric tension, light bodies are not sensibly attracted and the ordinary electrometer can no longer be of service to indicate what is going on in the body; nevertheless the electromotive action continues; for if, for example, water, or an acid or an alkali or a saline solution forms part of the circuit, these bodies are decomposed, especially when the electromotive action is constant, as has been known for some time; and furthermore as M. Oersted has recently discovered, when the electromotive action is produced by the contact of metals, the magnetic needle is turned from its direction when it is placed near any portion of the circuit; but these effects cease, water is no longer decomposed, and the needle comes back to its ordinary position as soon as the circuit is broken, when the tensions are reestablished and light bodies are again attracted. This proves that the tensions are not the cause of the decomposition of water, or of the changes of direction of the magnetic needle discovered by M. Oersted. This second case is evidently the only one which can occur if the electromotive action is developed between the different parts of the same conducting body. The consequences deduced in this memoir from the experiments of M. Oersted will lead us to recognize the existence of this condition in the only case where there is need as yet to admit it.

2. Let us see in what consists the difference of these entirely different orders of phenomena, one of which consists in the tension and attractions or repulsions which have been long known, and the other, in decomposition of water and a great many other substances, in the changes of direction of the needle, and in a sort of attractions and repulsions entirely different from the ordinary electric attractions and repulsions; which I believe I have first discovered and which I have named *voltaic attractions* and *repulsions* to distinguish them from the others. When there is not conducting continuity from one of the bodies, or systems of bodies, in which the electromotive action develops, to the other, and when these bodies are themselves conductors, as in Volta's pile, we can only conceive this action as constantly carrying positive electricity into the one body and negative electricity into the other: in the first moment, when there is nothing opposed to the effect that it tends to produce, the two electricities accumulate, each in the part of the whole system to which it is carried, but this effect is checked as soon as the difference of electric tensions gives to their mutual attraction, which tends to reunite them, a force sufficient to make equilibrium with the electromotive action. Then everything remains in this state, except for the leakage of electricity, which may take place little by little across the non-conducting body, the air, for example, which interrupts the circuit; for it appears that there are no bodies which are perfect insulators. As this leakage takes place the tension diminishes, but since when it diminishes, the mutual attraction of the two electricities no longer makes equilibrium with the electromotive action, this last force, in case it is constant, carries anew positive electricity on one side and negative electricity on the other, and the tensions are reestablished. It is this state of a system of electromotive and conducting bodies that I called *electric tension*. We know that it exists in the two halves of this system when we separate them or even in case they remain in contact after the electromotive action has ceased, provided

that then it arose by pressure or friction between bodies which are not both conductors. In these two cases the tension is gradually diminished because of the leakage of electricity of which we have recently spoken.

But when the two bodies or the two systems of bodies between which the electromotive action arises are also connected by conducting bodies in which there is no other electromotive action equal and opposite to the first, which would maintain the state of electrical equilibrium, and consequently the tensions which result from it, these tensions would disappear or at least would become very small and the phenomena occur which have been pointed out as characterizing this second case. But as nothing is otherwise changed in the arrangement of the bodies between which the electromotive action develops, it cannot be doubted that it continues to act, and as the mutual attraction of the two electricities, measured by the difference in the electric tensions, which has become nothing or has considerably diminished, can no longer make equilibrium with this action, it is generally admitted that it continues to carry the two electricities in the two senses in which it carried them before; in such a way that there results a double current, one of positive electricity, the other of negative electricity, starting out in opposite senses from the points where the electromotive action arises, and going out to reunite in the parts of the circuits remote from these points. The currents of which I am speaking are accelerated until the inertia of the electric fluids and the resistance which they encounter because of the imperfection of even the best conductors make equilibrium with the electromotive force, after which they continue indefinitely with constant velocity so long as this force has the same intensity, but they always cease on the instant that the circuit is broken. It is this state of electricity in a series of electromotive and conducting bodies which I name, for brevity, the *electric current*; and as I shall frequently have to speak of the two opposite senses in which the two electricities move, I shall understand every time that the question arises, to avoid tedious repetition, after the words "*sense of the electric current*" these words, *of positive electricity*; so that if we are considering, for example, a voltaic pile, the expression: *direction of the electric current in the pile*, will designate the direction from the end where hydrogen is disengaged in the decomposition of water to that end where oxygen is obtained; and this expression, *direction of the electric current in the conductor which makes connection between the two ends of the pile*, will designate the direction which goes, on the contrary, from the end where oxygen appears to that where the hydrogen develops. To include these two cases in a single definition we may say that what we may call the direction of the electric current is that followed by hydrogen and the bases of the salts when water or some saline substance is a part of the circuit, and is decomposed by the current, whether, in the voltaic pile, these substances are a part of the conductor or are interposed between the pairs of which the pile is constructed.

[Several paragraphs in which it is pointed out that the electric tensions cannot be the cause of chemical or magnetic actions are omitted.]

3. The ordinary electrometer indicates tension and the intensity of the tension; there was lacking an instrument which would enable us to recognize the presence of the electric current in a pile or a conductor and which would indicate the energy and the direction of it. This instrument now exists; all that is needed is that the pile, or any portion of the conductor, should be placed horizontally, approximately in the direction of the magnetic meridian, and that an apparatus similar to a compass, which, in fact, differs from it only in the use that is made of it, should be placed above the pile or either above or below a portion of the conductor. So long as the circuit is interrupted, the magnetic needle remains in its ordinary position, but it departs from this position as soon as the current is established, so much the more as the energy is greater, and it determines the direction of the current from this general fact, that if one places oneself in thought in the direction of the current in such a way that it is directed from the feet to the head of the observer and that he has his face turned toward the needle; the action of the current will always throw toward the left that one of the ends of the needle which points toward the north and which I shall always call the austral pole of the magnetic needle, because it is the pole similar to the southern pole of the earth. I express this more briefly by saying, that the austral pole of the needle is carried to the left of the current which acts on the needle. I think that to distinguish this instrument from the ordinary electrometer we should give it the name of *galvanometer* and that it should be used in all experiments on electric currents, as we habitually use an electrometer on electric machines, so as to see at every instant if a current exists and what is its energy.

   The first use that I have made of this instrument is to employ it to show that the current in the voltaic pile, from the negative end to the positive end, has the same effect on the magnetic needle as the current in the conductor which goes on the contrary from the positive end to the negative end.

   It is well to have for this experiment two magnetic needles, one of them placed on the pile and the other above or below the conductor; we see the austral pole of each needle move to the left of the current near which it is placed; so that when the second is above the conductor it is turned to the opposite side from that toward which the needle turns which has been placed on the pile, because the currents have opposite directions in these two portions of the circuit; the two needles, on the contrary, are turned toward the same side, remaining nearly parallel with each other, when the one is above the pile and the other below the conductor. As soon as the circuit is broken they come back at once in both cases to their ordinary position.

4. Such are the differences already recognized in the effects produced by electricity in the two states which I have described, of which the one consists, if not in rest, at least in a movement which is slow and only produced because of the difficulty of completely insulating the bodies in which the electric tension exhibits itself, the other, in a double current of positive and negative electricity along a continuous circuit of conducting bodies. In the ordinary theory of electricity, we suppose that the two fluids of which we consider it composed are unceasingly separated one from the other in a part of a circuit and carried rapidly in contrary senses into

another part of the same circuit, where they are continually reunited. Although the electric current thus defined can be produced with an ordinary machine by arranging it in such a way as to develop the two electricities and by joining by a conductor the two parts of the apparatus where they are produced, we cannot, unless we use a very large machine, obtain the current with an appreciable energy except by the use of the voltaic pile, because the quantity of electricity produced by a frictional machine remains the same in a given time whatever may be the conducting power of the rest of the circuit, whereas that which the pile sets in motion during a given time increases indefinitely as we join the two extremities by a better conductor.

But the differences which I have recalled are not the only ones which distinguish these two states of electricity. I have discovered some more remarkable ones still by arranging in parallel directions two straight parts of two conducting wires joining the ends of two voltaic piles; the one was fixed and the other, suspended on points and made very sensitive to motion by a counterweight, could approach the first or move from it while keeping parallel with it. I then observed that when I passed a current of electricity in both of these wires at once they attracted each other when the two currents were in the same sense and repelled each other when they were in opposite senses. Now these attractions or repulsions of electric currents differ essentially from those that electricity produces in the state of repose; first, they cease, as chemical decompositions do, as soon as we break the circuit of the conducting bodies; secondly, in the ordinary electric attractions and repulsions the electricities of opposite sort attract and those of the same name repel; in the attractions and repulsions of electric currents we have precisely the contrary; it is when the two conducting wires are placed parallel in such a way that their ends of the same name are on the same side and very near each other that there is attraction, and there is repulsion when the two conductors, still always parallel, have currents in them in opposite senses, so that the ends of the same name are as far apart as possible. Thirdly, in the case of attraction, when it is sufficiently strong to bring the movable conductor into contact with the fixed conductor, they remain attached to one another like two magnets and do not separate after a while, as happens when two conducting bodies which attract each other because they are electrified, one positively and the other negatively, come to touch. Finally, and it appears that this last circumstance depends on the same cause as the preceding, two electric currents attract or repel in vacuum as in air, which is contrary to that which we observe in the mutual action of two conducting bodies charged with ordinary electricity. It is not the place here to explain these new phenomena; the attractions and repulsions which occur between two parallel currents, according as they are directed in the same sense or in opposite senses, are facts given by an experiment which is easy to repeat. It is necessary in this experiment, in order to prevent the motions which would be given to the movable conductor by agitation of the air, to place the apparatus under a glass cover within which we introduce, through the base which carries it, those parts of the conductors which can be joined to the two ends of the pile. The most convenient arrangement of these conductors is to place one of them on two supports in a horizontal position,

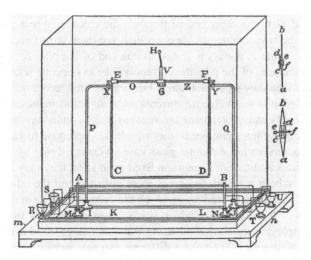

Fig. 8.1 Ampère's apparatus for measuring the force between two current-carrying wires.—[*K.K.*]

(see Fig. 8.1) in which it is fixed, and to hang up the other by two metallic wires, which are joined to it, on a glass rod which is above the first conductor and which rests on two other metal supports by very fine steel points; these points are soldered to the two ends of the metallic wires of which I have spoken, in such a way that connection is established through the supports by the aid of these points.

The two conductors are thus parallel and one beside the other in a horizontal plane; one of them is movable, because of the oscillations which it can make about the horizontal line passing through the ends of the two steel points and when it thus moves it necessarily remains parallel to the fixed conductor.

There is introduced above and in the middle of the glass rod a counterweight, to increase the mobility of the oscillating part of the apparatus, by raising its center of gravity.

I first thought that it would be necessary to set up the electric current in the two conductors by means of two different piles; but this is not necessary. The conductors may both make parts of the same circuit; for the electric current exists everywhere with the same intensity. We may conclude from this observation that the electric tensions of the two ends of the pile have nothing to do with the phenomena with which we are concerned; for there is certainly no tension in the rest of the circuit. This view is confirmed by our being able to move the magnetic needle at a great distance away from the pile by means of a very long conductor, the middle of which is curved over in the direction of the magnetic meridian above or below the needle. This experiment was suggested to me by the illustrious savant to whom the physico-mathematical sciences owe so much of the great progress that they have made in our time: it has fully succeeded.

Designate by *A* and *B* the two ends of the fixed conductor, by *C* the end of the movable conductor which is on the side of *A* and by *D* that of the same conductor which is on the side of *B*; it is plain that if one of the ends of the pile is joined to

A, B to C, and D to the other end of the pile, the electric current will be in the same sense in the two conductors; then we shall see them attract each other; if on the other hand, while A always is joined to one end of the pile, B is joined to D and C to the other end of the pile, the current will be in opposite senses in the two conductors and then they repel each other. Further, we may recognize that since the attractions and repulsions of electric currents act at all points in the circuit we may, with a single fixed conductor, attract and repel as many conductors and change the direction of as many magnetic needles as we please. I propose to have made two movable conductors within the same glass case so arranged that by making them parts of the same circuit, with a common fixed conductor, they may be alternately both attracted or both repelled, or one of them attracted and the other repelled at the same time, according to the way in which the connections are made. Following up the success of the experiment which was suggested to me by the Marquis de Laplace, by employing as many conducting wires and magnetized needles as there are letters, by fixing each letter on a different magnet, and by using a pile at a distance from these needles, which can be joined alternately by its own ends to the ends of each conductor, we may form a sort of telegraph, by which we can write all the matters which we may wish to transmit, across whatever obstacles there may be, to the person whose duty it is to observe the letters carried by the needles. By setting up above the pile a key-board of which the keys carry the same letters and by making connection by pressing them down, this method of correspondence could be managed easily and would take no more time than is necessary to touch the keys at one end and to read off each letter at the other.

If the movable conductor, instead of being adjusted so as to move parallel to the fixed conductor, can only turn in a plane parallel to the fixed conductor about a common perpendicular passing through their centers, it is clear, from the law that we have discovered of the attractions and repulsions of electric currents, that each half of the two conductors will attract or repel at the same time, according as the currents are in the same sense or in opposite senses; and consequently that the movable conductor will turn until it becomes parallel to the fixed conductor, in such a way that the currents are directed in the same sense: from which it follows that in the mutual action of two electric currents the directive action and the attractive or repulsive action depend on the same principle and are only different effects of one and the same action. It is no longer necessary, therefore, to set up between these two effects the distinction which it is so important to make, as we shall see very soon, when we are dealing with the mutual action of an electric current and of a magnet considered, as we ordinarily do, with respect to its axis, because in this action the two bodies tend to place themselves perpendicular to each other.

We now turn to the examination of this last action and of the action of two magnets on each other and we shall see that they both come under the law of the mutual action of two electric currents, if we conceive one of these currents as set up at every point of a line drawn on the surface of a magnet from one pole to the other, in planes perpendicular to the axis of the magnet, so that from the simple comparison of facts it seems to me impossible to doubt that there are really such currents about the axis of a magnet, or rather that magnetization consists in a process by which we give to the particles of steel the property of producing, in the sense of the currents of which we have spoken, the same electromotive action as is shown by the voltaic pile, by the oxidized zinc of the mineralogists, by heated tourmaline, and even in a pile made up of damp cardboard and discs of the same metal at two different temperatures. However, since this electromotive action is set up in the case of a magnet between the different particles of the same body, which is a good conductor, it can never, as we have previously remarked, produce any electric tension, but only a continuous current similar to that which exists in a voltaic pile re-entering itself in a closed curve. It is sufficiently evident from the preceding observations that such a pile cannot produce at any of its points either electric tensions or attractions or repulsions or chemical phenomena, since it is then impossible to insert a liquid in the circuit; but that the current which is immediately established in this pile will act to direct it or to attract or repel it either by another electric current or by a magnet, which, as we shall see, is only an assemblage of electric currents.

It is thus that we come to this unexpected result, that the phenomena of the magnet are produced by electricity and that there is no other difference between the two poles of a magnet than their positions with respect to the currents of which the magnet is composed, so that the austral pole is that which is to the right of these currents and the boreal pole that which is to the left.

[The rest of the memoir in the same volume of the Annales contains descriptions of more elaborate experiments by which the discoveries announced in the first part of the memoir are confirmed.]

8.3   Study Questions

QUES. 8.1. What is the nature of an electric current?

a) What is electric tension, and how can two bodies be placed in a state of electric tension? What happens when two bodies in electric tension are connected by a conducting body?
b) How does Ampère apply the Newtonian concepts of the action, inertia and resistance to the establishment of the state of electric current?
c) Of what does electric current consist? What does Ampère mean by the *sense of the electric current*? And what is the sense of the electric current *inside* a voltaic pile? Is it the same as in a conductor connecting the ends of the pile?
d) How is electric tension measured? How is electric current measured? And what benefit does a voltaic pile enjoy over, say, von Guericke's "frictional machine"?[4]

QUES. 8.2. Can Ampère's experiments on the force between currents be understood as a manifestation of the Coulomb force?

---

[4] See Chap. 3 of the present volume.

a) Describe Ampère's apparatus for measuring the force between electric currents.
b) In what four ways is the effect Ampère measured different than that produced by electric tension?
c) What precautions did Ampère take in order to ensure precision in his measurements?
d) What ingenious variation of the experiment was suggested to Ampère by the Marquis de Laplace? What invention was inspired by this suggestion?
e) Consider two square loops of wire; what is the tendency of these when a current is running through each?

QUES. 8.3. What, according to Ampère, is the cause of magnetism within a bar magnet? Why does he suggest this? Do you see any problems with his hypothesis? And is he right?

8.4   Exercises

EX. 8.1 (FORCE BETWEEN CURRENTS LABORATORY). A current balance, such as the one employed by Ampère, consists of two straight parallel conducting rods—one fixed and the other moveable—separated by a small distance. When electrical currents are driven through these rods, the attractive (or repulsive) force acting on the moveable rod may be determined by measuring the force required to keep it stationary. This is typically carried out by counterbalancing it with tiny milligram weights (or in some cases a torsion wire).[5] Using an available current balance, experimentally determine how the force between two parallel current-carrying rods, F, depends on both the current flowing through the wires, i, and the distance between the wires, r. You will need to figure out how to carry out your experiments so as to keep each one of these variables constant while changing the other. Then make a plot of $F(r)$ for constant i, and $F(i)$ for constant r. What is the mathematical relationship between the force, current and rod separation?

EX. 8.2 (AMPÈRE'S EXPERIMENT AND THE LORENTZ FORCE LAW). The force acting on each of two parallel current-carrying rods in a current balance is given empirically by the equation[6]

$$f = K \frac{i_a i_b}{r}. \tag{8.1}$$

---

[5] A relatively simple current balance (Model CP23530-50) is available from Sargent Welch, Chicago, IL. In addition to the balance itself, you will need a laser pointer (Model WL3677P-10), a DC power supply capable of delivering 10–20 A of current (Model WLS30972-70), a multimeter for measuring currents (Model WLS30712-53), a rheostat for regulating the electrical current (Air-Cooled Laboratory Rheostat, Model CP32179-10) and some 14AWG solid core hook-up wire with spade connectors for assembling the required circuit (available from Digi-Key Corporation, Thief River Falls, MN).

[6] See laboratory Ex. 8.1.

Here, f is the force (per unit length) acting on each of the rods, i_a and i_b are the electrical currents carried by the respective rods, and r is the perpendicular distance separating their central axes. The constant, K, appearing in Eq. 8.1 depends on both the nature of the medium surrounding the rods and the measurement units chosen.

On the one hand, these two current carrying rods may be thought to act directly on one another—so-called *action-at-a-distance*. On the other hand, they can be thought to act on one another *via* an intermediate agent—a *magnetic field*—which is called into existence in the space separating the wires.[7] This magnetic field communicates, or *mediates*, the force between the rods. According to this conception, the current in each rod generates a magnetic field in its vicinity; this magnetic field, in turn, exerts a force on the other current-carrying rod.[8]

To better understand the concept of mediated action, we must introduce the concept of the *lorentz force*, which is a force exerted on any electrical current which flows across magnetic field lines. According to this law, the force acting on a unit length of wire carrying an electrical current i perpendicular to a magnetic field of strength B is given by

$$f = iB. \tag{8.2}$$

In which direction does the lorentz force push the wire? Perpendicular to the plane defined by the orientation of the electrical current and the magnetic field in which the current is immersed.[9]

Now as an exercise, suppose that a straight wire carrying a current i_a is immersed in the magnetic field produced by a nearby parallel wire carrying a current i_b.

a) First, use Eq. 7.1 to find the strength of the magnetic field caused by i_b at the location of i_a.
b) Next, use the lorentz force law to find the force acting on i_a due to the magnetic field of i_b. Is your answer consistent with Eq. 8.1? (ANSWER: $f = \mu i_a i_b / 2\pi r_{ab}$.)
c) Suppose $i_a = 10$ and $i_b = 5$ A and that the wires are separated by 10 cm. What is the strength and direction of the force acting on each wire if the currents are in (i) opposite directions and (ii) the same direction.

EX. 8.3 (MAGNETIC TORQUE). Suppose that a square coil of wire has an area of $1\ \mathrm{cm}^2$. It is mounted so that it can rotate around an axis as shown in Fig. 8.2. The current running through the wire is 1 mA. This coil is then immersed in a uniform magnetic field whose strength is 100 mT. Find the maximum torque exerted by the

---

[7] Inspired by the work of Oersted, Michael Faraday suggested the existence of magnetic fields; see Chap. 28 of the present volume.

[8] The question of whether one object can, in fact, exert a force on another object "immediately", so to speak, across empty space will be explored in more detail when we come to James Clerk Maxwell's discussion of *action-at-a-distance* and *mediated action*; see Chap. 30 of the present volume.

[9] This will be formulated more precisely using the methods of vector algebra—specifically the cross-product—in Appendix A; see Ex. A.4.

Fig. 8.2 A coil of current
carrying wire placed in an
external magnetic field

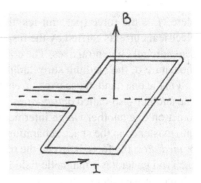

Fig. 8.3 Quantities used
when calculating the magnetic
field near a curved wire

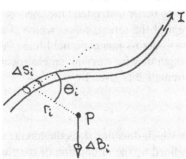

magnetic field on the coil. Briefly, you will need to calculate the force on each
section of the loop using the lorentz force law. Then you will need to compute the
torque caused by each force.[10] In what orientation with respect to the magnetic field
does the situation of maximum net torque occur? Is there any orientation of the loop
in which it experiences no net torque?

EX. 8.4 (BIOT–SAVART LAW). It is sometimes necessary to calculate the magnetic
field strength at a point, P, near an arbitrarily shaped wire carrying a constant electrical current, i. To do so, the following general procedure is used. First, the wire
carrying the current is conceptually divided into a large number, N, of very small
straight segments, Δs. Each of these segments produces a small magnetic field at
point P whose strength may be computed using the *Biot–Savart* law (see Eq. 8.3,
below). Finally, the components of these small magnetic fields, ΔB, are summed so
as to give the total magnetic field at point P.

The strength of the magnetic field due to the ith segment, Δs_i, is given by the
Biot–Savart law:

$$\Delta B_i = \frac{\mu}{4\pi} \frac{i \,\Delta s_i \sin \theta_i}{r_i^2} \tag{8.3}$$

---

[10] The torque acting on an object is the product of a force and a lever arm; see, for example Ex. 6.5
in Chap. 6 of volume II.

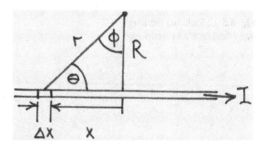

Fig. 8.4 Calculating the magnetic field near a straight wire

Here, r_i is the straight-line distance between Δs_i and the point P; θ_i is the angle between a tangent to segment Δs_i and a line connecting Δs_i and point P (see Fig. 8.3). The situation is of course complicated by the fact that the various magnetic fields, ΔB_i, may point in different directions, so the x, y, and z-components of the ΔB_i must be added (or subtracted) separately.[11]

In some cases, this summation may be performed analytically; the sum can then be converted into an integral by taking the limit

$$\lim_{N \to \infty} \sum_{i=0}^{N} \Delta B_i = \int dB$$

As an example of such a calculation, consider a long straight wire carrying a time-independent current, i (see Fig. 8.4). For exercise, you should fill in any missing steps of the following calculation of the magnetic field at point P located at a distance R from the wire. The distance R is very small compared to the length of the wire; this allows us to assume that the wire is effectively of infinite length. Using Eq. 8.3, the magnetic field due to a wire segment of length Δx which is located at distance x from the perpendicular connecting point P to the wire is given by

$$\Delta B(x) = \frac{\mu}{4\pi} \frac{i \Delta x \sin \theta}{r^2}.$$

Recall that the variables θ and r are themselves functions of the position, x, of the wire segment of length Δx. Converting finite quantities, such as ΔB and Δx, to infinitesimal quantities, dB and dx, and using a bit of trigonometry, one may write dB in terms of a single variable ϕ, which is the complement of the angle, θ:

$$dB(\phi) = \frac{\mu}{4\pi} \frac{i}{R} \cos \phi \, d\phi$$

---

[11] The direction of the magnetic field produced by each line segment is given by the right-hand rule, as mentioned previously in Ex. 7.4; see also the discussion of vector addition in Ex. A.2 of the present volume.

Fig. 8.5 Calculating the magnetic field near a kinked wire

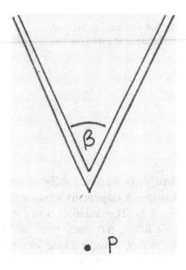

This equation may then be integrated between the (infinitely) far ends of the wire,
located at $\phi = -\pi/2$ and $\phi = \pi/2$, so as to obtain

$$B = \frac{\mu}{2\pi} \frac{i}{R} \tag{8.4}$$

Notice that our calculation of the magnetic field strength near a long straight current-
carrying wire using the Biot–Savart law is consistent with our previous Eq. 7.1.

EX. 8.5 (MAGNETIC FIELD NEAR A BENT WIRE). Suppose that a long straight wire
carrying a current i is bent sharply into a V-shape with an interior angle β, as shown
in Fig. 8.5. Find a mathematical expression for the strength of the magnetic field at
a distance R directly below the tip of the V and in the same plane as the wire.
(ANSWER: $\mu i \tan(\beta/2)/2\pi R$.)

8.5   Vocabulary

1. Electromagnetic
2. Electrodynamic
3. Electrostatic
4. Tourmaline
5. Electrometer
6. Alkali
7. Decomposition
8. Continuity
9. Meridian
10. Austral

11. Galvanometer
12. Agitation
13. Soldered
14. Savant
15. Telegraph
16. Magnetization

Chapter 9
Work and Weight

> *There is a kind, I might almost say, of artistic satisfaction, when*
> *we are able to survey the enormous wealth of Nature as a*
> *regularly-ordered whole—a Kosmos, an image of the logical*
> *thought of our own mind.*
>
> —Hermann von Helmholtz

9.1 Introduction

In our previous readings, we explored what are nowadays called *electrostatics* and *magnetostatics*. The former is the study of stationary electrical charges; the later is the study of static, or time-independent, magnetic fields. We began in the year 1600 with William Gilbert's theory of permanent magnets, and ended with Ampère's study of electro-magnetic forces in the 1820s. Shortly, we will step back in time once again to look at the concurrent development of our understanding of light, beginning in the late 1600s with Newton and Huygens' competing particle and wave theories, and ending with the experimental work of Young, Arago and Tyndall. After this historical look at electrostatics, magnetostatics and light as separate and distinct areas of study, we will finally see how they fit together into what became known as the *electromagnetic theory of light*, which was developed primarily by Faraday and Maxwell during the second half of the nineteenth century.

Before commencing our study of light, however, let us take a short detour, for the next three chapters, to look at one of the most significant organizing principles in all of science—the *conservation of energy*—which was articulated in its modern form in the mid 1800s by Hermann von Helmholtz (1821–1894). Hermann Helmholtz was born in the city of Potsdam in the Kingdom of Prussia.[1] Very early he developed a love of books—especially physics and mathematics books—but for economic reasons his parents encouraged him to study medicine. In 1837, he obtained a government grant to study at the Royal Friedrich-Wilhelm Institute of Medicine and Surgery in Berlin—with the stipulation that he would subsequently serve as a doctor in the Prussian army. So after graduating he was assigned to a regiment in Potsdam,

---

[1] The nobiliary particle *von* was added to Helmholtz's name in 1882, when German Emperor Wilhelm I bestowed on him the rank of hereditary nobility.

© Springer International Publishing Switzerland 2016

K. Kuehn, *A Student's Guide Through the Great Physics Texts*,
Undergraduate Lecture Notes in Physics, DOI 10.1007/978-3-319-21816-8_9

where he spent his free time doing experimental research on muscle physiology. At the time, living organisms were thought to possess a *vital principle*, as it were, which could initiate forces within the organism. How else, after all, could one explain the directed growth and willful action of plants and animals? Helmholtz was uncomfortable with this philosophy of nature, which seemed to suggest that living bodies were themselves capable of *perpetual motion*. So in 1847, informed by his studies of heat produced during muscle action, he published an essay entitled *Über die Erhaltung der Kraft*. In this essay, Helmholtz aimed to provide a "critical investigation and arrangement of the facts for the benefit of physiologists."[2] The facts to which he was referring—drawn from a wide range of mechanical, thermodynamic, and electrical systems—pertain to the question of whether natural forces are in fact capable of producing perpetual motion. Helmholtz thinks not, and proceeds to lay out his general theory of the conservation of force, which is today known as the *conservation of energy*.[3] Consequently, Helmholtz was relieved of his military obligation and was instead appointed professor at the University of Königsberg in 1849, where he lectured on general pathology and physiology. He would later serve as professor of anatomy and physiology at the University of Bonn, professor of physiology at the University of Heidelberg, and professor of physics at the University of Berlin. While preparing his lectures in Königsberg, Helmholtz conceived of the ophthalmoscope, and became the first to obtain a clear image of the human retina. He also measured the rate at which electrical signals travel through nerve fibers, established a number of significant mathematical theorems regarding topology and vortex motion in fluids,[4] wrote a comprehensive treatise on musical theory and the physics of sound perception,[5] and carried out investigations of vibrations in electronic circuits; these electromagnetic vibrations would later be employed by Helmholtz's student, Heinrich Hertz, to study high-frequency radio waves and confirm Maxwell's electromagnetic theory of light.[6]

Edmund Atkinson's 1885 English translation of Helmholtz's lecture, *On the Conservation of Force*, which is based on his famous 1847 essay, is reproduced in the next three chapters of the present volume. Helmholtz begins by offering some general thoughts on how the natural sciences are distinct from the mental sciences. Do you agree with his sentiments?

---

[2] Helmholtz provides this explanation in a speech he delivered on the occasion of his 70th birthday in 1891. See Chap. 14 of Helmholtz, H., *Science and Culture*, The University of Chicago Press, 1995.

[3] The term *energy* was introduced by the British scientist Thomas Young. See Lectures V and VIII in Young, T., *Course of lectures on natural philosophy and the mechanical arts*, vol. 1, Printed for Joseph Johnson, London, 1807.

[4] For a comprehensive review of Helmholtz's impact on the subsequent study of fluid dynamics, see Meleshko, V. V., Coaxial axisymmetric vortex rings: 150 years after Helmholtz, *Theor. Comput. Fluid Dyn.*, 24, 403–431, 2010.

[5] Helmholtz, H., *On the Sensation of Tone as a Physiological Basis for the Theory of Music*, fourth ed., Longmans, Green, and Co., New York, Bombay and Calcutta, 1912.

[6] Read Hertz's own description of his apparatus, method and discoveries in Chap. 33 of the present volume.

9.2   Reading: Helmholtz, *On the Conservation of Force*

Helmholtz, H., On the Conservation of Force, in *Popular Lectures on Scientific Subjects*, edited by E. Atkinson, D. Appleton and Company, New York, 1885. Pages 317–331.

9.2.1   *Introduction to a Series of Lectures Delivered at Carlsruhe in the Winter of 1862–1863*

As I have undertaken to deliver here a series of lectures, I think the best way in which I can discharge that duty will be to bring before you, by means of a suitable example, some view of the special character of those sciences to the study of which I have devoted myself. The natural sciences, partly in consequence of their practical applications, and partly from their intellectual influence on the last four centuries, have so profoundly, and with such increasing rapidity, transformed all the relations of the life of civilized nations; they have given these nations such increase of riches, enjoyment of life, of the preservation of health, of means of industrial and of social intercourse, and even such increase of political power, that every educated man who tries to understand the forces at work in the world in which he is living, even if he does not wish to enter upon the study of a special science, must have some interest in that peculiar kind of mental labour, which works and acts in the sciences in question.

On a former occasion I have already discussed the characteristic differences which exist between the natural and the mental sciences as regards the kind of scientific work. I then endeavoured to show that it is more especially in the thorough conformity with law which natural phenomena and natural products exhibit, and in the comparative ease with which laws can be stated, that this difference exists. Not that I wish by any means to deny, that the mental life of individuals and peoples is also in conformity with law, as is the object of philosophical, philological, historical, moral, and social sciences to establish. But in mental life, the influences are so interwoven, that any definite sequence can but seldom be demonstrated. In Nature the converse is the case. It has been possible to discover series of natural phenomena with such accuracy and completeness that we can predict their future occurrence with the greatest certainty; or in cases in which we have power over the conditions under which they occur, we can direct them just according to our will. The greatest of all instances of which the human mind can effect by means of a well-recognised law of natural phenomena is that afforded by modern astronomy. The one simple law of gravitation regulates the motions of the heavenly bodies not only of our planetary system, but also of the far more distant double stars; from which, even the ray of light, the quickest of all messengers, needs years to reach our eye; and, just on account of this simple conformity with law, the motions of bodies in question can be accurately predicted and determined both for the past and for future years and centuries to a fraction of a minute.

On this exact conformity with law depends also the certainty with which we know how to tame the impetuous force of steam, and to make it the obedient servant of our wants. On this conformity depends, moreover, the intellectual fascination which chains the physicist to his subjects. It is an interest of quite a different kind to that which mental and moral sciences afford. In the latter it is man in the various phases of his intellectual activity who chains us. Every great deed of which history tells us, every mighty passion which art can represent, every picture of manners, of civic arrangements, of the culture of peoples of distant lands or of remote times, seizes and interests us, even if there is no exact scientific connection among them. We continually find points of contact and comparison in our conceptions and feelings; we get to know the hidden capacities and desires of the mind, which in the ordinary peaceful course of civilised life remain unawakened.

It is not to be denied that, in the natural sciences, this kind of interest is wanting. Each individual fact, taken by itself, can indeed arouse our curiosity or our astonishment, or be useful to us in its practical applications. But intellectual satisfaction we obtain only from a connection of the whole, just from its conformity with law. *Reason* we call that faculty innate in us of discovering laws and applying them with thought. For the unfolding of the peculiar forces of pure reason in their entire certainty and in their entire bearing, there is no more suitable arena than inquiry into Nature in the wider sense, the mathematics included. And it is not only the pleasure at the successful activity of one of our most essential mental powers; and the victorious subjections to the power of our thought and will of an external world, partly unfamiliar, and partly hostile, which is the reward of this labour; but there is a kind, I might almost say, of artistic satisfaction, when we are able to survey the enormous wealth of Nature as a regularly-ordered whole—a *kosmos*, an image of the logical thought of our own mind.

The last decades of scientific development have led us to the recognition of a new universal law of all natural phenomena, which from its extraordinarily extended range, and from the connection which it constitutes between natural phenomena of all kinds, even of the remotest times and the most distant places, is especially fitted to give us an idea of what I have described as the character of the natural sciences, which I have chosen as the subject of this lecture.

This law is *the Law of the Conservation of Force*, a term the meaning of which I must first explain. It is not entirely new; for individual domains of natural phenomena it was enunciated by Newton and Daniel Bernoulli; and Rumford and Humphry Davy have recognised distinct features of its presence in the laws of heat.

The possibility that it was of universal application was first stated by Dr. Julius Robert Mayer, a Schwabian physician (now living in Heilbronn), in the year 1842, while almost simultaneously with, and independently of him, James Prescot Joule, an English manufacturer, made a series of important and difficult experiments on the relation of heat to mechanical force, which supplied the chief points in which the comparison of the new theory with experience was still wanting.

The law in question asserts, that the *quantity of force which can be brought into action in the whole of Nature is unchangeable*, and can neither be increased nor diminished. My first object will be to explain to you what is understood by *quantity of force*; or, as the same idea is more popularly expressed with reference to its technical application, what we call *amount of work* in the mechanical sense of the word.

The idea of work for machines, or natural processes, is taken from comparison with the working power of a man; and we can therefore best illustrate with human labour the most important features of the question with which we are concerned. In speaking of the work of machines and of natural forces we must, of course, in this comparison eliminate anything in which activity of intelligence comes into play. The latter is also capable of the hard and intense work of thinking, which tries a man just as muscular exertion does. But whatever of the actions of intelligence is met with in the work of machines, of course is due to the mind of the constructor and cannot be assigned to the instrument at work.

Now, the external work of man is of the most varied kind as regards the force or ease, the form and rapidity, of the motions used on it, and the kind of work produced. But both the arm of the blacksmith who delivers his powerful blows with the heavy hammer, and that of the violinist who produces the most delicate variations in sound, and the hand of the lace-maker who works with threads so fine that they are on the verge of the invisible, all these acquire the force which moves them in the same manner and by the same organs, namely, the muscles of the arms. An arm the muscles of which are lamed is incapable of doing any work; the moving force of the muscle must be at work in it, and these must obey the nerves, which bring to them orders from the brain. That member is then capable of the greatest variety of motions; it can compel the most varied instruments to execute the most diverse tasks.

Just so it is with machines: they are used for the most diversified arrangements. We produce by their agency an infinite variety of movements, with the most various degrees of force and rapidity, from powerful steam hammers and rolling mills, where gigantic masses of iron are cut and shaped like butter, to spinning and weaving frames, the work of which rivals that of the spider. Modern mechanism has the richest choice of means of transferring the motion of one set of rolling wheels to another with greater or less velocity; of changing the rotating motion of wheels into the up-and-down motion of the piston rod, of the shuttle, of falling hammers and stamps; or, conversely, of changing the latter into the former; or it can, on the other hand, change movements of uniform into those of varying velocity, and so forth. Hence this extraordinarily rich utility of machines for so extremely varied branches of industry. But one thing is common to all these differences; they all need a *moving force*, which sets and keeps them in motion, just as the works of the human hand all need the moving force of the muscles.

Now, the work of the smith requires a far greater and more intense exertion of the muscles than that of the violin player; and there are in machines corresponding differences in the power and duration of the moving force required. These differences, which correspond to the different degree of exertion of the muscles in human labour, are alone what we have to think of when we speak of the *amount of work* of a machine. We have nothing to do here with the manifold character of the actions and arrangements which the machines produce; we are only concerned with an expenditure of force.

This very expression which we use so fluently, 'expenditure of force,' which indicates that the force applied has been expended and lost, leads us to a further characteristic analogy between the effects of the human arm and those of machines. The greater the exertion, and the longer it lasts, the more is the arm *tired*, and the more *is the store of its moving force for the time exhausted.* We shall see that this peculiarity of becoming exhausted by work is also met with in the moving forces of inorganic nature; indeed, that this capacity of the human arm of being tired is only one of the consequences of the law with which we are now concerned. When fatigue sets it, recovery is needed, and this can only be effected by rest and nourishment. We shall find that also in the inorganic moving forces, when their capacity for work is spend, there is spent a possibility of reproduction, although in general other means must be used to this end than in the case of the human arm.

From the feeling of exertion and fatigue in our muscles, we can form a general idea of what we understand by amount of work; but we must endeavour, instead of the indefinite estimate afforded by this comparison, to form a clear and precise idea of the standard by which we have to measure the amount of work. This we can do better by the simplest inorganic moving forces than by the actions of our muscles, which are a very complicated apparatus, acting in an extremely intricate manner.

Let us now consider that moving force which we know best, and which is simplest—gravity. It acts, for example, as such in those clocks which are driven by a weight. This weight, fastened to a string, which is wound round a pulley connected with the first toothed wheel of the clock, cannot obey the pull of gravity without setting the whole clockwork in motion. Now I must beg you to pay special attention to the following points: the weight cannot put the clock in motion without itself sinking; did the weight not move, it could not move the clock, and its motion can only be such a one as obeys the action of gravity. Hence, if the clock is to go, the weight must continually sink lower and lower, and it must at length sink so far that the string which supports it is run out. The clock then stops. The usual effect of its weight is for the present exhausted. Its gravity is not lost or diminished; it is attracted by the earth as before, but the capacity of this gravity to produce the motion of the clock work is lost. It can only keep the weight at rest in the lowest point of its path, it cannot farther put it in motion.

But we can wind up the clock by the power of the arm, by which the weight is again raised. When this has been done, it has regained its former capacity, and can again set the clock in motion.

We learn from this that a raised weight possesses a *moving force*, but that it must necessarily sink if this force is to act; that by sinking, this moving force is exhausted, but by using another extraneous moving force—that of the arm—its activity can be restored.

The work which the weight has to perform in driving the clock is not indeed great. It has continually to overcome the small resistances which the friction of the axles and teeth, as well as the resistance of the air, opposed to the motion of the wheels, and it has to furnish the force for the small impulses and sounds which the pendulum produces at each oscillation. If the weight is detached from the clock, the pendulum swings for a while before coming to a rest, but its motion becomes each moment feebler, and ultimately ceases entirely, being gradually used up by the small hindrances I have mentioned. Hence, to keep the clock going, there must be a moving force, which, though small, must be continually at work. Such a one is the weight.

We get, moreover, from this example, a measure for the amount of work. Let us assume that a clock is driven by a weight of a pound, which falls 5 ft in 24 h. If we fix ten such clocks, each with a weight of 1 pound, then ten clocks will be driven 24 h; hence as each has to overcome the same resistances in the same time as the others, ten times as much work is performed for 10 pounds through 5 ft. Hence, we conclude that the height of the fall being the same, the work increases directly as the weight.

Now, if we increase the length of the string so that the weight runs down 10 ft, the clock will go 2 days instead of one; and with double the height of the fall, the weight will overcome on the second day the same resistances as the first, and will therefore do twice as much work as when it can only run down 5 ft. The weight being the same, the work increases as the height of fall. Hence, we may take the product of the weight into the height of fall as a measure of work, at any rate, in the present case. The application of this measure is, in fact, not limited to the individual case, but the universal standard adopted in manufactures for measuring magnitude of work is a *foot-pound*—that is, the amount of work which a pound raised through a foot can produce.[7]

We may apply this measure of work to all kinds of machines, for we should be able to set them all in motion by means of a weight sufficient to turn a pulley. We could thus always express the magnitude of any driving force, for any given machine, by the magnitude and height of fall of such a weight as would be necessary to keep the machine going with its arrangements until it had performed a certain work. Hence it is that the measurement of work by foot pounds is universally applicable. The use of such a weight as a driving force would not indeed be practically advantageous in those cases in which we were compelled to raise it by the power of our own arm; it would in that case be simpler to work the machine by the direct action of that arm. In the clock we use a weight so that we need not stand the whole day that the clockwork, as we should have to do to move it directly. By winding up the clock we accumulate a store of working capacity in it, which is sufficient for the expenditure of the next 24 h.

The case is somewhat different when Nature herself raises the weight, which then works for us. She does not do this with solid bodies, at least not with such regularity as to be utilised; but she does it abundantly with water, which, being raised to the tops of mountains by meteorological processes, returns in streams from them. The gravity of the water we use a moving force, the most direct application being in

---

[7] This is the *technical* measure of work; to convert it into scientific measure it must be multiplied by the intensity of gravity.

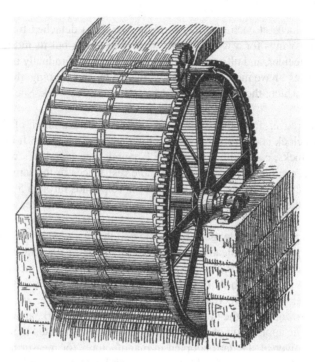

Fig. 9.1 An overshot wheel.—[*K.K.*]

what are called *overshot* wheels, one of which is represented in Fig. 9.1. Along the circumference of such a wheel are a series of buckets, which act as receptacles for the water, and, on the side turned to the observer, have the tops uppermost; on the opposite side the tops of the buckets are upside-down. The water flows in at *M* into the buckets of the front of the wheel, and at *F*, where the mouth begins to incline downwards, it flows out. The buckets on the circumference are filled on the side turned to the observer, and empty on the other side. Thus the former are weighted by the water contained in them, the latter not; the weight of the water acts continuously on only one side of the wheel, draws this down, and thereby turns the wheel; the other side of the wheel offers no resistance, for it contains no water. It is thus the weight of the falling water which turns the wheel, and furnishes the motive power. But you will at once see that the mass of water which turns the wheel must necessarily fall in order to do so, and that though, when it reaches the bottom, it has lost none of its gravity, it is no longer in a position to drive the wheel, if it is not restored to its original position, either by the power of the human arm or by means of some other natural force. If it can flow from the mill-stream to still lower levels, if may be used to work other wheels. But when it has reached its lowest level, the sea, the last remainder of the moving force is used up, which is due to gravity—that is, to the attraction of the earth, and it cannot act by its weight until it has been again raised to a high level. As this is actually effected by meteorological processes, you will at once observe that these are to be considered as sources of moving force.

Water power was the first inorganic force which man learnt to use instead of his own labour or that of domestic animals. According to Strabo, it was known that King Mithradates of Pontus, who was also celebrated for his knowledge of Nature; near his palace there was a water wheel. Its use was first introduced among the Romans, in the time of the first Emperors. Even now we find water mills in all mountains, valleys, or wherever there are rapidly-flowing regularly-filled brooks and streams. We find water power used for all purposes which can possibly be effected by machines. It drives mills which grind corn, sawmills, hammers and oil presses, spinning frames and looms, and so forth. It is the cheapest of all motive powers, it flows spontaneously from the inexhaustible stores of Nature; but it is restricted to a particular place, and only in mountainous countries is it present in any quantity; in level countries extensive reservoirs are necessary for damming the rivers to produce any amount of water power.

Before passing to the discussion of other motive forces I must answer an objection which may readily suggest itself. We all know that there are numerous machines, systems of pulleys, levers and cranes, by the aid of which heavy burdens may be lifted by a comparatively small expenditure of force. We have all of us often seen one or two workmen hoist heavy masses of stones to great heights, which they would be quite unable to do directly; in like manner, one or two men, by means of a crane, can transfer the largest and heaviest chests from a ship to the quay. Now, it may be asked, If a large, heavy weight had been used for driving a machine, would it not be very easy, by means of a crane or system of pulleys, to raise it anew, so that it could again be used as a motor, and thus acquire a motive power, without being compelled to use a corresponding exertion in raising the weight?

The answer to this is, that all these machines, in that degree in which for the moment they facilitate the exertion, also prolong it, so that by their help no motive power is ultimately gained. Let us assume that four labourers have to raise a load of four hundredweight by means of a rope passing over a single pulley. Every time the rope is pulled down through 4 ft, the load is also raised through 4 ft. But now, for the sake of comparison, let us suppose the same load hung to a block of four pulleys, as represented in Fig. 9.2. A single labourer would now be able to raise the load by the same exertion of force as each one of the four put forth. But when he pulls the rope through 4 ft, the load only rises 1 ft, for the length through which he pulls the rope, at A, is uniformly distributed in the block over four ropes, so that each of these is only shortened by a foot. To raise the load, therefore, to the same height, the one man must necessarily work four times as long as the four together did. But the total expenditure of work is the same, whether four labourers work for a quarter of an hour or one works for an hour.

If, instead of human labour, we introduce the work of a weight, and hang to the block a load of 400, and at A, where otherwise the labourer works, a weight of 100 pounds, the block is then in equilibrium, and, without any appreciable exertion of the arm, may be set in motion. The weight of 100 pounds sinks, that of 400 rises. Without any measurable expenditure of force, the heavy weight has been raised

Fig. 9.2 A weight suspended
by four pullies.—[*K.K.*]

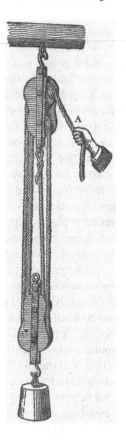

by the sinking of the smaller one. But observe that the smaller weight will have
sunk through four times the distance that the greater one has risen. But a fall of
100 pounds through 4 ft is just as much as 400 foot-pounds as a fall of 400 pounds
through 1 ft.

The action of levers in all their various modifications is precisely similar. Let *ab*,
Fig. 9.3, be a simple lever, supported at *c*, the arm *cb* being four times as long as the
other arm *ac*. Let a weight of 1 pound be hung at *b*, and a weight of 4 pounds be at
a, the lever is then in equilibrium, and the least pressure of the finger is sufficient,
without any appreciable exertion of force, to place it in the position *a'b'*, in which
the heavy weight of 4 pounds has been raised, while the 1-pound weight has sunk.
But here, also, you will observe no work has been gained, for while the heavy weight
has been raised through 1 in., the lighter one has fallen through 4 in.; and 4 pounds
through 1 in. is, as work, equivalent to the product of 1 pound through 4 in.

Most other fixed parts of machines may be regarded as modified and compound
levers; a toothed-wheel, for instance as a series of levers, the ends of which are represented by the individual teeth, and one after the other of which is put in activity

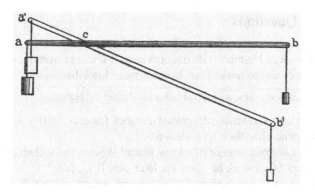

Fig. 9.3 A simple lever.—[*K.K.*]

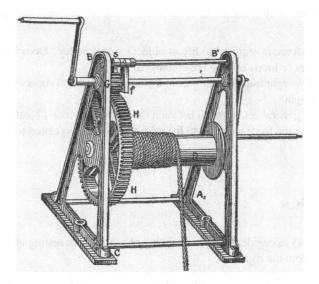

Fig. 9.4 A crabwinch.—[*K.K.*]

in the degree in which the tooth in question seizes or is seized by the adjacent pinion. Take, for instance, the crabwinch, represented in Fig. 9.4. Suppose the pinion on the axis of the barrel of the winch has 12 teeth, and the toothed-wheel, HH, 72 teeth, that is, six times as many as the former. The winch must now be turned round six times before the toothed-wheel, H, and the barrel, D, have made one turn, and before the rope which raises the load has been lifted by a length equal to the circumference of the barrel. The workman thus requires six times the time, though to be sure only one-sixth of the exertion, which he would have to use if the handle were directly applied to the barrel, D. In all these machines, and parts of machines, we find it confirmed that in proportion as the velocity of the motion increases its power diminishes, and that when the power increases the velocity diminishes, but that the amount of work is never thereby increased.

9.3   Study Questions

QUES. 9.1. How does Helmholtz distinguish mental sciences from natural sciences? Do they both conform to laws? If so, how are these laws discovered?

QUES. 9.2. What is the new universal law introduced by Helmholtz?

a) Which previous scientists recognized distinct features of this law? How did Helmholtz generalize their formulations?
b) What is the common feature of all machines? What is the technical meaning of the term *work*? Can work be saved for later use? If so, how?
c) How much work is accomplished by a 1-pound weight which falls by 5 ft, under the influence of gravity? More generally, what are the units of work?

QUES. 9.3. How much weight can be lifted with a machine such as a pulley or a lever?

a) How much force is required to lift an object using a pulley? Does it matter how many pullies or loops of rope are employed?
b) How much weight must be hung from a balance in order to support, or even lift, another weight?
c) What is the general relationship between the distance that an object is lifted by a machine and the force required to lift it? And how is this related to Helmholtz's new law?

9.4   Exercises

EX. 9.1 (LEVER). Consider a lever consisting of a 5 ft beam resting atop a fulcrum which is 2 ft from the right side.

a) If the left end of the lever is depressed at a speed of 1 in./s, with what speed does the right end ascend? (ANSWER: ⅔ in./s)
b) What force must be applied to the left end in order to lift a weight of 8 pounds placed at the right end of the lever at a constant velocity? What happens if a greater force is applied?
c) Is the action of the lever consistent with Helmholtz's universal law? Explain.

EX. 9.2 (HYDROELECTRIC POWER). One hundred and fifty thousand gallons of water plunge over Niagara Falls each second. The height of Niagara Falls is 176 ft. In principle, how many watts (Joules per second) of electrical energy could a hydroelectric plant produce from Niagara falls? (ANSWER: 300 MW)

EX. 9.3 (KOSMOS ESSAY). What does Helmholtz mean when he says that Nature is a "regularly-ordered whole—a *kosmos*"? Is he right? If so, what are the implications of this claim?

EX. 9.4 (VITAL PRINCIPLE ESSAY). Why might scientists have believed that motion and force can be initiated by a vital principle which resides in living things? Consider: how does a person's *desire* to move his body give rise to his body's subsequent *motion*? How might Helmholtz address this "mind-body" problem?

9.5   Vocabulary

1. Conformity
2. Phenomena
3. Impetuous
4. Conception
5. Kosmos
6. Manifold
7. Inorganic
8. Intricate
9. Extraneous
10. Motive
11. Appreciable
12. Equilibrium
13. Crabwinch
14. *Vis viva*

Chapter 10
Kinetic and Potential Energy

The velocity of a moving mass can act as motive force.
—Hermann von Helmholtz

10.1 Introduction

In the first part of his lecture *On the Conservation of Force*, Helmholtz defines the *work* done by a moving force as the product of the moving force and the distance over which the force acts:

$$W = F \cdot d. \tag{10.1}$$

Thus, lifting a 5-pound body through a vertical distance of 1 ft requires an agent to do 5 ft-pounds of work.[1] A lifting force which happens to be larger than 5 pounds would simply rise at a constant velocity. This is important, since up until now Helmholtz has been focusing his attention on bodies which are lifted (or which fall) at a constant velocity. What, then, is the connection between the work done on (or by) a body and the body's change in velocity? It is this question to which Helmholtz now turns.

10.2 Reading: Helmholtz, *On the Conservation of Force*

Helmholtz, H., On the Conservation of Force, in *Popular Lectures on Scientific Subjects*, edited by E. Atkinson, D. Appleton and Company, New York, 1885, pp. 331–342.

In the overshot mill wheel, described above, water acts by its weight. But there is another form of mill wheels, what is called the *undershot wheel*, in which it only acts by its impact, as represented in Fig. 10.1. These are used where the height from

---

[1] Moving the same weight horizontally at a constant speed does not require the agent to do any work at all, since the force which supports the weight is perpendicular to its direction of motion.

© Springer International Publishing Switzerland 2016
K. Kuehn, *A Student's Guide Through the Great Physics Texts*,
Undergraduate Lecture Notes in Physics, DOI 10.1007/978-3-319-21816-8_10

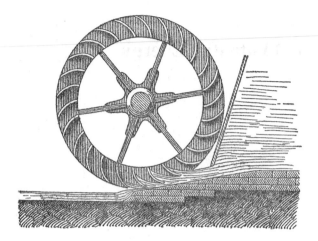

Fig. 10.1 An undershot wheel.—[*K.K.*]

which the water comes is not great enough to flow on the upper part of the wheel. The lower part of undershot wheels dips in the flowing water which strikes against their float-boards and carries them along. Such wheels are used in swift-flowing streams which have a scarcely perceptible fall, as, for instance, on the Rhine. In the immediate neighborhood of such a wheel, the water need not necessarily have a great fall if it only strikes with considerable velocity. It is the velocity of the water exerting an impact against the float-boards, which acts in this case, and which produces the motive power.

Windmills, which are used in the great plains of Holland and North Germany to supply the want of falling water, afford another instance of the action of velocity. The sails are driven by air in motion—by wind. Air at rest could just as little drive a windmill as water at rest a water wheel. The driving force depends here on the velocity of moving masses.

A bullet resting in the hand is the most harmless thing in the world; by its gravity it can exert no great effect; but when fired and endowed with great velocity it drives through all obstacles with the most tremendous force.

If I lay the head of a hammer gently on a nail, neither its small weight nor the pressure of my arm is quite sufficient to drive the nail into the wood; but if I swing the hammer and allow it to fall with great velocity, it acquires new force, which can overcome far greater hindrances.

These examples teach us that the velocity of a moving mass can act as motive force. In mechanics, velocity in so far as it is motive force, and can produce work, is called *vis viva*. The name is not well chosen; it is too apt to suggest to us the force of living beings. Also in this case, you will see, from the instances of the hammer and of the bullet, that velocity is lost, as such, when it produces working power. In the case of the water mill, or of the windmill, a more careful investigation of the moving masses of water and air is necessary to prove that part of their velocity has been lost by the work which they have performed.

Fig. 10.2 A pendulum.
—[K.K.]

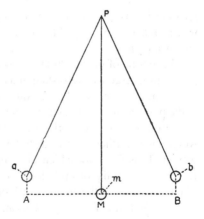

The relation of velocity to working power is most simply and clearly seen in a simple pendulum, such as can be constructed by any weight which we suspend to a cord. Let M (see Fig. 10.2) be such a weight, of a spherical form; AB, a horizontal line drawn through the center of the sphere; P the point at which the cord is fastened. If now I draw the weight M on one side towards A, it moves in the arc Ma, the end of which, a is somewhat higher than the point A in the horizontal line. The weight is thereby raised to the height Aa. Hence my arm must exert a certain force to bring the weight to a. Gravity resists this motion, and endeavours to bring back the weight to M, the lowest point which it can reach.

Now, if after I have brought the weight to a I let it go, it obeys this force of gravity and returns to M, arrives there with a certain velocity, and no longer remains quietly hanging at M as it did before, but swings beyond M towards b, where its motion stops as soon as it has traversed on the side of B an arc equal in length to that on the side of A, and after it has risen to a distance Bb above the horizontal line, which is equal to the height Aa, to which my arm had previously raised it. In b the pendulum returns, swings the same way back through M towards a, and so on, until its oscillations are gradually diminished, and ultimately annulled by the resistance of the air and by friction.

You see here that the reason why the weight, when it comes from a to M, and does not stop there, but ascends to b, in opposition to the action of gravity, is only to be sought in its velocity. The velocity which it has acquired in moving from the height Aa is capable of again raising it to Bb. The velocity of the moving mass, M, is thus capable of raising this mass; that is to say, in the language of mechanics, of performing work. This would also be the case if we had imparted such a velocity to the suspended weight by a blow.

From this we learn further how to measure the working power of velocity—or, what is the same thing, the *vis viva* of the *moving mass*. It is equal to the work, expressed in foot-pounds, which the same mass can exert after its velocity has been used to raise it, under the most favourable circumstances, to as great a height as possible.[2] This does not depend on the direction of the velocity; for if we swing a weight attached to a thread in a circle, we can even change a downward motion into an upward one.

The motion of the pendulum shows us very distinctly how the forms of working power hitherto considered—that of a raised weight and that of a moving mass—may merge into one another. In the points *a* and *b*, Fig. 10.2, the mass has no velocity; at the point *M* it has fallen as far as possible, but possesses velocity. As the weight goes from *a* to *m* the work of the raised weight is changed into *vis viva*; as the weight goes further from *m* to *b* the *vis viva* is changed into the work of a raised weight. Thus the work which the arm originally imparted to the pendulum is not lost in these oscillations provided we may leave out of consideration the influence of the resistance of the air and of friction. Neither does it increase, but it continually changes the form of its manifestation.

Let us now pass to other mechanical forces, those of elastic bodies. Instead of the weights which drive our clocks, we find in timepieces and watches, steel springs which are coiled in winding up the clock, and are uncoiled by the working of the clock. To coil up the spring we consume the force of the arm; this has to overcome the resisting elastic force of the spring as we wind it up, just as in the clock we have to overcome the force of gravity which the weight exerts. The coiled up spring can, however, perform work; it gradually expends this acquired capability in driving the clockwork.

If I stretch a crossbow and afterwards let it go, the stretched string moves the arrow; it imparts to it force in the form of velocity. To stretch the cord my arm must work for a few seconds; this work is imparted to the arrow at the moment it is shot off. Thus the crossbow concentrates into an extremely short time the entire work which the arm had communicated in the operation of stretching; the clock, on the contrary, spreads it over one or several days. In both cases, no work is produced which my arm did not originally impart to the instrument; it is only expended more conveniently.

The case is somewhat different if by any other natural process I can place an elastic body in a state of tension without having to exert my arm. This is possible and is most easily observed in the case of gasses.

If, for instance, I discharge a firearm loaded with gunpowder, the greater part of the mass of the powder is converted into gasses at a very high temperature, which have a powerful tendency to expand, and can only be retained in the narrow space in which they are formed, by the exercise of the most powerful pressure. In expanding with enormous force they propel the bullet, and impart to it a great velocity, which we have already seen is a form of work.

In this case, then, I have gained work which my arm has not performed. Something, however, has been lost—, the gunpowder, that is to say, whose constituents

---

[2] The measure of *vis viva* in theoretical mechanics is half the product of the weight into the square of the velocity. To reduce it to the technical measure of the work we must divide it by the intensity of gravity; that is, by the velocity at the end of the first second of a freely falling body.

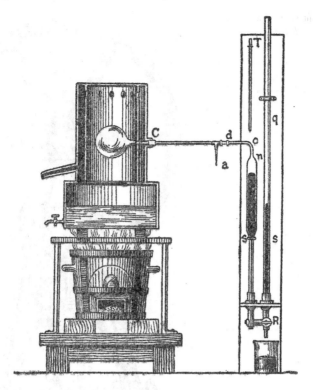

Fig. 10.3 A boiler.—[*K.K.*]

have changed into other chemical compounds, from which they cannot, without further ado, be restored to their original condition. Here, then, a chemical change has taken place, under the influence of which work has been gained.

Elastic forces are produced in gases by the aid of heat, on a far greater scale.

Let us take, as the most simple instance, atmospheric air. In Fig. 10.3 an apparatus is represented such as Regnault used for measuring the expansive force of heated gasses. If no great accuracy is required in the measurement, the apparatus may be arranged more simply. At *C* is a glass globe filled with dry air, which is placed in a metal vessel, in which it can be heated by steam. It is connected with the U-shaped tube *Ss*, which contains a liquid, and the limbs of which communicate with each other when the stopcock *R* is closed. If the liquid is in equilibrium in the tube *Ss* when the globe is cold, it rises in the leg *s*, and ultimately overflows when the globe is heated. If, on the contrary, when the globe is heated, equilibrium be restored by allowing some of the liquid to flow out at *R*, as the globe cools it will be drawn up towards *n*. In both cases liquid is raised, and work thereby produced.

The same experiment is continuously repeated on the largest scale in steam engines, though, in order to keep up a continual disengagement of compressed gases

Fig. 10.4 Front view of a
steam engine.—[*K.K.*]

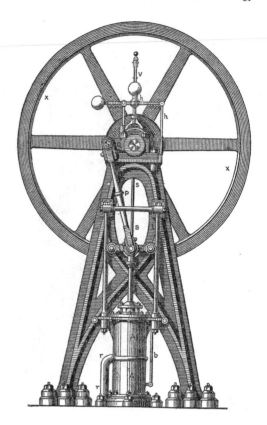

from the boiler, the air in the globe in Fig. 10.3, which would soon reach the maximum of its expansion, is replaced by water, which is gradually changed into steam by the application of heat. But steam, so long as it remains such, is an elastic gas which endeavours to expand exactly like atmospheric air.

And instead of the column of liquid which was raised in our last experiment, the machine is caused to drive a solid piston which imparts its motion to other parts of the machine. Figure 10.4 represents a front view of the working parts of a high-pressured engine, and Fig. 10.5 a section. The boiler in which steam is generated is not represented; the steam passes through the tube zz, Fig. 10.5, to the cylinder AA, in which moves a tightly fitting piston C. The parts between the tube zz and the cylinder AA, that is the slide valve in the valve-chest KK, and the two tubes d and e allow the steam to pass first below and then above the piston, while at the same time the steam has free exit from the other half of the cylinder. When the steam passes under the piston, it forces it upward; when the piston has reached the top of its course the position of the valve in KK changes, and the steam passes above the piston and forces it down again. The piston rod acts by means of the connecting rod P, on the crank Q of the flywheel X and sets this in motion. By means of the rod s, the motion of the rod regulates the opening and closing of the valve. But we need

Fig. 10.5 Side view of a
steam engine.—[*K.K.*]

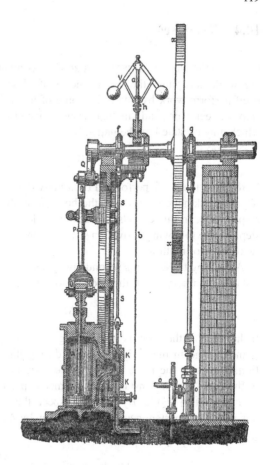

not here enter into those mechanical arrangements, however ingeniously they have
been devised. We are only interested in the manner in which heat produces elastic
vapour, and how this vapour, in its endeavour to expand, is compelled to move the
solid parts of the machine, and furnish work.

10.3   Study Questions

QUES. 10.1. Can a moving object, such as a projectile, perform work? If so, how
much?

QUES. 10.2. How is steam like an elastic substance? Can steam accomplish work?
If so, under what circumstances?

10.4   Exercises

EX. 10.1 (WORK AND KINETIC ENERGY).   As mentioned by Helmholtz in the reading, work can change the kinetic energy of a body. Conversely, a moving body can itself perform work by giving up some of its kinetic energy. The precise relationship between work and kinetic energy is given by the *work-kinetic energy theorem*, which may be stated mathematically as

$$W = K_2 - K_1. \tag{10.2}$$

Here, W is the work performed on a body during some time interval, and K_1 and K_2 are the kinetic energies of the body at the beginning and end of this interval, respectively. In this exercise, you should clearly write out (and fill in the missing steps of) the following derivation so as to show how Eq. 10.2 follows from Newton's second law of motion.[3]

According to Newton's second law, when a constant force, F, acts on a body for a short time interval, Δt, the body's momentum, p, changes by an amount

$$\Delta p = F \, \Delta t. \tag{10.3}$$

If the mass of the body is unchanging then we can write $\Delta p = m\Delta v$, where Δv is the change in the body's velocity during the interval in which the force acts. Furthermore, if the force acting on the body is unchanging, then the acceleration of the body is also constant, and the mean speed during the time interval is given by $v_m = (v_1 + v_2)/2$. By the mean speed theorem,[4] the distance traveled during this period of constant acceleration is given by $\Delta x = v_m \Delta t$. Multiplying Δp by v_m gives

$$\Delta p \cdot v_m = m(v_2 - v_1) \cdot \frac{(v_1 + v_2)}{2}.$$

But since $\Delta p = F\Delta t$ and $v_m = \Delta x / \Delta t$, this can be written as

$$F \cdot \Delta x = \frac{1}{2} m \, v_2^2 - \frac{1}{2} m \, v_1^2.$$

Now the work done by a constant force acting over a distance Δx is given by $W = F \cdot \Delta x$. So if we identify the kinetic energy of a body as $K = \frac{1}{2}mv^2$, then we have Eq. 10.2.

---

[3] Newton's three laws of motion are presented as the *Axioms* in Book I of his *Principia*; see Chap. 21 of volume II.

[4] The mean speed theorem states that the distance traveled during a given time interval by a uniformly accelerating body is the same as the distance traveled by a second body moving at the average speed of the first body during the same time interval. This theorem is employed by Galileo's to determine the distance traveled by falling bodies in Book 3 his *Dialogues*; see Chap. 9 of volume II.

Fig. 10.6 Computing the
work required to compress a
spring which obeys Hooke's
law

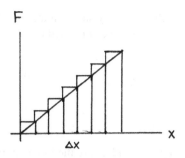

But what if the force is not constant? In this case, it helps to conceptually divide
the total distance, d, over which the force acts into a large number, N, of tiny distance intervals, Δx, during each of which the force is *approximately* constant. Then
the total change in kinetic energy of the body over the distance d can be computed
by summing the work done during each of these tiny distance intervals.[5]

EX. 10.2 (PENDULUM SPEED).   A tiny one 1-pound lead ball is suspended from the
ceiling by a 6 foot ft cord. If the cord is initially drawn back by an angle of 5° from
the vertical, and then released from a state of rest, what will be the maximum speed
which the lead ball acquires? (ANSWER: About 1.2 ft/s)

EX. 10.3 (WORK AND POTENTIAL ENERGY).   If a constant force acts on a body over
a distance, d, then the work given by the product in Eq. 10.1 may be understood
geometrically as the area of a rectangle whose height and width represent the force
and distance, respectively. In this problem, we will consider the work done by a
force which is *not* constant. This will allow us to predict the speed of a slingshot
bullet using the concepts of work, potential energy and kinetic energy. Let us begin
by considering how much work it takes to compress an ordinary spring by a distance
d. The force which an agent must overcome in attempting to stretch (or compress)
a spring is given by *Hooke's law*:

$$F = kx. \tag{10.4}$$

Here, the spring constant k, characterizes the stiffness of the spring and x represents
the amount of compression of the spring. Notice that the strength of the force is not
constant—it changes as the spring's length changes. Thus, one cannot simply multiply the spring force, Eq. 10.4, by the total distance, d. How then can one calculate
the amount of work required to compress the spring? To solve this problem, we
conceptually divide the total compression distance, d, into a large number of tiny
compression steps, each of size Δx, as depicted in Fig. 10.6. During the ith compression step, the force, $F(x_i)$, is approximately constant and is equal to kx_i. The
work, ΔW_i, during the ith step is then given by the product $F(x_i) \cdot \Delta x$. Then the
total work done during the compression is found by summing the work done during

---

[5] This procedure will be explained in more detail in Ex. 10.3, below; see Eq. 10.5.

each of the tiny steps. In the limit that $\Delta x \to 0$ and the number of steps $N \to \infty$, the sum can be converted to an integral

$$W = \lim_{N \to \infty} \sum_{i=0}^{N} F(x_i) \cdot \Delta x$$

$$= \int_{x_1}^{x_2} F(x) \cdot dx.$$

(10.5)

Here, x_1 and x_2 are the initial and final values of the spring's compression. Having compressed the spring, one may say that the work done on the spring in the act of compression is "stored" inside the spring in the form of *potential energy*. In practice, this means only that the compressed spring may now deliver precisely this amount of work by expanding back to its original size.

Now here is the problem. Assume that the cord of a slingshot, when pulled back, obeys Hooke's law. (a) First, find the spring constant of the elastic cord such that the slingshot can fire a 10-g lead ball at a speed of 200 m/s when drawn back 20 in. (b) What is the kinetic energy of this projectile? (c) By how much can the speed of the projectile be increased by drawing back the sling just one more inch? (d) Is the kinetic energy of the projectile simply proportional to the distance of stretch?

10.5   Vocabulary

1. Perceptible
2. Endow
3. Hindrance
4. Endeavor
5. Traverse
6. Annul
7. Oscillation
8. Impart
9. Manifest
10. Equilibrium

Chapter 11
Conservation of Energy

> *Work is wealth. A machine which could produce work from nothing was as good as one which made gold.*
>
> —Hermann von Helmholtz

11.1 Introduction

In his *On the Conservation of Force*, Helmholtz argued that by performing work on a body (or collection of bodies) one can either change its *configuration* or its *speed*. In the first case, the work is stored, so to speak, as *potential energy*.[1] Examples of this are the lifting of a body to a new height or the stretching of a spring. The precise amount of work required—and the potential energy stored—depends on the force which must be overcome in order to change its configuration from an initial to a final state.[2] In the second case, the work is stored as *kinetic energy*.[3] The kinetic energy of a moving object is a measure of how much work the moving object can perform on another object (such as a target) before coming to a complete stop. Generally speaking, what we today call the *energy* of an object (or a configuration of objects) is just the amount of work which it is capable of performing. Moreover—and this is Helmholtz's main point—this quantity of work seems to be conserved: when work is done on an object, the object can, in turn, do precisely the same amount of work on another object. This is what Helmholtz means when he states that "the quantity of force which can be brought into action in the whole of Nature is unchangeable, and can neither be increased nor diminished."

Thus far, Helmholtz has focused on mechanical systems, such as water-wheels, pendulums, windmills, pulleys and levers. Now, in the remainder of the essay, Helmholtz generalizes his previous analysis to consider steam engines, chemical affinity, batteries and electrical machines. As you read this text, see if you can detect any flaws in his reasoning. Specifically, what experimental observations might falsify Helmholtz's new universal law of nature?

---

[1] Helmholtz did not use this term. The term *energy* derives from the Greek word "erg" which means "work"—so the energy of an object is, etymologically speaking, the "work inside" the object.

[2] See Ex. 10.3 , and especially Eq. 10.5, in the previous chapter of the present volume.

[3] Helmholtz reluctantly refers to this quantity as the *vis viva*—or living force—of a moving mass.

© Springer International Publishing Switzerland 2016
K. Kuehn, *A Student's Guide Through the Great Physics Texts*,
Undergraduate Lecture Notes in Physics, DOI 10.1007/978-3-319-21816-8_11

11.2   Reading: Helmholtz, *On the Conservation of Force*

Helmholtz, H., On the Conservation of Force, in *Popular Lectures on Scientific Subjects*, edited by E. Atkinson, D. Appleton and Company, New York, 1885, pp. 342–362.

You all know how powerful and varied are the effects of which steam engines are capable; with them has really begun the great development of industry which has characterised our century before all others. Its most essential superiority over motive powers formerly known is that it is not restricted to a particular place. The store of coal and the small quantity of water which are the sources of its power can be brought everywhere, and steam engines can even by made movable, as is the case with steam ships and locomotives. By means of these machines we can develop motive power to almost an indefinite extent at any place on the earth's surface, in deep mines and even on the middle of the ocean; while water and windmills are bound to special parts of the surface of the land. The locomotive transports travellers and goods over the land in numbers and with a speed which must have seemed an incredible fable to our forefathers, who looked upon the mail coach with its six passengers in the inside, and its 10 miles an hour, as an enormous progress. Steam engines traverse the ocean independently of the direction of the wind, and, successfully resisting storms which would drive sailing vessels far away, reach their goal at the appointed time. The advantages which the concourse of numerous and variously skilled workmen in all branches offers in large towns where wind and water power are wanting, can be utilised, for steam engines find place everywhere, and supply the necessary crude force; thus the more intelligent human force may be spared for better purposes; and, indeed, wherever the nature of the ground or the neighbourhood of suitable lines of communications present a favourable opportunity for the development of industry, the motive power is also present in the form of steam engines.

We see, then, that heat can produce mechanical power; but in the cases which we have discussed we have seen that the quantity of force which can be produced by a given measure of a physical process is always accurately defined, and that the further capacity for work of the natural forces is either diminished or exhausted by the work which has been performed. How is it now with *Heat* in this respect?

This question was of decisive importance in the endeavour to extend the law of Conservation of Force to all natural processes. In the answer lay the chief difference between the older and newer views in these respects. Hence it is that many physicists designate that view of Nature corresponding to the law of conservation of force with the name of *Mechanical Theory of Heat.*

The older view of the nature of heat was that it is a substance, very fine and imponderable indeed, but indestructible, and unchangeable in quantity, which is an essential fundamental property of all matter. And, in fact, in a large number of natural processes, the quantity of heat which can be demonstrated by the thermometer is unchangeable.

By conduction and radiation, it can indeed pass from hotter to colder bodies; but the quantity of heat which the former lose can be shown by the thermometer to have reappeared in the latter. Many processes, too, were known, especially in the passage of bodies from the solid to the liquid and gaseous states, in which heat disappeared—at an rate as regards the thermometer. But when the gaseous body was restored to the liquid, and the liquid to the solid state, exactly the same quantity of heat reappeared which formerly seemed to have been lost. Heat was said *to have become latent.* On this view, liquid water differed from solid ice in containing a certain quantity of heat bound, which just because it was bound, could not pass to the thermometer, and therefore was not indicated by it. Aqueous vapour contains a far greater quantity of heat thus bound. But if the vapour be precipitated, and the liquid water restored to the state of ice, exactly the same amount of heat is liberated as had become latent in the melting of the ice and in the vaporisation of the water.

Finally, heat is sometimes produced and sometimes disappears in chemical processes. But even here it might be assumed that the various chemical elements and chemical compounds contain certain constant quantities of latent heat, which, when they changed their composition, are sometimes liberated and sometimes must be supplied from external sources. Accurate experiments have shown that the quantity of heat which is developed by a chemical process—for instance, in burning a pound of pure carbon into carbonic acid—is perfectly constant, whether the combustion is slow or rapid, whether it takes place all at once or by intermediate stages. This also agreed very well with the assumption, which was the basis of the theory of heat, that heat is a substance entirely unchangeable in quantity. The natural processes which have here been briefly mentioned, were the subject of extensive experimental and mathematical investigations, especially of the great French physicists in the last decade of the former, and the first decade of the present, century; and a rich and accurately-worked chapter of physics has been developed, in which everything agreed excellently with the hypothesis—that heat is a substance. On the other hand, the invariability in the quantity of heat in all these processes could at that time be explained in no other manner than that heat is a substance.

But one relation of heat—namely, that to mechanical work—had not been accurately investigated. A French engineer, Sadi Carnot, son of the celebrated War Minister of the Revolution, had indeed endeavoured to deduce the work which heat performs, by assuming that the hypothetical caloric endeavoured to expand like a gas; and from this assumption he deduced in fact a remarkable law as to the capacity of heat for work, which even now, though with an essential alteration introduced by Clausius, is among the bases of the modern mechanical theory of heat, and the practical conclusions from which, so far as they could at that time be compared with experiments, have held good.

But it was already known that whenever two bodies in motion rubbed against each other, heat was developed anew, and it could not be said whence it came.

The fact is universally recognised; the axle of a carriage which is badly greased and where the friction is great, becomes hot—so hot, indeed, that it may take fire; machine wheels with iron axles going at a great rate may become so hot that they weld to their sockets.

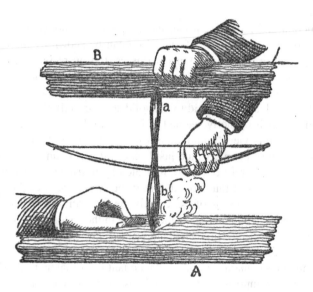

Fig. 11.1 Using friction to start a fire.—[*K.K.*]

A powerful degree of friction is not, indeed, necessary to disengage an appreciable degree of heat; thus, a lucifer match, which by rubbing is so heated that the phosphoric mass ignites, teaches this fact. Nay, it is enough to rub the dry hands together to feel the heat produced by friction, and which is far greater than the heating which takes place when the hands lied gently on each other. Uncivilised people use the friction of two pieces of wood to kindle a fire. With this view, a sharp spindle of hard wood is made to revolve rapidly on a base of soft wood in the manner represented in Fig. 11.1.

So long as it was only a question of the friction of solids, in which particles from the surface become detached and compressed, it might be supposed that some changes in structure of the bodies rubbed might here liberate latent heat, which would thus appear as heat of friction.

But heat can also be produced by the friction of liquids, in which there could be no question of changes in structure, or of the liberation of latent heat. The first decisive experiment of this kind was made by Sir Humphry Davy in the commencement of the present century. In a cooled space he made two pieces of ice rub against each other, and thereby caused them to melt. The latent heat which the newly formed water must have here assimilated could not have been conducted to it by the cold ice, or have been produced by a change of structure; it could have come from no other cause than from friction, and must have been created by friction.

Heat can also be produced by the impact of imperfectly elastic bodies as well as by friction. This is the case, for instance, when we produce fire by striking flint against steel, or when an iron bar is worked for some time by powerful blows of the hammer.

If we inquire into the mechanical effects of friction and of inelastic impact, we find at once that these are the processes by which all terrestrial movements are brought to rest. A moving body whose motion was not retarded by any resisting force would continue to move to all eternity. The motions of the planets are an instance of this. This is apparently never the case with the motion of the terrestrial bodies, for they are always in contact with other bodies which are at rest, and rub against them. We can, indeed, very much diminish their friction, but never completely annul it. A wheel which turns about a well-worked axle, once set in motion, continues it for a long time; and the longer, the more truly and smoother the axle is made to turn, the better it is greased, and the less the pressure it has to support. Yet the *vis viva* of the motion which we have imparted to such a wheel when we started it, is gradually lost in consequence of friction. It disappears, and if we do not carefully consider the matter, it seems as if the *vis viva* which the wheel had possessed had been simply destroyed without any substitute.

A bullet which is rolled on a smooth horizontal surface continues to roll until its velocity is destroyed by friction on the path, caused by the very minute impacts on its little roughnesses.

A pendulum which has been put in vibration can continue to oscillate for hours if the suspension is good, without being driven by a weight; but by the friction against the surrounding air, and by that at its place of suspension, it ultimately comes to rest.

A stone which has fallen from a height has acquired a certain velocity on reaching the earth; this we know is the equivalent of a mechanical work; so long as this velocity continues as such, we can direct it upwards by means of suitable arrangements, and thus utilise it to raise the stone again. Ultimately the stone strikes against the earth and comes to rest; the impact has destroyed its velocity, and therewith apparently also the mechanical work which this velocity could have affected.

If we review the results of all these instances, which each of you could easily add to from your own daily experience, we shall see that friction and inelastic impact are processes in which mechanical work is destroyed, and heat produced in its place.

The experiments of Joule, which have been already mentioned, lead us a step further. He has measured in foot pounds the amount of work which is destroyed by the friction of solids and by friction of liquid; and, on the other hand, he has determined the quantity of heat which is thereby produced, and has established a definite relation between the two. His experiments show that when heat is produced by the consumption of work, a definite quantity of work is required to produce that amount of heat which is known to physicists as the *unit of heat*; the heat, that is to say, which is necessary to raise 1 g of water through 1 °C. The quantity of work necessary for this is, according to Joule's best experiments, equal to the work which a gramme would perform in falling through a height of 425 m.

In order to show how closely concordant are his numbers, I will adduce the results of a few series of experiments which he obtained after introducing the latest improvements in his methods.

Table 11.1 Calculated
mechanical equivalents of
heat disengaged from various permanent gasses when
compressed

With atmospheric air	426.0 m
With oxygen	425.7 m
With nitrogen	431.3 m
With hydrogen	425.3 m

1. A series of experiments in which water was heated by friction in a brass vessel. In the interior of this vessel a vertical axle provided with 16 paddles was rotated, the eddies thus produced being broken by a series of projecting barriers, in which parts were cut out large enough for the paddles to pass through. The value of the equivalent was 424.9 m.
2. Two similar experiments, in which mercury in an iron vessel was substituted for water in a brass one, gave 425 and 426.3 m.
3. Two series of experiments, in which a conical ring rubbed against another, both surrounded by mercury, gave 426.7 and 425.6 m.

Exactly the same relations between heat and work were also found in the reverse process—that is, when work was produced by heat. In order to execute this process under physical conditions that could be controlled as perfectly as possible, permanent gases and not vapours were used, although the latter are, in practice, more convenient for producing large quantities of work, as in the case of the steam engine. A gas which is allowed to expand with moderate velocity becomes cooled. Joule was the first to show the reason of this cooling. For the gas has, in expanding, to overcome the resistance which the pressure of the atmosphere and the slowly yielding side of the vessel oppose to it: or, if it cannot of itself overcome this resistance, it supports the arm of the observer which does it. Gas thus performs work, and this work is produced at the cost of its heat. Hence the cooling. If, on the contrary, the gas is suddenly allowed to issue into a perfectly exhausted space where it finds no resistance, it does not become cool, as Joule has shown; or if individual parts of it become cool, others become warm; and, after the temperature has become equalised, this is exactly as much as before the sudden expansion of the gaseous mass.

How much heat the various gases disengage when they are compressed, and how much work is necessary for their compression; or, conversely, how much heat disappears when they expand under a pressure equal to their own counterprbessure, was partly known from the older physical experiments of Regnault by extremely perfect methods. Calculations with the best data of this kind give us the value of the thermal equivalent from experiments (Table 11.1.):—

Comparing these numbers with those which determine the equivalence of heat and mechanical work in friction, as close an agreement is seen as can at all be expected from numbers which have been obtained by such varied investigations of different observers.

Thus then: a certain quantity of heat may be changed into a definite quantity of work; this quantity of work can also be retransformed into heat, and, indeed, into exactly the same quantity of heat as that from which it originated; in a mechanical point of view, they are exactly equivalent. Heat is a new form in which a quantity of work may appear.

These facts no longer permit us to regard heat as a substance, for its quantity is not unchangeable. It can be produced anew from the *vis viva* of motion destroyed; it can be destroyed, and then produces motion. We must rather conclude from this that heat itself is a motion, an internal invisible motion of the smallest elementary particles of bodies. If, therefore, motion seems lost in friction and impact, it is not actually lost, but only passes from the great visible masses to their smallest particles; while in steam engines the internal motion of the heated gaseous particles is transferred to the piston of the machine, accumulated in it, and combined in a resultant whole.

But what is the nature of this internal motion can only be asserted with any degree of probability in the case of gases. Their particles probably cross one another in rectilinear paths in all directions, until, striking another particle, or against the side of the vessel, they are reflected in another direction. A gas would thus be analogous to a swarm of gnats, consisting, however, of particles infinitely small and infinitely more closely packed. This hypothesis, which has been developed by Krönig, Clausius, and Maxwell, very well accounts for all the phenomena of gases.

What appeared to the earlier physicists to be the constant quantity of heat is nothing more than the whole motive power of the motion of heat, which remains constant so long as it is not transformed into other forms of work, or results afresh from them.

We turn now to another kind of natural forces which can produce work—I mean the chemical. We have today already come across them. They are the ultimate cause of the work which gunpowder and the steam engine produce; for the heat which is consumed in the latter, for example, originates in the combustion of carbon—that is to say, in a chemical process. The burning of coal is the chemical union of carbon with the oxygen of the air, taking place under the influence of the chemical affinity of the two substances.

We may regard this force as an attractive force between the two, which, however, only acts through them with extraordinary power, if the smallest particles of the two substances are in closest proximity to each other. In combustion this force acts; the carbon and oxygen atoms strike against each other and adhere firmly, inasmuch as they form a new compound—carbonic acid—a gas known to all of you as that which ascends from all fermenting and fermented liquids—from beer and champagne. Now this attraction between the atoms of carbon and of oxygen performs work just as much as that which the earth in the form of gravity exerts upon a raised weight. When the weight falls to the ground, it produces an agitation, which is partly transmitted to the vicinity as sound waves, and partly remains as the motion of heat. The same result we must expect from chemical action. When carbon and oxygen atoms have rushed against each other, the newly-formed particles of carbonic acid must be in the most violent molecular motion—that is, in the motion of heat. And this is so. A pound of carbon burned with oxygen to form carbonic acid, gives as much heat as is necessary to raise 80.9 pounds of water from the freezing to the boiling point; and just as the same amount of work is produced when a weight falls, whether it falls slowly or fast, so also the same quantity of heat is produced by the combustion of carbon, whether this is slow or rapid, whether it takes place all at once, or by successive steps.

When the carbon is burned, we obtain in its stead, and in that of the oxygen, the gaseous product of combustion—carbonic acid. Immediately after combustion it is incandescent. When it has afterwards imparted heat to the vicinity, we have in the carbonic acid the entire quantity of carbon that the entire quantity of oxygen, and also the force of affinity quite as strong as before. But the action of the latter is now limited to holding the atoms of carbon and oxygen firmly united; they can no longer produce either heat or work any more than a fallen weight can do work if it has not been again raised by some extraneous force. When the carbon has been burnt we take no further trouble to retain the carbonic acid; it can do no more service, we endeavour to get it out of the chimneys of our houses as fast as we can.

Is it possible, then, to tear asunder the particles of carbonic acid, and give to them once more the capacity of work which they had before they were combined, just as we can restore the potentiality of a weight by raising it from the ground? It is indeed possible. We shall afterwards see how it occurs in the life of plants; it can also be effected by inorganic processes, though in roundabout ways, the explanation of which would lead us too far from our present course.

This can, however, be easily and directly shown for another element, hydrogen, which can be burnt just like carbon. Hydrogen with carbon is a constituent of all combustible vegetable substances, among others, it is also an essential constituent of the gas which is used for lighting our streets and rooms; in the free state it is also a gas, the lightest of all, and burns when ignited with a feebly luminous blue flame. In this combustion—that is, in the chemical combination of hydrogen and oxygen, a very considerable quantity of heat is produced; for a given weight of hydrogen, four times as much heat as in the combustion of the same weight of carbon. The product of combustion is water, which therefore, is not of itself further combustible, for the hydrogen in it is completely saturated with oxygen. The force of affinity, therefore of hydrogen for oxygen, like that of carbon for oxygen, performs work in combustion, which appears in the form of heat. In the water which has been formed during combustion, the force of affinity is exerted between the elements as before, but its capacity for work is lost. Hence the two elements must be again separated, their atoms torn apart, if new effects are to be produced from them.

This we can do by the aid of currents of electricity. In the apparatus depicted in Fig. 11.2, we have two glass vessels filled with acidulated water a and a_1, which are separated in the middle by a porous plate moistened with water. In both sides are fitted platinum wires, k, which are attached to platinum plates, i and i_1. As soon as a galvanic current is transmitted through the water by the platinum wires, k, you see bubbles of gas ascend from the plates i and i_1. These bubbles are the two elements of water, hydrogen on the one hand, and oxygen on the other. The gases emerge through the tubes g and g_1. If we wait until the upper part of the vessels and the tubes have been filled with it, we can inflame hydrogen at one side; it burns with a blue flame. If I bring a glimmering spill near the mouth of the other tube, it bursts into flame, just as happens with oxygen gas, in which the processes of combustion

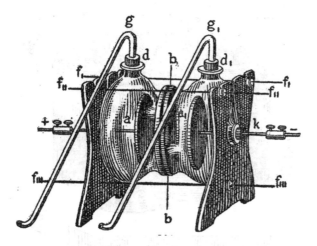

Fig. 11.2 Electrolysis apparatus.—[*K.K.*]

are far more intense than in atmospheric air, where the oxygen mixed with nitrogen is only one-fifth of the whole volume.

If I hold a glass filled with water over the hydrogen flame, the water, newly formed in combustion, condenses upon it.

If a platinum wire be held in the almost non-luminous flame, you see how intensely it is ignited; in a plentiful current of a mixture of the gases, hydrogen and oxygen, which have been liberated in the above experiment, the almost infusible platinum might even be melted. The hydrogen which has here been liberated from the water by the electrical current has regained the capacity of producing large quantities of heat by a fresh combination with oxygen; its affinity for oxygen has regained for its capacity for work.

We here become acquainted with a new source of work, the electric current which decomposes water. This current is itself produced by a galvanic battery, Fig. 11.3. Each of the four vessels contains nitric acid, in which there is a hollow cylinder of very compact carbon. In the middle of the carbon cylinder is a cylindrical porous vessel of white clay, which contains dilute sulphuric acid; in this dips a zinc cylinder. Each zinc cylinder is connected by a metal ring with the carbon cylinder of the next vessel, the last zinc cylinder, *n* is connected with one platinum plate, and the first carbon cylinder, *p*, with the other platinum plate of the apparatus for the decomposition of water.

If now the conducting circuit of this galvanic apparatus is completed, and the decomposition of water begins, a chemical process takes place simultaneously in the cells of the voltaic battery. Zinc takes oxygen from the surrounding water and undergoes a slow combustion. The product of combustion thereby produced, oxide of zinc, unites further with sulphuric acid, for which it has a powerful affinity, and sulphate of zinc, a saline kind of substance, dissolves in the liquid. The oxygen, moreover, which is withdrawn from it is taken by the water from the nitric acid

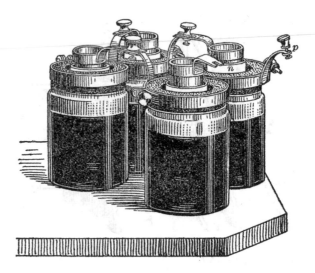

Fig. 11.3 A galvanic battery.—[*K.K.*]

surrounding the cylinder of carbon, which is very rich in it, and readily gives it up. Thus, in the galvanic battery, zinc burns to sulphate of zinc at the cost of the oxygen of nitric acid.

Thus, while one product of combustion, water, is again separated, a new combustion is taking place—that of zinc. While we there reproduce chemical affinity which is capable of work, it is here lost. The electric current is, as it were, only the carrier which transfers the chemical force of the zinc uniting with oxygen and acid to water in the decomposing cell, and uses it for overcoming the chemical force of hydrogen and oxygen.

In this case, we can restore work which has been lost, but only by using another force, that of oxidising zinc.

Here we have overcome chemical forces by chemical forces, through the instrumentality of the electric current. But we can attain the same object by mechanical forces, if we produce the electrical current by a magneto-electric machine, Fig. 11.4. If we turn the handle, the anker RR', on which is coiled copper-wire, rotates in front of the poles of the horseshoe magnet, and in these coils electrical currents are produced, which can be led from the points a and b. If the ends of these wires are connected with the apparatus for decomposing water, we obtain hydrogen and oxygen, though in far smaller quantity than by the aid of the battery which we used before. But this process is interesting, for the mechanical force of the arm which turns the wheel produces the work which is required for separating the combined chemical elements. Just as the steam engine changes chemical into mechanical force, the magneto-electric machine transforms mechanical force into chemical.

The application of electrical currents opens out a large number of relations between the various natural forces. We have decomposed water into its elements

Fig. 11.4 A magneto-electric machine.—[*K.K.*]

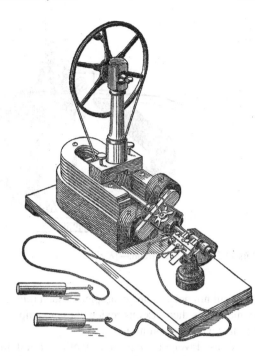

by such currents, and should be able to decompose a large number of other chemical compounds. On the other hand, in ordinary galvanic batteries electrical currents are produced by chemical forces.

In all conductors through which electrical currents pass they produce heat; I stretch a thin platinum wire between the ends n and p of the galvanic battery, Fig. 11.3; it becomes ignited and melts. On the other hand, electrical currents are produced by heat in what are called thermo-electrical elements.

Iron which is brought near a spiral copper wire, traversed by an electrical current, becomes magnetic, and then attracts other pieces of iron, or a suitably placed steel magnet. We thus obtain mechanical actions which meet with extended applications in the electrical telegraph, for instance. Figure 11.5 represents a Morse's telegraph in one third of the natural size. The essential part is a horseshoe shaped iron core, which stands in the copper spirals bb. Just over the top of this is a small steel magnet cc, which is attracted the moment an electrical current, arriving by the telegraph wire, traverses the spirals bb. The magnet cc is rigidly fixed in the lever dd, at the other end of which is a style; this makes a mark on a paper band, drawn by a clockwork, as often and as long as cc is attracted by the magnetic action of the electrical current. Conversely, by reversing the magnetism in the iron core of the spirals bb, we should obtain in them an electrical current just as we have obtained such currents in the magneto-electric machine, Fig. 11.4; in the spirals of that machine there is an iron core which, by being approached to the poles of the large horseshoe magnet, is sometimes magnetised in one and sometimes in the other direction.

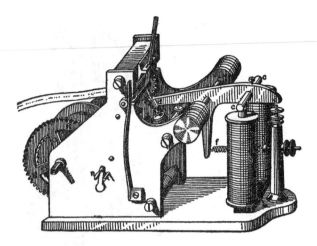

Fig. 11.5 Morse's telegraph.—[*K.K.*]

I will not accumulate examples of such relations; in subsequent lectures we shall come across them. Let us review these examples once more, and recognise in them the law which is common to all.

A raised weight can produce work, but in doing so it must necessarily sink from its height, and, when it has fallen as deep as it can fall, its gravity remains as before, but it can no longer do work.

A stretched spring can do work, but in doing so it becomes loose. The velocity of a moving mass can do work, but in doing so it comes to rest. Heat can perform work; it is destroyed in the operation. Chemical forces can perform work, but they exhaust themselves in the effort.

Electrical currents can perform work, but to keep them up we must consume either chemical or mechanical forces, or heat.

We may express this generally. *It is a universal character of all known natural forces that their capacity for work is exhausted in the degree in which they actually perform work.*

We have seen, further, that when a weight fell without performing any work, it *either* acquired velocity or produced heat. We might also drive a magneto-electrical machine by a falling weight; it would then furnish electrical currents.

We have seen that chemical forces, when they come into play, produce either heat or electrical currents or mechanical work.

We have seen that heat may be changed into work; there are apparatus (thermo-electric batteries) in which electrical currents are produced by it. Heat can directly separate chemical compounds; thus, when we burn limestone, it separates carbonic acid from lime.

Thus, whenever the capacity for work of one natural force is destroyed, it is transformed into another kind of activity. Even within the circuit of inorganic natural forces, we can transform each of them into an active condition by the aid of any other natural force which is capable of work. The connections between the various natural forces which modern physics has revealed, are so extraordinarily numerous that several entirely different methods may be discovered for each of these problems.

I have stated how we are accustomed to measure mechanical work, and how the equivalent in work of heat may be found. The equivalent in work of chemical processes is again measured by the heat which they produce. By similar relations, the equivalent in work of the other natural forces may be expressed in terms of mechanical work.

If, now, a certain quantity of mechanical work is lost, there is obtained, as experiments made with the object of determining this point show, an equivalent quantity of heat, or, instead of this, of chemical force; and, conversely, when heat is lost, we gain an equivalent quantity of chemical or mechanical force; and, again, when chemical force disappears, an equivalent of heat or work; so that in all these interchanges between various inorganic natural forces working force may indeed disappear in one form, but then it reappears in exactly equivalent quantity in some other form; it is thus neither increased nor diminished, but always remains in exactly the same quantity. We shall subsequently see that the same law holds good also for processes in organic nature, so far as the facts have been tested. It follows thence *that the total quantity of all the forces* capable of work *in the whole universe remains eternal and unchanged throughout all their changes.* All change in nature amounts to this, that force can change its form and locality without its quantity being changed. The universe possesses, once for all, a store of force which is not altered by any change of phenomena, can neither be increased nor diminished, and which maintains any change which takes place on it.

You see how, starting from considerations based on the immediate practical interests of technical work, we have been led up to a universal natural law, which, as far as all previous experience extends, rules and embraces all natural processes; which is no longer restricted to the practical objects of human utility, but expresses a perfectly general and particularly characteristic property of all natural forces, and which, as regards generality, is to be placed by the side of the laws of the unalterability of mass, and the unalterability of the chemical elements.

At the same time, it also decides a great practical question which has been much discussed in the last two centuries, to the decision of which an infinity of experiments has been made and an infinity of apparatus constructed—that is, the question of the possibility of a perpetual motion. By this was understood a machine which was to work continuously without the aid of any external driving force. The solution of this problem promised enormous gains. Such a machine would have all the advantages of steam without requiring the expenditure of fuel. Work is wealth. A machine which could produce work from nothing was as good as one which made gold. This problem had thus for a long time occupied the place of gold making, and had confused many a pondering brain. That a perpetual motion could not be produced by the aid of the then known mechanical forces could be demonstrated in the last century by the aid of mathematical mechanics which had at that time been developed. But to show also that it is not possible even if heat, chemical forces, electricity and magnetism were made to co-operate, could not be done without a knowledge of our law in all its generality. The possibility of a perpetual motion was first finally negatived by the law of the conservation of force, and this law might also be expressed in the practical form that no perpetual motion is possible, that force cannot be produced from nothing; something must be consumed.

You will only be ultimately able to estimate the importance and the scope of our law when you have before your eyes a series of its applications to individual processes in nature.

What I have today mentioned as to the origin of the moving forces which are at our disposal, directs us to something beyond the narrow confines of our laboratories and or manufactories, to the great operations at work in the life of the earth and of the universe. The force of falling water can only flow down from the hills when rain and snow bring it to them. To furnish these, we must have aqueous vapour in the atmosphere, which can only be effected by the aid of heat, and this heat comes from the sun. The steam engine needs the fuel which the vegetable life yields, whether it be the still active life of the surrounding vegetation, or the extinct life which has produced the immense coal deposits in the depths of the earth. The forces of man and animals must be restored by nourishment; all nourishment comes ultimately from the vegetable kingdom, and leads us back to the same source.

You see then that when we inquire into the origin of the moving forces which we take into our service, we are thrown back upon the meteorological processes in the earth's atmosphere, on the life of plants in general, and on the sun.

11.3 Study Questions

QUES. 11.1. Which is true, the caloric theory of heat or the mechanical theory of heat?

a) What is the caloric theory of heat? And what is the strongest evidence in favor of the caloric theory?
b) What is the relationship between resistive forces and heat? What experiments most clearly demonstrate this relationship?
c) What happens to the temperature of a gas which expands while doing work? What about when it expands without doing work? Does it depend upon the type of gas under consideration?
d) What, exactly, does the mechanical theory of heat claim? What is the strongest evidence for the mechanical theory?

QUES. 11.2. Can chemical reactions accomplish work?

a) Carefully describe the process of combustion of coal.
b) In what sense is chemical affinity similar to gravitational attraction? And how is spent fuel analogous to a weight which has fallen to the bottom of a pendulum clock?
c) How can electrical current be used to revitalize spent fuel? How, in turn, can chemical affinity be used to create electrical currents?

QUES. 11.3. Is the law of conservation of force true?

a) What are some of the numerous examples which Helmholtz provides in which work is exhausted by one system in order to produce the capacity for work in another system?
b) What does Helmholtz mean by *perpetual motion*? What would be the benefit of constructing a perpetual motion machine?
c) Does the law of the conservation of force *prove* the impossibility of perpetual motion, or does it *assume* the impossibility of perpetual motion?
d) What, according to Helmholtz, is the origin of all the moving forces on the earth? Does this sound reasonable? Are there any philosophical—or even theological—implications of accepting Helmholtz's law?

11.4 Exercises

EX. 11.1 (COMBUSTION).   How many ounces of carbon must be burned so as to heat a liter of water by $10\,°F$? What is the maximum height through which the heat produced by such a combustion might be capable of lifting 1 L of water? (ANSWER: 0.024 oz.)

EX. 11.2 (CALORIC THEORY ESSAY).   Are Helmholtz's arguments in favor of the mechanical theory of heat convincing? In particular, how might a committed proponent of the caloric theory of heat consistently maintain this theory despite Helmholtz's arguments?

EX. 11.3 (CONSERVATION OF ENERGY ESSAY).   Is Helmholtz's generalized principle of the conservation of energy, in fact, true? To sharpen your thinking, you might consider the following questions. Does this principle follow from Newton's second law of motion, as did the *work-kinetic energy theorem*?[4] More generally, does Helmholtz argue deductively or inductively for the conservation of energy? Is there any empirical evidence which, in fact, contradicts the generalized principle of conservation of energy? How might such evidence be explained if encountered in the laboratory? And what type of evidence would constitute a definitive refutation of the principle of conservation of energy?

11.5 Vocabulary

1. Fly-wheel
2. Conduction
3. Radiation

---

[4] See Ex. 10.1 in the previous chapter of the present volume.

4. Latent
5. Lucifer-match
6. Inelastic
7. Terrestrial
8. Concordant
9. Adduce
10. Rectilinear
11. Affinity
12. Proximity
13. Adhere
14. Fermentation
15. Combustion
16. Incandescent
17. Galvanic
18. Magneto-electrical
19. Thermo-electric

Chapter 12
Geometric Optics

*The Point from which Rays diverge or to which they converge
may be called their Focus.*

—Isaac Newton

12.1 Introduction

Chronologically, our last chapter took us into the latter half of the nineteenth century with Helmholtz's famous statement of the conservation of energy. Now let us return, as it were, to the seventeenth century to follow the concurrent development of the theory—or more precisely competing theories—of light. We begin with the work of Isaac Newton (1642–1727), who set forth his corpuscular, or particle, theory of light in 1704 in a book called *Opticks*.[1] Unlike his more famous *Mathematical Principles of Natural Philosophy*, the *Opticks* was written in his native English. It consists of three Books in which the reader is presented with a litany of experimental observations touching upon the nature of light. The first book, from which the reading in this chapter is extracted, begins with a succinct overview of the laws of optics pertaining to mirrors and lenses. The reader who has studied Newton previously will perhaps notice that, like the first book of his famous *Principia*, the first book of his *Opticks* proceeds deductively from carefully articulated definitions and axioms.[2] It should be noted, however, that unlike in his *Principia*, the axioms presented in his *Opticks* were generally agreed upon as having been empirically verified at the time of his writing. The following, then, should be seen as an "Introduction to Readers of quick Wit and good Understanding not yet versed in Opticks."

---

[1] For biographical notes on Newton, see Chap. 19 of volume II.

[2] See, for instance, Newton's definitions and laws of motion presented in Chaps. 19–21 of volume II.

© Springer International Publishing Switzerland 2016
K. Kuehn, *A Student's Guide Through the Great Physics Texts,*
Undergraduate Lecture Notes in Physics, DOI 10.1007/978-3-319-21816-8_12

12.2   Reading: Newton, *Opticks*

Newton, I., *Opticks: or A Treatise of the Reflections, Refractions, Inflections &
Colours of Light*, 4th ed., William Innes at the West-End of St. Pauls, London, 1730.
Book I.

12.2.1   Definitions

DEFINITION I.   *By the Rays of Light I understand its least Parts, and those as well
Successive in the same Lines, as Contemporary in Several Lines.*

For it is manifest that Light consists of Parts, both Successive and Contemporary;
because in the same place you may stop that which comes one moment, and let pass
that which comes presently after; and in the same time you may stop it in any one
place, and let it pass in any other. For that part of Light which is stopp'd cannot be
the same with that which is let pass. The least Light or part of Light, which may be
stopp'd alone without the rest of the Light, or propagated alone, or do or suffer any
thing alone, which the rest of the Light doth not or suffers not, I call a Ray of Light.

DEFINITION II.   *Refrangibility of the Rays of Light, is their Disposition to be
refracted or turned out of their Way in passing out of one transparent Body or
Medium into another. And a greater or less Refrangibility of Rays, is their Disposition to be turned more or less out of their Way in like Incidences on the same
Medium.*

Mathematicians usually consider the Rays of Light to be Lines reaching from the
luminous Body to the Body illuminated, and the refraction of those Rays to be bending or breaking of those lines in their passing out of one Medium into another.
And thus may Rays and Refractions be considered, if Light be propagated in an
instant. But by an Argument taken from the Æquations of the times of the Eclipses
of *Jupiter's Satellites*, it seems that Light is propagated in time, spending in its passage from the Sun to us about 7 min of time: And therefore I have chosen to define
Rays and Refractions in such general terms as may agree to Light in both cases.

DEFINITION III.   *Reflexibility of Rays, is their Disposition to be reflected or turned
back into the same Medium from any other Medium upon whose Surface they fall.
And Rays are more or less reflexible, which are turned back more or less easily.*

As if Light pass out of Glass into Air, and by being inclined more and more to the
common Surface of the Glass and Air, begins at length to be totally reflected by that
Surface; those sorts of Rays which at like Incidences are reflected most copiously,
or by inclining the Rays begin soonest to be totally reflected, are most reflexible.

DEFINITION IV.   *The Angle of Incidence is that Angle, which the Line described
by the incident Ray contains with the Perpendicular to the reflecting or refracting
Surface at the Point of Incidence.*

DEFINITION V. *The Angle of Reflexion or Refraction, is the Angle which the line described by the reflected or refracted Ray containeth with the Perpendicular to the reflecting or refracting Surface at the Point of Incidence.*

DEFINITION VI. *The Sines of Incidence, Reflexion, and Refraction, are the Sines of the Angles of Incidence, Reflexion, and Refraction.*

DEFINITION VII. *The Light whose Rays are all alike Refrangible, I call Simple, Homogeneal and Similar; and that whose Rays are some more Refrangible than others, I call Compound, Heterogeneal and Dissimilar.*

The former Light I call Homogeneal, not because I would affirm it so in all respects, but because the Rays which agree in Refrangibility, agree at least in all those their other Properties which I consider in the following Discourse.

DEFINITION VIII. *The Colours of Homogeneal Lights, I call Primary, Homogeneal and Simple; and those of Heterogeneal Lights, Heterogeneal and Compound.*

For these are always compounded of the colours of Homogeneal Lights; as will appear in the following discourse.

12.2.2 Axioms

AXIOM I. *The Angles of Reflexion and Refraction, lie in one and the same Plane with the Angle of Incidence.*

AXIOM II. *The Angle of Reflexion is equal to the Angle of Incidence.*

AXIOM III. *If the refracted Ray be returned directly back to the Point of Incidence, it shall be refracted into the Line before described by the incident Ray.*

AXIOM IV. *Refraction out of the rarer Medium into the denser, is made towards the Perpendicular; that is, so that the Angle of Refraction be less than the Angle of Incidence.*

AXIOM V. *The Sine of Incidence is either accurately or very nearly in a given Ratio to the Sine of Refraction.*

Whence if that Proportion be known in any one Inclination of the incident Ray, 'tis known in all the Inclinations, and thereby the Refraction in all cases of Incidence on the same refracting Body may be determined. Thus if the Refraction be made out of Air into Water, the Sine of Incidence of the red Light is to the Sine of its Refraction as 4 to 3. If out of Air into Glass, the Sines are as 17 to 11. In Light of other Colours the Sines have other Proportions: but the difference is so little that it need seldom be considered.

Suppose therefore, that RS (in Fig. 12.1) represents the Surface of stagnating Water, and that C is the point of Incidence in which any Ray coming in the Air from A in the Line AC is reflected or refracted, and I would know whither this Ray shall go after Reflexion or Refraction: I erect upon the Surface of the Water from the point of Incidence the Perpendicular CP and produce it downwards to Q, and

Fig. 12.1 The bending of a
light ray by the surface of
water according to Snell's law
of refraction.—[*K.K.*]

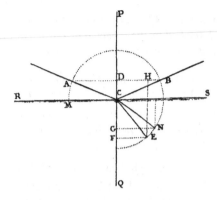

Fig. 12.2 Refraction of a light
ray as it passes through a glass
prism.—[*K.K.*]

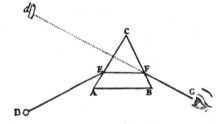

conclude by the first Axiom, that the Ray after Reflexion and Refraction, shall be
found somewhere in the Plane of the Angle of Incidence ACP produced. I let fall
therefore upon the Perpendicular CP the Sine of Incidence AD; and if the reflected
Ray be desired, I produce AD to B so that DB be equal to AD, and draw CB.
For this Line CB shall be the reflected Ray; the Angle of Reflexion BCP and its
Sine BD being equal to the Angle and Sine of Incidence, as they ought to be by the
second Axiom, But if the refracted Ray be desired, I produce AD to H, so that DH
may be to AD as the Sine of Refraction to the Sine of Incidence, that is, (if the Light
be red) as 3 to 4; and about the Center C and in the Plane ACP with Radius CA
describing a Circle ABE, I draw a parallel to the Perpendicular CPQ, the Line HE
cutting the Circumference in E, and joining CE, this Line CE shall be the Line of
the refracted Ray. For if EF be let fall perpendicularly on the Line PQ, this Line
EF shall be the Sine of Refraction of the Ray CE, the Angle of Refraction being
ECQ; and this Sine EF is equal to DH, and consequently in Proportion to the Sine
of Incidence AD as 3 to 4.

In like manner, if there be a Prism of Glass (that is, a Glass bounded with two
Equal and Parallel Triangular ends, and three plain and well polished Sides, which
meet in three Parallel Lines running from the three Angles of one end to the three
Angles of the other end) and if the Refraction of the Light in passing cross this Prism
be desired: Let ACB (in Fig. 12.2) represent a Plane cutting this Prism transversely
to its three Parallel lines or edges there where the Light passeth through it, and let
DE be the Ray incident upon the first side of the Prism AC where the Light goes
into the Glass; and by putting the Proportion of the Sine of Incidence to the Sine of

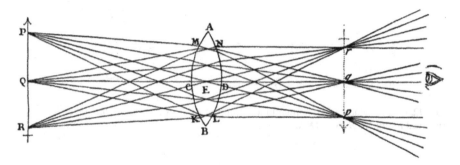

Fig. 12.3 The Rays of light from any point on an object which pass through a double-convex lens are focused in a plane on the opposite side of the lens.—[*K.K.*]

Refraction as 17 to 11 find *EF* the first refracted Ray. Then taking this Ray for the Incident Ray upon the second side of the Glass *BC* where the Light goes out, find the next refracted Ray *FG* by putting the Proportion of the Sine of Incidence to the Sine of Refraction as 11 to 17. For if the Sine of Incidence out of Air into Glass be to the Sine of Refraction as 17 to 11, the Sine of Incidence out of Glass into Air must on the contrary be to the Sine of Refraction as 11 to 17 by the third Axiom.

Much after the same manner, if *ACBD* (in Fig. 12.3) represent a Glass spherically convex on both sides (usually called a *Lens*, such as is a Burning-glass, or Spectacle-glass, or an Object-glass of a Telescope) and it be required to known how Light falling upon it from any lucid point *Q* shall be refracted, let *QM* represent a Ray falling upon any point *M* of its first spherical Surface *ACB*, and by erecting a Perpendicular to the Glass at the point *M*, find the first refracted Ray *MN* by the Proportion of the Sines 17 to 11. Let that Ray in going out of the Glass be incident upon *N*, and then find the second refracted Ray *Nq* by the Proportion of the Sines 11 to 17. And after the same manner may the Refraction be found when the Lens is convex on one side and plane or concave on the other, or concave on both sides.

AXIOM VI.   *Homogeneal Rays which flow from several Points of any Object, and fall perpendicularly or almost perpendicularly on any reflecting or refracting Plane or spherical Surface, shall afterwards diverge from so many other Points, or be parallel to so many other Lines, or converge to so many other Points, either accurately or without any sensible Error. And the same thing will happen, if the Rays be reflected or refracted successively by two or three or more Plane or Spherical Surfaces.*

The Point from which Rays diverge or to which they converge may be called their *Focus*.[3] And the Focus of the incident Rays being given, that of the reflected or

---

[3] What Newton calls the *focus*, many contemporary textbooks call the *image* (or *object*) *distance*. This is not to be confused with the *focal length* of a lens or mirror. The focal length coincides with the image distance in the special case of paraxial rays; see Ex. 12.6.—[*K.K.*]

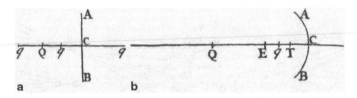

Fig. 12.4 Points along a line normal to a flat mirror ACB (**a**), and along the principle axis of spherical mirror ACB (**b**).—[K.K.]

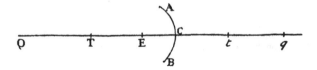

Fig. 12.5 The principle axis of a spherical refracting surface ACB.—(K.K.)

refracted ones may be found by finding the Refraction of any two Rays, as above; or more readily thus.

CASE 1. Let ACB (in Fig. 12.4a) be a reflecting or refracting Plane, and Q the Focus of the incident Rays, and QqC a Perpendicular to that Plane.

And if this perpendicular be produced to q, so that qC be equal to QC, the Point q shall be the Focus of the reflected Rays: Or if qC be taken on the same side of the Plane with QC, and in proportion to QC as the Sine of Incidence to the Sine of Refraction, the Point q shall be the Focus of the refracted Rays.

CASE 2. Let ACB (in Fig. 12.4b) be the reflecting Surface of any Sphere whose Centre is E. Bisect any Radius thereof, (suppose EC) in T, and if in that Radius on the same side the Point T you take the Points Q and q, so that TQ, TE, and Tq, be continual Proportionals,[4] and the Point Q be the Focus of the incident Rays, the Point q shall be the Focus of the reflected ones.

CASE 3. Let ACB (in Fig. 12.5) be the refracting Surface of any Sphere whose Centre is E. In any Radius thereof EC produced both ways take TE and Ct equal to one another and severally in such Proportion to that Radius as the lesser of the Sines of Incidence and Refraction hath to the difference of those Sines. And then if in the same Line you find any two Points Q and q, so that TQ be to TE as Et to tq, taking tq the contrary way from t which TQ lieth from T, and if the Point Q be the Focus of any incident Rays, the Point q shall be the Focus of the refracted ones.

And by the same means the Focus of the Rays after two or more Reflexions or Refractions may be found.

CASE 4. Let $ACBD$ (in Fig. 12.6) be any refracting Lens, spherically Convex or Concave or Plane on either side, and let CD be its Axis (that is, the Line which cuts

---

[4] Since they are continual proportionals, TE is a mean proportion of TQ and Tq. That is, $TQ/TE = TE/Tq$. See Ex. 12.7 for a geometrical proof of this Newtonian equation and its relationship to the *thin lens equation*.—[K.K.]

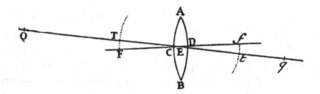

Fig. 12.6 A ray of light passing through a double-concave lens.—[*K.K.*]

both its Surfaces perpendicularly, and passes through the Centres of the Spheres,) and in this Axis produced let F and f be the Foci of the refracted Rays found as above, when the incident Rays on both sides the Lens are parallel to the same Axis; and upon the Diameter Ff bisected in E, describe a Circle. Suppose now that any Point Q be the Focus of any incident Rays. Draw QE cutting the said Circle in T and t, and therein take tq in such proportion to tE as tE or TE hath to TQ. Let tq lie the contrary way from t which TQ doth from T, and q shall be the Focus of the refracted Rays without any sensible Error, provided the Point Q be not so remote from the Axis, nor the Lens so broad as to make any of the Rays fall too obliquely on the refracting Surfaces.[5]

And by the like Operations may the reflecting or refracting Surfaces be found when the two Foci are given, and thereby a Lens be formed, which shall make the Rays flow towards or from what Place you please.[6]

So then the Meaning of this Axiom is, that if Rays fall upon any Plane or Spherical Surface or Lens, and before their Incidence flow from or towards any Point Q, they shall after Reflexion or Refraction flow from or towards the Point q found by the foregoing Rules. And if the incident Rays flow from or towards several points Q, the reflected or refracted Rays shall flow from or towards so many other Points q found by the same Rules. Whether the reflected and refracted Rays flow from or towards the Point q is easily known by the situation of that Point. For if that point be on the same side of the reflecting or refracting Surface or Lens with the Point Q, and the incident Rays flow from the Point Q, the reflected flow towards the Point q and the refracted from it; and if the incident Rays flow towards Q, the reflected flow from q, and the refracted towards it. And the contrary happens when q is on the other side of the Surface.

AXIOM VII.   *Wherever the Rays which come from all the Points of any Object meet again in so many Points after they have been made to converge by Reflection or Refraction, there they will make a Picture of the Object upon any white Body on which they fall.*

---

[5] In our Author's *Lectiones Opticæ*, Part I. Sect. IV. Prop. 29, 30, there is an elegant Method of determining these *Foci*; not only in spherical Surfaces, but likewise in any other curved Figure whatever: And in Prop. 32, 33, the same thing is done for any Ray lying out of the Axis.

[6] *Ibid.* Prop. 34.

So if *P R* (in Fig. 12.3) represent any Object without Doors, and *A B* be a Lens
placed at a hole in the Window-shut of a dark Chamber, whereby the Rays that
come from any Point *Q* of that Object are made to converge and meet again in the
Point *q*; and if a Sheet of white Paper be held at *q* for the Light there to fall upon
it, the Picture of that Object *P R* will appear upon the Paper in its proper shape
and Colours. For as the Light which comes from the Point *Q* goes to the Point *q*,
so the Light which comes from other Points *P* and *R* of the Object, will go to so
many other correspondent Points *p* and *r* (as is manifest by the sixth Axiom;) so
that every Point of the Object shall illuminate a correspondent Point of the Picture,
and thereby make a Picture like the Object in Shape and Colour, this only excepted,
that the Picture shall be inverted. And this is the Reason of that vulgar Experiment
of casting the Species of Objects from abroad upon a Wall or Sheet of white Paper
in a dark Room.

In like manner, when a Man views any Object *P Q R*, (in Fig. 12.7) the Light
which comes from the several Points of the Object is so refracted by the transparent
skins and humors of the Eye, (that is, by the outward coat *E F G*, called the *Tunica
Cornea*, and by the crystalline humor *A B* which is beyond the Pupil *mk*) as to
converge and meet again in so many Points in the bottom of the Eye, and there to
paint the Picture of the Object upon that skin (called the *Tunica Retina*) with which
the bottom of the Eye is covered. For Anatomists, when they have taken off from
the bottom of the Eye that outward and most thick Coat called the *Dura Matter*, can
then see through the thinner Coats, the Pictures of Objects lively painted thereon.
And these Pictures, propagated by Motion along the Fibres of the Optick Nerves
into the Brain, are the Cause of Vision. For accordingly as these Pictures are perfect
or imperfect, the Object is seen perfectly or imperfectly. If the Eye be tinged with
any colour (as in the Disease of the *Jaundice*) so as to tinge the Pictures in the
bottom of the Eye with that Colour, then all Objects appear tinged with the same
Colour. If the Humours of the Eye by old Age decay, so as by shrinking to make the
Cornea and Coat of the *Crystalline Humour* grow flatter than before, the Light will
not be refracted enough, and for want of a sufficient Refraction will not converge
to the bottom of the Eye but to some place beyond it, and by consequence paint
in the bottom of the Eye a confused Picture, and according to the Indistinctness
of this Picture the Object will appear confused. This is the reason of the decay of
sight in old Men, and shows why their Sight is mended by Spectacles. For those
Convex glasses supply the defect of plumpness in the Eye, and by increasing the
Refraction make the Rays converge sooner, so as to convene distinctly at the bottom
of the Eye if the Glass have a due degree of convexity. And the contrary happens
in short-sighted Men whose Eyes are too plump. For the Refraction being now too
great, the Rays converge and convene in the Eyes before they come at the bottom;
and therefore the Picture made in the bottom and the Vision caused thereby will not
be distinct, unless the Object be brought so near the Eye as that the place where the
converging Rays convene may be removed to the bottom, or that the plumpness of
the Eye be taken off and the Refractions diminished by a Concave-glass of a due
degree of Concavity, or lastly that by Age the Eye grow flatter till it come to a due

Fig. 12.7 Rays of light from an object form an image on retina after being refracted by the lens *AB*.—[*K.K.*]

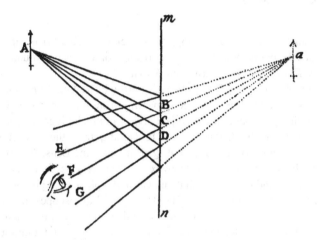

Fig. 12.8 When light rays from an object, *A*, reflect from a plane mirror, *mn*, an upright virtual image of the object is formed behind the surface of the mirror at *a*, from where the light rays appear to diverge.—[*K.K.*]

Figure: for short-sighted Men see remote Objects best in Old Age, and therefore they are accounted to have the most lasting Eyes.

AXIOM VIII.   *An Object seen by Reflexion or Refraction, appears in that place from whence the Rays after their last Reflexion or Refraction diverge in falling on the Spectator's Eye.*

If the Object *A* (in Fig. 12.8) be seen by Reflexion on a Looking-glass *mn*, it shall appear, not in its proper place *A*, but behind the Glass at *a*, from whence any Rays *AB*, *AC*, *AD*, which flow from one and the same Point of the Object, do after their Reflexion made in the Points, *B*, *C*, *D*, diverge in going from the Glass to *E*, *F*, *G*, where they are incident on the Spectator's Eyes. For these Rays do make the same Picture in the bottom of the Eyes as if they had come from the Object really placed at *a* without the Interposition of the Looking-glass; and all Vision is made according to the place and shape of that Picture.

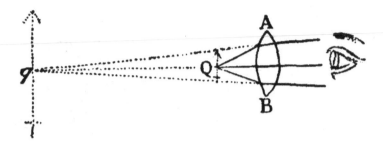

Fig. 12.9 Light from an object, Q, placed just behind a magnifying glass is refracted, forming a magnified image at q.—[*K.K.*]

In like manner the Object D (in Fig. 12.2) seen through a Prism, appears not in its proper place D, but is thence translated to some other place d situated in the last refracted Ray FG drawn backward from F to d.

And so the Object Q (in Fig. 12.9) seen through the Lens AB, appears at the place q from whence the Rays diverge in passing from the Lens to the Eye. Now it is to be noted, that the Image of the Object at q is so much bigger or lesser than the Object it self at Q, as the distance of the Image at q from the Lens AB is bigger or less than the distance of the Object at Q from the same Lens. And if the Object be seen through two or more such Convex or Concave-glasses, every Glass shall make a new Image, and the Object shall appear in the place of the bigness of the last Image. Which consideration unfolds the Theory of Microscopes and Telescopes. For that Theory consists in almost nothing else than the describing such Glasses as shall make the last Image of any Object as distinct and large and luminous as it can conveniently be made.

I have now given in Axioms and their Explications the sum of what hath hitherto been treated of in Opticks. For what hath been generally agreed on I content my self to assume under the notion of Principles, in order to what I have farther to write. And this may suffice for an Introduction to Readers of quick Wit and good Understanding not yet versed in Opticks: Although those who are already acquainted with this Science, and have handled Glasses, will more readily apprehend what followeth.

12.3   Study Questions

QUES. 12.1. What is a ray of light? Is a ray of light divisible into parts? Is it transmitted instantaneously? If not, how much time does it take to travel? And are all rays of light the same?

QUES. 12.2. What is the difference between the reflexion and refraction of a ray of light? In particular, for Fig. 12.1, which lines represent the incident, reflected, and refracted rays? How can one compute the various angles in this diagram? And for the law of refraction, do the ratio of the Sines depend upon the color of the ray of light?

QUES. 12.3. How does a ray of light behave when passing through a glass prism? In particular, for the prism depicted in Fig. 12.2, is light refracted identically at both surfaces, AC and CB? And in which direction must one look through the prism in order to see an image of the object D?

QUES. 12.4. How does light behave when passing through a glass lens? For the lens depicted in Fig. 12.3, where are the light rays which are emitted from point P of the object PQR focused by the lens? What about from points Q and R?

QUES. 12.5. What are the rules for locating the focus, q, of a light ray emitted from a point, Q, on a source object after it (i) strikes a plane mirror, (ii) passes through a plane refracting surface, (iii) strikes a spherical mirror, (iv) passes through a single convex spherical refracting surface, and (v) passes through a spherically convex lens.

QUES. 12.6. How does the eye work? And what optical defects lead to far- and short-sightedness?

a) What are the parts of the eye depicted in Fig 12.7? How is this similar to Fig. 12.3?
b) What happens if the eye is flattened or the lens is not curved enough? Where will the lens focus the image of a nearby object? What is this effect called?
c) What happens if the eye is too plump or the lens is too curved? Where will the lens focus the image of a distant object?
d) How can far- and short-sightedness be corrected?
e) Is there a minimum object distance such that the rays of light from an object can form a focused image on the retina?

QUES. 12.7. When an object is placed in front of a planar mirror, where does its image appear? In front of or behind the mirror? Why is this? Is the image upright or inverted? And is the image magnified or is it the same size as the object?

QUES. 12.8. When an object is viewed through a double-convex lens (a magnifying glass) where does its image appear? Is the image upright or inverted? Is it magnified? If so, by how much?

12.4 Exercises

EX. 12.1 (RELFECTION PROOF). In CASE 1 OF AXIOM VI, Newton argues that when a ray of light is emitted from an object at point Q, and subsequently reflects from a flat surface (ACB), another point, q—which lies on the *opposite* side of the surface as Q—acts as the focus of the reflected light ray (see Figs. 12.4 and 12.10). What, according to Newton, is the ratio of the distances qC and QC? Can you deduce this ratio geometrically by starting from the law of reflection (AXIOM II)? Is Newton correct? Does the distance of the object's image behind the mirror make sense?

Fig. 12.10 A light ray from
an object reflects from a flat
surface

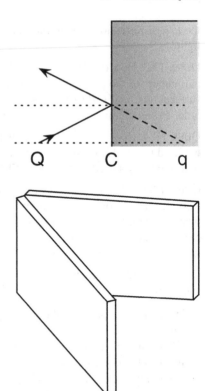

Q   C   q

Fig. 12.11 Mirrors erected in
a V-configuration on a table-
top

EX. 12.2 (REFLECTION FROM MIRRORS IN A V-CONFIGURATION). Suppose that two
flat mirrors are arranged in a V-configuration on a table-top with a 60° angle between
their mirrored surfaces (see Fig. 12.11). A horizontal ray of light from a candle
enters the region between the mirrors such that there is a 30° angle between the ray
and each surface. If the incoming ray does not strike at the junction of the mirrors,
then how many times does it bounce before escaping? What if the angle is 15°
instead of 30°?

EX. 12.3 (REFRACTION PROOF). In CASE 1 OF AXIOM VI, Newton argues that when
a ray of light is emitted from an object at point Q, and subsequently passes through
a flat surface (ACB) into a different medium, a different point, q—which lies on
the *same* side of the surface as Q—acts as the focus of the refracted light ray (see
Figs. 12.4 and 12.12). What, according to Newton, is the ratio of the distances qC
and QC? Can you find this ratio geometrically by starting from the law of refraction
(AXIOM V)? Is Newton's ratio correct?

EX. 12.4 (BOW FISHING). A bow fisherman spies a fish at the bottom of a stagnant
two-foot deep pool of water; the fish appears to be resting 5 ft. away from the base
of the rock on which he stands. If he fires an arrow from a height of 4 ft. above the
water's surface directly toward the apparent location of the fish, by how much will

Fig. 12.12 A light ray from an object refracts when passing through a flat surface

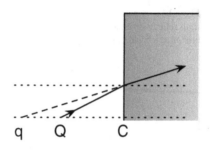

q   Q   C

he miss his target? Toward what point on the water's surface must he instead aim in order to hit the fish? (ANSWER: He overshoots by almost 7 in.)

EX. 12.5 (TOTAL INTERNAL REFLECTION). When a ray of light passing through a dense material (such as glass) strikes a surface leading into a less dense material (such as air) the ray may not be able to escape the more dense material at all. This occurs if the angle of refraction in the less dense material would be greater than or equal to 90°. In such cases, the ray is said to be *totally internally reflected* back into the denser material.[7] As an example, suppose that a ray of light is incident upon the face of a glass prism (see Fig. 12.2). A plane cutting transversely through the prism forms an equilateral triangle ABC. If the refractive index of glass is 1.5, then what is the maximum angle of incidence upon the (external) face AC so as to ensure total internal reflection of the light ray from the opposite (internal) face CB? Will the ray emerge from the third (internal) face, AB, or will it once again be totally internally reflected? So at which face of the prism must the eye look in order to see object D?

EX. 12.6 (FOCAL LENGTH OF A SPHERICAL MIRROR). In this exercise, we will attempt to prove that the focal length of a spherical mirror is approximately equal to half the radius of curvature of the mirror. A mirror's focal length is defined as the distance from the surface of the mirror to its *focal point*—the point at which rays of light traveling initially parallel to the axis of the mirror converge to a focus after reflecting from the mirror. Consider Fig. 12.13, which depicts a spherical mirror $ADCB$. The radius of curvature of the mirror is EC. A paraxial ray of light, traveling along GD, strikes the mirror at a short distance, DF, from the axis of the mirror, $ETFC$.[8] The ray reflects from the mirror along line DT.

a) First, prove that $\angle GDE = \angle EDT$. Which of Newton's AXIOMS must you invoke?
b) Next, what is the mathematical relationship between $\angle DTF$ and $\angle EDT$. Why, geometrically, must this be true?

---

[7] Strictly speaking, the relevant property is not the relative densities of the materials, but rather their relative refractive indices.

[8] A *paraxial ray* is a ray of light which lies very near the axis of the mirror.

Fig. 12.13 A paraxial ray of
light reflects from a spherical
mirror to a focal point

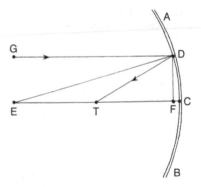

c) Finally, prove that if DF is small, then $TC \simeq EC/2$. (Hint: if $\angle DEF$ is small,
   then $\tan(\angle DEF) \simeq \angle DEF$.) Thus, the focal length of a spherical mirror is
   equal to one-half its radius of curvature.[9]

EX. 12.7 (SPHERICAL MIRRORS AND THE THIN-LENS EQUATION). When an object is
placed in front of a spherical mirror, rays of light from any point on the object reflect
from the mirror and converge at a point in front of the mirror, forming an image of
the object. In this exercise we will find a mathematical relationship between the
object distance, the *image distance* and the *radius of curvature* of the mirror. Consider the spherical mirror $ACDB$ depicted in Fig. 12.14. An object of height HQ
is placed along the axis $QETC$, at a distance QC from the mirror's surface. In this
case, QC is chosen to be larger than the radius of curvature, of the mirror, EC. To
carry out our proof, we now consider three paraxial rays of light emitted from an
arbitrary point, H, on the object: HB, HD, and HC.

a) If HB passes through point E before striking the mirror at B, prove that it
   reflects directly back to point H. From which of Newton's AXIOMS does this
   follow?
b) If HC strikes the axis of the mirror at C, prove that the angle between the
   incident and reflected rays are such that $\angle HCQ = \angle hCQ$.
c) If HD passes through point T, where $TC = EC/2$, prove that the reflected ray
   is parallel to the mirror's axis. Why must this be? (Hint: T is the focal point; see
   Ex. 12.6.)
d) Prove that the reflected rays HC and HD converge to a single point, h. This is
   the location at which an image of an arbitrary object point, H, will be formed.
e) We have just shown that an image of the object is formed at a distance Tq from
   the mirror's focal point. Let us now prove that TE is a *mean proportion* of Tq
   and TQ, as claimed by Newton in CASE 2 of AXIOM VI. Begin by proving that

---

[9] Notice, however, that when an object is placed at an arbitrary distance from the mirror, the object's
image will not typically form at this distance from the mirror. The image distance equals the focal
length only for objects placed very far—technically an infinite distance—from the mirror's surface.
In Ex. 12.7, we will find the image distance for an object placed closer to the mirror's surface.

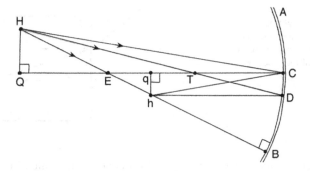

Fig. 12.14 Three rays of light emitted from the tip of an object converge and form an image after reflecting from the surface of a spherical mirror

$$\frac{QC}{QH} = \frac{qC}{qh} \quad \text{and that} \quad \frac{TC}{qh} = \frac{TQ}{QH}.$$

Combine these to show that

$$\frac{QC}{qC} = \frac{TQ}{TC}.$$

Next, show that

$$\frac{TQ + TE}{Tq + TE} = \frac{TQ}{TE},$$

and finally that TE is a mean proportion of Tq and TQ:

$$\frac{Tq}{TE} = \frac{TE}{TQ}.$$

f) If we express the *focal length* of the mirror as f, the *object distance* from the mirror as d_o, and the *image distance* from the mirror as d_i, show that the previous equation may be written as

$$(d_o - f)(d_i - f) = f^2. \tag{12.1}$$

g) Prove that the above "Newtonian" form of the *thin-lens equation* may be re-written as[10]

$$\frac{1}{f} = \frac{1}{d_o} + \frac{1}{d_i} \tag{12.2}$$

---

[10] The thin lens equation, as suggested by its name, applies not only to spherical mirrors, but also to paraxial rays passing through a thin lens, as depicted in Figs. 12.6, 12.7 and 12.9.

Fig. 12.15 The entire trajectory of a ray of light passing through a glass block may be reconstructed from two points on each side of the block, as described in Ex. 12.9

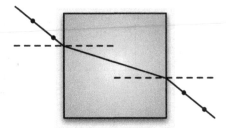

h) The *magnification* of an optical system is defined as the ratio of the image and object sizes. Prove that the magnification may be written in terms of the image and object distances as

$$M = \frac{d_i}{d_o} \qquad (12.3)$$

EX. 12.8 (REFLECTION LABORATORY). Stand up two flat hand-held mirrors in a V-configuration on a table-top (see Fig. 12.11). Erect a small object, such as a pen cap, in the space between the mirrors. Then place your eye behind the object so that you can see the image(s) of the object in the mirrors. If the interior angle between the mirrors is 180°, then a single image of the object should appear at the junction of the two mirrors. As you reduce this angle, θ, you will notice that the lone image at the junction splits into two images. At a certain angle a new image appears at the junction and this in turn splits into two more. Suppose we agree to label the angle between the mirrors at which the nth new image appears at the junction as θ_n . Thus $n = 1$ refers to the first image appearing at the junction, with θ_1 being 180°. First, verify that $\theta_2 = 90$. Next, measure $\theta_3, \theta_4 \ldots$. Is there a mathematical relationship between n and θ_n? For each n, note whether the newly created image is true or reversed (right to left). Finally, for each of the cases, $n = 1 \ldots 4$, construct a ray-diagram depicting (from above) the mirrors, the object, the image(s), and the path of the light rays which leave the object and arrive at your eye. How many reflections are needed to produce each of the virtual images? And what does this have to do with whether the virtual image is true or reversed? (Fig. 12.15)

EX. 12.9 (REFRACTION LABORATORY). The goal of this laboratory exercise is to explore refraction. Lay a transparent block (such as a 5 cm $\times$ 5 cm $\times$ 1 cm glass plate, face down) on a piece of white paper which is itself taped to a sheet of cork-board or card-board. If a pin stuck in the cardboard behind the block is viewed through the block, it will appear somewhat displaced from its actual position. The goal is to trace the path of a ray of light from this original pin to the viewer's eye. To do so, while viewing the original pin through the block, erect a second pin in line with—but just behind—the original pin on the opposite side of the block. Then erect two more pins *on the viewer's side* of the block which appear to lie in line with the first two. The trajectory of the light ray from the pin to the eye through the block may now be reconstructed, using a pencil and straight-edge, from the locations of the four stick pins and the outline of the block traced on the paper. Using this method, verify Newton's fourth and fifth AXIOMS over a wide range of incidence angles. What is the ratio of the Sines for your interface? Next, repeat the above procedure, but use a prism instead of a block. Rather than verifying Newton's AXIOMS, attempt to determine the *angle of minimum deviation* of a ray of light passing through the prism. The angle of deviation of the outgoing ray of light is measured from the initial trajectory of the incident ray. Does the angle of minimum deviation depend upon the angle of incidence?

12.5 Vocabulary

1. Successive
2. Contemporary
3. Propagate
4. Refrangibility
5. Reflexibility
6. Homogeneal
7. Heterogeneal
8. Discourse
9. Consequent
10. Prism
11. Transverse
12. Convex
13. Converge
14. Bisect
15. Concave
16. Axis
17. Oblique
18. Manifest
19. Vulgar
20. Cornea
21. Humor
22. Retina
23. Pupil
24. Jaundice
25. Tinge
26. Interposition
27. Explication
28. Hitherto
29. Apprehend

Chapter 13
The Wave Theory of Light

Now there is no doubt at all that light also comes from the
luminous body to our eyes by some movement impressed on the
matter which is between the two.

—Christiaan Huygens

13.1 Introduction

At the outset of Book I of his *Opticks*, Newton provided a concise introduction to
what is now known as *geometric optics*. As opposed to *wave optics* (which we will
consider shortly) geometric optics treats light very simply as a set of rays whose
trajectories are governed by well-defined rules, such as the laws of reflection and
refraction.[1] Before exploring Newton's theory of light in greater detail, let us turn to
the work of his famous contemporary, Christiaan Huygens. Huygens (1629–1695),
was born into a prominent Dutch family in The Hague. He was homeschooled by
his father, who was himself a distinguished mathematician and musician. Huygens
went on to attend the University of Leiden, where he studied law and mathematics, before attending the College of Orange in Breda. After taking his degree in law
and serving briefly as a diplomat, Huygens returned to The Hague in 1654. Thereafter, he devoted much of his time to research and writing. His work spanned the
fields of geometry, probability theory, optics, gravity, astronomy, cosmology and
mechanics. He was also very active in devising and constructing useful scientific
devices. For example, in an effort to understand how telescopes work, Huygens
and his brother Constantijn began to grind their own lenses in 1655. This facilitated his subsequent discovery of the Orion Nebula and Saturn's moon, Titan.
In 1659 Huygens published a book entitled *Systema Saturnium*, or *The System of
Saturn*; in it he advanced the novel theory that Saturn's recently observed protuber-
ances were in fact a solid ring surrounding the planet. During this time, Huygens
also devised a new means for accurate time-keeping—the pendulum clock—which
would prove useful in astronomical observations and (to a lesser extent) in sea-faring
navigation. In 1673 he published a book entitled *Horologium Oscillatorium*, in

---

[1] See Newton's optical AXIOMS in Chap. 12 of the present volume.

© Springer International Publishing Switzerland 2016
K. Kuehn, *A Student's Guide Through the Great Physics Texts*,
Undergraduate Lecture Notes in Physics, DOI 10.1007/978-3-319-21816-8_13

which he described the construction and the theory of operation of the pendulum clock.

Huygens' interest in astronomy is evident in his book on optics, the *Traité de lumière*, which was originally written in French in 1678 while Huygens lived in Paris. Several chapters from a 1912 English translation of this work by Silvanus P. Thompson are included in the next several chapters of the present volume. Huygens begins this *Treatise on Light* by reminding the reader of some of the well-known properties of light rays. What are these properties, and how does he attempt to explain them? Perhaps most interestingly, do you think that these properties of light *demand* an explanation? What does Huygens think?

13.2 Reading: Huygens, *Treatise on Light*

Huygens, C., *Treatise on Light*, Macmillan, London, 1912.

13.2.1 Preface

I wrote this Treatise during my sojourn in France 12 years ago, and I communicated it in the year 1678 to the learned persons who then composed the Royal Academy of Science, to the membership of which the King had done me the honour of calling me. Several of that body who are still alive will remember having been present when I read it, and above the rest those amongst them who applied themselves particularly to the study of Mathematics; of whom I cannot cite more than the celebrated gentlemen Cassini, Römer, and De la Hire. And, although I have since corrected and changed some parts, the copies which I had made of it at that time may serve for proof that I have yet added nothing to it save some conjectures touching the formation of Iceland Crystal, and a novel observation on the refraction of Rock Crystal. I have desired to relate these particulars to make known how long I have meditated the things which now I publish, and not for the purpose of detracting from the merit of those who, without having seen anything that I have written, may be found to have treated of like matters: as has in fact occurred to two eminent Geometricians, Messieurs Newton and Leibnitz, with respect to the Problem of the figure of glasses for collecting rays when one of the surfaces is given.

One may ask why I have so long delayed to bring this work to the light. The reason is that I wrote it rather carelessly in the Language in which it appears, with the intention of translating it into Latin, so doing in order to obtain greater attention to the thing. After which I proposed to myself to give it out along with another Treatise on Dioptrics, in which I explain the effects of Telescopes and those things which belong more to that Science. But the pleasure of novelty being past, I have put off from time to time the execution of this design, and I know not when I shall ever come to an end if it, being often turned aside either by business or by some new study. Considering which I have finally judged that it was better worth while to publish this writing, such as it is, than to let it run the risk, by waiting longer, of remaining lost.

There will be seen in it demonstrations of those kinds which do not produce as great a certitude as those of Geometry, and which even differ much therefrom, since whereas the Geometers prove their Propositions by fixed and incontestable Principles, here the Principles are verified by the conclusions to be drawn from them; the nature of these things not allowing of this being done otherwise.

It is always possible to attain thereby to a degree of probability which very often is scarcely less than complete proof. To wit, when things which have been demonstrated by the Principles that have been assumed correspond perfectly to the phenomena which experiment has brought under observation; especially when there are a great number of them, and further, principally, when one can imagine and fore-see new phenomena which ought to follow from the hypotheses which one employs, and when one finds that therein the fact corresponds to our prevision. But if all these proofs of probability are met with in that which I propose to discuss, as it seems to me they are, this ought to be a very strong confirmation of the success of my inquiry; and it must be ill if the facts are not pretty much as I represent them. I would believe then that those who love to know the Causes of things and who are able to admire the marvels of Light, will find some satisfaction in these various speculations regarding it, and in the new explanation of its famous property which is the main foundation of the construction of our eyes and of those great inventions which extend so vastly the use of them.

I hope also that there will be some who by following these beginnings will pen-etrate much further into this question than I have been able to do, since the subject must be far from being exhausted. This appears from the passages which I have indi-cated where I leave certain difficulties without having resolved them, and still more from matters which I have not touched at all, such as Luminous Bodies of several sorts, and all that concerns Colours; in which no one until now can boast of having succeeded. Finally, there remains much more to be investigated touching the nature of Light which I do not pretend to have disclosed, and I shall owe much in return to him who shall be able to supplement that which is here lacking to me in knowledge.

The Hague. The 8 January 1690.

13.2.2   Chapter I: On Rays Propagated in Straight Lines

As happens in all the sciences in which Geometry is applied to matter, the demon-strations concerning Optics are founded on truths drawn from experience. Such are that the rays of light are propagated in straight lines; that the angles of reflexion and of incidence are equal; and that in refraction the ray is bent according to the law of sines, now so well known, and which is no less certain than the preceding laws.

The majority of those who have written touching the various parts of Optics have contented themselves with presuming these truths. But some, more inquiring, have desired to investigate the origin and the causes, considering these to be in themselves wonderful effects of Nature. In which they advanced some ingenious things, but not however such that the most intelligent folk do not wish for better and more satisfactory explanations. Wherefore I here desire to propound what I have meditated on the subject, so as to contribute as much as I can to the explanation of this department of Natural Science, which, not without reason, is reputed to be one of its most difficult parts. I recognize myself to be much indebted to those who were the first to begin to dissipate the strange obscurity in which these things were enveloped, and to give us hope that they might be explained by intelligible reasoning. But, on the other hand I am astonished also that even here these have often been willing to offer, as assured and demonstrative, reasonings which were far from conclusive. For I do not find that any one has yet given a probable explanation of the first and most notable phenomena of light, namely why it is not propagated except in straight lines, and how visible rays, coming from an infinitude of diverse places, cross one another without hindering one another in any way.

I shall therefore essay in this book, to give, in accordance with the principles accepted in the Philosophy of the present day, some clearer and more probable reasons, firstly of these properties of light propagated rectilinearly; secondly of light which is reflected on meeting other bodies. Then I shall explain the phenomena of those rays which are said to suffer refraction on passing through transparent bodies of different sorts; and in this part I shall also explain the effects of the refraction of the air by the different densities of the Atmosphere.

Thereafter I shall examine the causes of the strange refraction of a certain kind of Crystal which is brought from Iceland. And finally I shall treat of the various shapes of transparent and reflecting bodies by which rays are collected at a point or are turned aside in various ways. From this it will be seen with what facility, following our new Theory, we find not only the Ellipses, Hyperbolas, and other curves which Mr. Des Cartes has ingeniously invented for this purpose; but also those which the surface of a glass lens ought to possess when its other surface is given as spherical or plane, or of any other figure that may be.

It is inconceivable to doubt that light consists in the motion of some sort of matter. For whether one considers its production, one sees that here upon the Earth it is chiefly engendered by fire and flame which contain without doubt bodies that are in rapid motion, since they dissolve and melt many other bodies, even the most solid; or whether one considers its effects, one sees that when light is collected, as by concave mirrors, it has the property of burning as a fire does, that is to say it disunites the particles of bodies. This is assuredly the mark of motion, at least in the true Philosophy, in which one conceives the causes of all natural effects in terms of mechanical motions. This, in my opinion, we must necessarily do, or else renounce all hopes of ever comprehending anything in Physics.

And as, according to this Philosophy, one holds as certain that the sensation of sight is excited only by the impression of some movement of a kind of matter which acts on the nerves at the back of our eyes, there is here yet one reason more for

Fig. 13.1 Descartes' construction for measuring the speed of light during a lunar eclipse.—[*K.K.*]

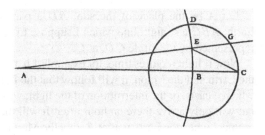

believing that light consists in a movement of the matter which exists between us and the luminous body.

Further, when one considers the extreme speed with which light spreads on every side, and how, when it comes from different regions, even from those directly opposite, the rays traverse one another without hindrance, one may well understand that when we see a luminous object, it cannot be by any transport of matter coming to us from this object, in the way in which a shot or an arrow traverses the air; for assuredly that would too greatly impugn these two properties of light, especially the second of them. It is then in some other way that light spreads; and that which can lead us to comprehend it is the knowledge which we have of the spreading of Sound in the air.

We know that by means of the air, which is an invisible and impalpable body, Sound spreads around the spot where it has been produced, by a movement which is passed on successively from one part of the air to another; and that the spreading of this movement, taking place equally rapidly on all sides, ought to form spherical surfaces ever enlarging and which strike our ears. Now there is no doubt at all that light also comes from the luminous body to our eyes by some movement impressed on the matter which is between the two; since, as we have already seen, it cannot be by the transport of a body which passes from one to the other. If, in addition, light takes time for its passage—which we are now going to examine—it will follow that this movement, impressed on the intervening matter, is successive; and consequently it spreads, as Sound does, by spherical surfaces and waves: for I call them waves from their resemblance to those which are seen to be formed in water when a stone is thrown into it, and which present a successive spreading as circles, though these arise from another cause, and are only in a flat surface.

To see then whether the spreading of light takes time, let us consider first whether there are any facts of experience which can convince us to the contrary. As to those which can be made here on the Earth, by striking lights at great distances, although they prove that light takes no sensible time to pass over these distances, one may say with good reason that they are too small, and that the only conclusion to be drawn from them is that the passage of light is extremely rapid. Mr. Des Cartes, who was of opinion that it is instantaneous, founded his views, not without reason, upon a better basis of experience, drawn from the Eclipses of the Moon; which, nevertheless, as I shall show, is not at all convincing. I will set it forth, in a way a little different from his, in order to make the conclusion more comprehensible (see Fig. 13.1).

Let A be the place of the sun, BD a part of the orbit or annual path of the Earth: ABC a straight line which I suppose to meet the orbit of the Moon, which is represented by the circle CD, at C.

Now if light requires time, for example 1 h, to traverse the space which is between the Earth and the Moon, it will follow that the Earth having arrived at B, the shadow which it casts, or the interruption of the light, will not yet have arrived at the point C, but will only arrive there an hour after. It will then be 1 h after, reckoning from the moment when the Earth was at B, that the Moon, arriving at C, will be obscured: but this obscuration or interruption of the light will not reach the Earth till after another hour. Let us suppose that the Earth in these 2 h will have arrived at E. The Earth then, being at E, will see the Eclipsed Moon at C, which it left an hour before, and at the same time will see the sun at A. For it being immovable, as I suppose with Copernicus, and the light moving always in straight lines, it must always appear where it is. But one has always observed, we are told, that the eclipsed Moon appears at the point of the Ecliptic opposite to the Sun; and yet here it would appear in arrear of that point by an amount equal to the angle GEC, the supplement of AEC. This, however, is contrary to experience, since the angle GEC would be very sensible, and about 33°. Now according to our computation, which is given in the Treatise on the causes of the phenomena of Saturn, the distance BA between the Earth and the Sun is about 12,000 diameters of the Earth, and hence 400 times greater than BC the distance of the Moon, which is 30 diameters. Then the angle ECB will be nearly 400 times greater than BAE, which is 5′; namely, the path which the earth travels in 2 h along its orbit; and thus the angle BCE will be nearly 33°; and likewise the angle CEG, which is greater by 5′.

But it must be noted that the speed of light in this argument has been assumed such that it takes a time of 1 h to make the passage from here to the Moon. If one supposes that for this it requires only 1 min of time, then it is manifest that the angle CEG will only be 33′; and if it requires only 10 s of time, the angle will be less than 6′. And then it will not be easy to perceive anything of it in observations of the Eclipse; nor, consequently, will it be permissible to deduce from it that the movement of light is instantaneous.

It is true that we are here supposing a strange velocity that would be a 100,000 times greater than that of Sound. For Sound, according to what I have observed, travels about 180 Toises in the time of 1 s, or in about one beat of the pulse. But this supposition ought not to seem to be an impossibility; since it is not a question of the transport of a body with so great a speed, but of a successive movement which is passed on from some bodies to others. I have then made no difficulty, in meditating on these things, in supposing that the emanation of light is accomplished with time, seeing that in this way all its phenomena can be explained, and that in following the contrary opinion everything is incomprehensible. For it has always seemed to me that even Mr. Des Cartes, whose aim has been to treat all the subjects of Physics intelligibly, and who assuredly has succeeded in this better than any one before him, has said nothing that is not full of difficulties, or even inconceivable, in dealing with Light and its properties.

Fig. 13.2 Römer's method for
measuring the speed of light
using eclipses of Jupiter's
moon.—[*K.K.*]

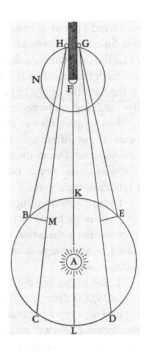

But that which I employed only as a hypothesis, has recently received great seemingness as an established truth by the ingenious proof of Mr. Römer which I am going here to relate, expecting him himself to give all that is needed for its confirmation. It is founded as is the preceding argument upon celestial observations, and proves not only that Light takes time for its passage, but also demonstrates how much time it takes, and that its velocity is even at least six times greater than that which I have just stated.

For this he makes use of the Eclipses suffered by the little planets which revolve around Jupiter, and which often enter his shadow: and see what is his reasoning (see Fig. 13.2). Let A be the Sun, $BCDE$ the annual orbit of the Earth, F Jupiter, GN the orbit of the nearest of his Satellites, for it is this one which is more apt for this investigation than any of the other three, because of the quickness of its revolution. Let G be this Satellite entering into the shadow of Jupiter, H the same Satellite emerging from the shadow.

Let it be then supposed, the Earth being at B some time before the last quadrature, that one has seen the said Satellite emerge from the shadow; it must needs be, if the Earth remains at the same place, that, after 42½ h, one would again see a similar emergence, because that is the time in which it makes the round of its orbit, and when it would come again into opposition to the Sun. And if the Earth, for instance, were to remain always at B during 30 revolutions of this Satellite, one would see it again emerge from the shadow after 30 times 42½ h. But the Earth having been carried along during this time to C, increasing thus its distance from Jupiter, it follows that if Light requires time for its passage the illumination of the little planet will be perceived later at C than it would have been at B, and that there must be added to this time of 30 times 42½ h that which the Light has required to traverse the space MC, the difference of the spaces CH, BH. Similarly at the other quadrature when the earth has come to E from D while approaching toward Jupiter, the immersions of the Satellite ought to be observed at E earlier than they would have been seen if the Earth had remained at D.

Now in quantities of observations of these Eclipses, made during 10 consecutive years, these differences have been found to be very considerable, such as 10 min and more; and from them it has been concluded that in order to traverse the whole diameter of the annual orbit KL, which is double the distance from here to the sun, Light requires about 22 min of time.

The movement of Jupiter in his orbit while the Earth passed from B to C, or from D to E, is included in this calculation; and this makes it evident that one cannot attribute the retardation of these illuminations or the anticipation of the eclipses, either to any irregularity occurring in the movement of the little planet or to its eccentricity.

If one considers the vast size of the diameter KL, which according to me is some 24,000 diameters of the Earth, one will acknowledge the extreme velocity of Light. For, supposing that KL is no more than 22,000 of these diameters, it appears that being traversed in 22 min this makes the speed a 1000 diameters in 1 min, that is 16⅔ diameters in 1 s or in one beat of the pulse, which makes more than 1100 times a 100,000 toises; since the diameter of the Earth contains 2865 leagues, reckoned at 25 to the degree, and each league is 2282 Toises, according to the exact measurement which Mr. Picard made by order of the King in 1669. But Sound, as I have said above, only travels 180 toises in the same time of 1 s: hence the velocity of Light is more than 600,000 times greater than that of Sound. This, however, is quite another thing from being instantaneous, since there is all the difference between a finite thing and an infinite. Now the successive movement of Light being confirmed in this way, it follows, as I have said, that it spreads by spherical waves, like the movement of Sound.

But if the one resembles the other in this respect, they differ in many other things; to wit, in the first production of the movement which causes them; in the matter in which the movement spreads; and in the manner in which it is propagated. As to that which occurs in the production of Sound, one knows that it is occasioned by the agitation undergone by an entire body, or by a considerable part of one, which shakes all the contiguous air. But the movement of the Light must originate as from each point of the luminous object, else we should not be able to perceive all the different parts of that object, as will be more evident in that which follows. And I do not believe that this movement can be better explained than by supposing that all those of the luminous bodies which are liquid, such as flames, and apparently the sun and the stars, are composed of particles which float in a much more subtle medium which agitates them with great rapidity, and makes them strike against the particles of the ether which surrounds them, and which are much smaller than they. But I hold also that in luminous solids such as charcoal or metal made red hot in the fire, this same movement is caused by the violent agitation of the particles of the metal or of the wood; those of them which are on the surface striking similarly against the ethereal matter. The agitation, moreover, of the particles which engender the light ought to be much more prompt and more rapid than is that of the bodies which cause sound, since we do not see that the tremors of a body which is giving out a sound are capable of giving rise to Light, even as the movement of the hand in the air is not capable of producing Sound.

Now if one examines what this matter may be in which the movement coming from the luminous body is propagated, which I call Ethereal matter, one will see that it is not the same that serves for the propagation of Sound. For one finds that the latter is really that which we feel and which we breathe, and which being removed from any place still leaves there the other kind of matter that serves to convey Light. This may be proved by shutting up a sounding body in a glass vessel from which the air is withdrawn by the machine which Mr. Boyle has given us, and with which he has performed so many beautiful experiments. But in doing this of which I speak, care must be taken to place the sounding body on cotton or on feathers, in such a way that it cannot communicate its tremors either to the glass vessel which encloses it, or to the machine; a precaution which has hitherto been neglected. For then after having exhausted all the air one hears no Sound from the metal, though it is struck.

One sees here not only that our air, which does not penetrate through glass, is the matter by which Sound spreads; but also that it is not the same air but another kind of matter in which Light spreads; since if the air is removed from the vessel the Light does not cease to traverse it as before.

And this last point is demonstrated even more clearly by the celebrated experiment of Torricelli, in which the tube of glass from which the quicksilver has withdrawn itself, remaining void of air, transmits Light just the same as when air is in it. For this proves that a matter different from air exists in this tube, and that this matter must have penetrated the glass or the quicksilver, either one or the other, though they are both impenetrable to the air. And when, in the same experiment, one makes the vacuum after putting a little water above the quicksilver, one concludes equally that the said matter passes through glass or water, or through both.

As regards the different modes in which I have said the movements of Sound and of Light are communicated, one may sufficiently comprehend how this occurs in the case of Sound if one considers that the air is of such a nature that it can be compressed and reduced to a much smaller space than that which it ordinarily occupies. And in proportion as it is compressed the more does it exert an effort to regain its volume; for this property along with its penetrability, which remains notwithstanding its compression, seems to prove that it is made up of small bodies which float about and which are agitated very rapidly in the ethereal matter composed of much smaller parts. So that the cause of the spreading of Sound is the effort which these little bodies make in collisions with one another, to regain freedom when they are a little more squeezed together in the circuit of these waves than elsewhere.

But the extreme velocity of Light, and other properties which it has, cannot admit of such a propagation of motion, and I am about to show here the way in which I conceive it must occur. For this, it is needful to explain the property which hard bodies must possess to transmit movement from one to another.

13.3 Study Questions

QUES. 13.1. What is Huygens' view of the nature of light?

a) What are the three laws of optics initially mentioned by Huygens? Does he question their validity? How is Huygens's approach to optics different (if at all) than that of many of his predecessors?
b) What was the opposing theory of light with which Huygens contends? What are two behaviors of light which this opposing theory cannot explain?
c) Regarding the first objection: How did Descartes propose to measure the speed of light? How did Römer actually measure the speed of light?
d) What, exactly, rendered Römer's method superior? And what conclusion(s), regarding the nature of light, does Huygens draw from Römer's measurements?

13.4 Exercises

EX. 13.1 (DESCARTES' METHOD OF MEASURING THE SPEED OF LIGHT). If the speed of light is not infinite, then the sun, earth and moon will not be perfectly lined up at the moment a lunar eclipse is observed from Earth. Based on the accepted value of the speed of light, by how much do they deviate from a straight line? Is this deviation practically measurable? Hints: you might set up a diagram like that of Huygens' Fig. 13.1 and then calculate time that light from the sun takes to pass the earth, reflect from the moon and return to the earth; the average orbital speed of the earth can be obtained from the diameter of its orbit and the duration of a year. (ANSWER: $\angle BAE \simeq 0.1$ arcseconds)

EX. 13.2 (RÖMER'S METHOD OF MEASURING THE SPEED OF LIGHT). In this exercise, we will explore how Römer calculated the speed of light by measuring the different number of eclipses of Jupiter's moon observed when the earth travels (i) away from Jupiter along BC, and (ii) towards Jupiter along DE (see Fig. 13.2).

Imagine, for a moment, running at speed u away from a paint ball gun that fires pellets with muzzle velocity v once every Δt seconds. How long will it be between when successive pellets strike you? When one of them hits you, the next one will still be a distance $\Delta x = v \Delta t$ behind you. So it will take $\Delta t' = \Delta x/(v - u)$ seconds for the next one to hit you, since its speed relative to you is $(v - u)$. Writing this in terms of Δt, we get $\Delta t' = \frac{v}{v-u} \Delta t$. Since the frequency at which the pellets strike you is $f' = 1/\Delta t'$, we can write the frequency of hits in terms of the frequency of pellets emitted by the gun as

$$f' = \frac{v - u}{v} f. \tag{13.1}$$

This effect is called the *doppler shift*: the frequency of "absorption events" counted by a moving target is different than the frequency of the "emission events" counted by the source. The doppler shift accounts for the difference in perceived pitch (frequency) of a train whistle when a train is approach and receding from an observer.

We can now use the doppler shift formula, Eq. 13.1, to calculate the number of eclipses seen by an observer traveling towards and away from Jupiter. Generally, we can find the number of events (*e.g.* eclipses) seen by an observer in motion using the formula

$$N = \left(\frac{\text{\# observed events}}{\text{time}}\right)(\text{time of travel}) = f'\frac{L}{u}. \tag{13.2}$$

Here we have assumed the observer is moving at speed u through a distance L, which we will take to be either BC or DE. By combining Eqs. 13.1 and 13.2 we obtain formulas for N_1 and N_2.

$$
\begin{aligned}
N_1 &= \frac{v - u}{v} f\frac{L}{u} = \frac{v - u}{vu} fL \\
N_2 &= \frac{v + u}{v} f\frac{L}{u} = \frac{v + u}{vu} fL.
\end{aligned}
\tag{13.3}
$$

The difference in signs ($\pm$) appearing in Eq. 13.3 is due to the different direction of motion of the earth. Now we can easily calculate the difference in the number of eclipses observed, ΔN, along the two routes BC and DE:

$$\Delta N = \frac{2fL}{v}. \tag{13.4}$$

Notice that Eq. 13.4 depends only on the frequency of eclipses, f, the length of travel, L, and the speed of light, v. Now using some reference data (*e.g.* the orbital periods of Io and Earth), estimate the difference in the observed number of eclipses of Jupiter's moon, Io, during two separate 1-month periods during which the earth is (i) approaching, and (ii) receding from Jupiter. Are your results consistent with those of Römer?

13.5 Vocabulary

1. Intelligible
2. Infinitude
3. Rectilinear
4. Refraction
5. Impalpable
6. Supplement
7. Emanation
8. Celestial
9. Quadrature
10. Toise
11. League
12. Ether
13. Quicksilver

Chapter 14
Huygens' Principle

Each little region of a luminous body, such as the Sun, a candle,
or a burning coal, generates its own waves of which that region
is the centre.

—Christiaan Huygens

14.1 Introduction

Christiaan Huygens began his famous *Treatise on Light* by reminding the reader of the peculiar properties exhibited by rays of light. They travel in straight lines, obey the law of reflection when striking a mirror, bend when traveling from one medium to another, travel at an extraordinary speed, and are able to pass through one another unhindered. Huygens then suggested that these properties of light might be explained by a wave theory, analogous to the well-established wave theory of sound. But if sound requires a medium such as air, then what type of medium does the propagation of light necessitate? And how can a wave theory of light explain the typically rectilinear (*i.e.* straight-line) motion of light rays when passing through apertures and past obstructions? These are the types of questions which Huygens must now address.

14.2 Reading: Huygens, *Treatise on Light*

Huygens, C., *Treatise on Light*, Macmillan, London, 1912.

14.2.1 Chapter I: On Rays Propagated in Straight Lines, Continued

When one takes a number of spheres of equal size, made of some very hard substance, and arranges them in a straight line, so that they touch one another, one finds, on striking with a similar sphere against the first of these spheres, that the

© Springer International Publishing Switzerland 2016
K. Kuehn, *A Student's Guide Through the Great Physics Texts*,
Undergraduate Lecture Notes in Physics, DOI 10.1007/978-3-319-21816-8_14

motion passes as in an instant to the last of them, which separates itself from the row, without one's being able to perceive that the others have been stirred. And even that one which was used to strike remains motionless with them. Whence one sees that the movement passes with an extreme velocity which is the greater, the greater the hardness of the substance of the spheres.

But it is still certain that this progression of motion is not instantaneous, but successive, and therefore must take time. For if the movement, or the disposition to movement, if you will have it so, did not pass successively through all these spheres, they would all acquire the movement at the same time, and hence would all advance together; which does not happen. For the last one leaves the whole row and acquires the speed of the one which was pushed. Moreover there are experiments which demonstrate that all the bodies which we reckon of the hardest kind, such as quenched steel, glass, and agate, act as springs and bend somehow, not only when extended as rods but also when they are in the form of spheres or of other shapes. That is to say they yield a little in themselves at the place where they are struck, and immediately regain their former figure. For I have found that on striking with a ball of glass or of agate against a large and quite thick thick piece of the same substance which had a flat surface, slightly soiled with breath or in some other way, there remained round marks, of smaller or larger size according as the blow had been weak or strong. This makes it evident that these substances yield where they meet, and spring back: and for this time must be required.

Now in applying this kind of movement to that which produces Light there is nothing to hinder us from estimating the particles of the ether to be of a substance as nearly approaching to perfect hardness and possessing a springiness as prompt as we choose. It is not necessary to examine here the causes of this hardness, or of that springiness, the consideration of which would lead us too far from our subject. I will say, however, in passing that we may conceive that the particles of the ether, notwithstanding their smallness, are in turn composed of other parts and that their springiness consists in the very rapid movement of a subtle matter which penetrates them from every side and constrains their structure to assume such a disposition as to give to this fluid matter the most overt and easy passage possible. This accords with the explanation which Mr. Des Cartes gives for the spring, though I do not, like him, suppose the pores to be in the form of round hollow canals. And it must not be thought that in this there is anything absurd or impossible, it being on the contrary quite credible that it is this infinite series of different sizes of corpuscles, having different degrees of velocity, of which Nature makes use to produce so many marvellous effects.

But though we shall ignore the true cause of springiness we still see that there are many bodies which possess this property; and thus there is nothing strange in supposing that it exists also in little invisible bodies like the particles of the Ether. Also if one wishes to seek for any other way in which the movement of Light is successively communicated, one will find none which agrees better, with uniform progression, as seems to be necessary, than the property of springiness; because if this movement should grow slower in proportion as it is shared over a greater quantity of matter, in moving away from the source of the light, it could not conserve this great velocity over great distances. But by supposing springiness in the ethereal

Fig. 14.1 Huygens' depiction
of the communication of
impulses by adjacent particles
of ether.—[*K.K.*]

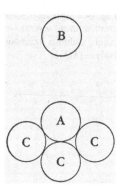

matter, its particles will have the property of equally rapid restitution whether they
are pushed strongly or feebly; and thus the propagation of Light will always go on
with an equal velocity.

And it must be known that although the particles of the ether are not ranged thus
in straight lines, as in our row of spheres, but confusedly, so that one of them touches
several others, this does not hinder them from transmitting their movement and from
spreading it always forward. As to this it is to be remarked that there is a law of
motion serving for this propagation, and verifiable by experiment (see Fig. 14.1). It
is that when a sphere, such as *A* here, touches several other similar spheres *CCC*,
if it is struck by another sphere *B* in such a way as to exert an impulse against all
the spheres *CCC* which touch it, it transmits to them the whole of its movement,
and remains after that motionless like the sphere *B*. And without supposing that the
ethereal particles are of spherical form (for I see indeed no need to suppose them
so) one may well understand that this property of communicating an impulse does
not fail to contribute to the aforesaid propagation of movement.

Equality of size seems to be more necessary, because otherwise there ought to be
some reflexion of movement backwards when it passes from a smaller particle to a
larger one, according to the *Laws of Percussion* which I published some years ago.

However, one will see hereafter that we have to suppose such an equality not so
much as a necessity for the propagation of light as for rendering that propagation
easier and more powerful; for it is not beyond the limits of probability that the
particles of the ether have been made equal for a purpose so important as that of
light, at least in that vast space which is beyond the region of atmosphere and which
seems to serve only to transmit the light of the Sun and the Stars.

I have then shown in what manner one may conceive Light to spread successively,
by spherical waves, and how it is possible that this spreading is accomplished with
as great a velocity as that which experiments and celestial observations demand.
Whence it may be further remarked that although the particles are supposed to be in
continual movement (for there are many reasons for this) the successive propagation
of the waves cannot be hindered by this; because the propagation consists nowise
in the transport of those particles but merely in a small agitation which they cannot
help communicating to those surrounding, notwithstanding any movement which
may act on them causing them to be changing positions amongst themselves.

Fig. 14.2 Spherical waves emitted from several points within a luminous body.—[*K.K.*]

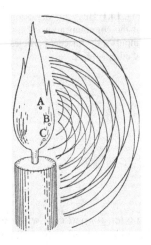

But we must consider still more particularly the origin of these waves, and the manner in which they spread. And, first, it follows from what has been said on the production of Light, that each little region of a luminous body, such as the Sun, a candle, or a burning coal, generates its own waves of which that region is the centre (see Fig. 14.2). Thus in the flame of a candle, having distinguished the points *A*, *B*, *C*, concentric circles described about each of these points represent the waves which come from them. And one must imagine the same about every point of the surface and of the part within the flame.

But as the percussions at the centres of these waves possess no regular succession, it must not be supposed that the waves themselves follow one another at equal distances: and if the distances marked in the figure appear to be such, it is rather to mark the progression of one and the same wave at equal intervals of time than to represent several of them issuing from one and the same centre.

After all, this prodigious quantity of waves which traverse one another without confusion and without effacing one another must not be deemed inconceivable; it being certain that one and the same particle of matter can serve for many waves coming from different sides or even from contrary directions, not only if it is struck by blows which follow one another closely but even for those which act on it at the same instant. It can do so because the spreading of the movement is successive. This may be proved by the row of equal spheres of hard matter (see Fig. 14.3), spoken of above. If against this row there are pushed from two opposite sides at the same time two similar spheres *A* and *D*, one will see each of them rebound with the same velocity which it had in striking, yet the whole row will remain in its place, although the movement has passed along its whole length twice over. And if these contrary movements happen to meet one another at the middle sphere, *B*, or at some other such as *C*, that sphere will yield and act as a spring at both sides, and so will serve at the same instant to transmit these two movements.

But what may at first appear full strange and even incredible is that the undulations produced by such small movements and corpuscles, should spread to such

Fig. 14.3 A row of equal spheres of hard matter struck simultaneously from both ends.—[*K.K.*]

immense distances; as for example from the Sun or from the Stars to us. For the force of these waves must grow feeble in proportion as they move away from their origin, so that the action of each one in particular will without doubt become incapable of making itself felt to our sight. But one will cease to be astonished by considering how at a great distance from the luminous body an infinitude of waves, though they have issued from different points of this body, unite together in such a way that they sensibly compose one single wave only, which, consequently, ought to have enough force to make itself felt. Thus this infinite number of waves which originate at the same instant from all points of a fixed star, big it may be as the Sun, make practically only one single wave which may well have force enough to produce an impression on our eyes. Moreover from each luminous point there may come many thousands of waves in the smallest imaginable time, by the frequent percussion of the corpuscles which strike the Ether at these points: which further contributes to rendering their action more sensible.

There is the further consideration in the emanation of these waves, that each particle of matter in which a wave spreads, ought not to communicate its motion only to the next particle which is in the straight line drawn from the luminous point, but that it also imparts some of it necessarily to all the others which touch it and which oppose themselves to its movement. So it arises that around each particle there is made a wave of which that particle is the centre. Thus if *DCF* is a wave emanating from the luminous point *A*, which is its centre (see Fig. 14.4), the particle *B*, one of those comprised within the sphere *DCF*, will have made its particular or partial wave *KCL*, which will touch the wave *DCF* at *C* at the same moment that the principal wave emanating from the point *A* has arrived at *DCF*; and it is clear that it will be only the region *C* of the wave *KCL* which will touch the wave *DCF*, to wit, that which is in the straight line drawn through *AB*. Similarly the other particles of the sphere *DCF*, such as *bb*, *dd*, etc., will each make its own wave. But each of these waves can be infinitely feeble only as compared with the wave *DCF*, to the composition of which all the others contribute by the part of their surface which is most distant from the centre *A*.

One sees, in addition, that the wave *DCF* is determined by the distance attained in a certain space of time by the movement which started from the point *A*; there being no movement beyond this wave, though there will be in the space which it encloses, namely in parts of the particular waves, those parts which do not touch the sphere *DCF*. And all this ought not to seem fraught with too much minuteness or subtlety, since we shall see in the sequel that all the properties of Light, and everything pertaining to its reflexion and its refraction, can be explained in principle by this means. This is a matter which has been quite unknown to those who hitherto have begun

Fig. 14.4 Huygens' principle illustrating how a principle wave, emitted from a point source, gives rise to partial waves, and hence regions of light and shadow.—[*K.K.*]

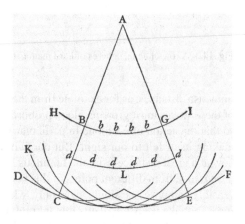

to consider the waves of light, amongst whom are Mr. Hooke in his *Micrographia*, and Father Pardies, who, in a treatise of which he let me see a portion, and which he was unable to complete as he died shortly afterward, had undertaken to prove by these waves the effects of reflexion and refraction. But the chief foundation, which consists in the remark I have just made, was lacking in his demonstrations; and for the rest he had opinions very different from mine, as may be will appear some day if his writing has been preserved.

To come to the properties of Light. We remark first that each portion of a wave ought to spread in such a way that its extremities lie always between the same straight lines drawn from the luminous point. Thus the portion *BG* of the wave, having the luminous point *A* as its centre, will spread into the arc *CE* bounded by the straight lines *ABC*, *AGE*. For although the particular waves produced by the particles comprised within the space *CAE* spread also outside this space, they yet do not concur at the same instant to compose a wave which terminates the movement, as they do precisely at the circumference *CE*, which is their common tangent.

And hence one sees the reason why light, at least if its rays are not reflected or broken, spreads only by straight lines, so that it illuminates no object except when the path from its source to that object is open along such lines.

For if, for example, there were an opening *BG*, limited by opaque bodies *BH*, *GI*, the wave of light which issues from the point *A* will always be terminated by the straight lines *AC*, *AE*, as has just been shown; the parts of the partial waves which spread outside the space *ACE* being too feeble to produce light there.

Now, however small we make the opening *BG*, there is always the same reason causing the light there to pass between straight lines; since this opening is always large enough to contain a great number of particles of the ethereal matter, which are of an inconceivable smallness; so that it appears that each little portion of the wave necessarily advances following the straight line which comes from the luminous point. Thus then we may take the rays of light as if they were straight lines.

It appears, moreover, by what has been remarked touching the feebleness of the particular waves, that it is not needful that all the particles of the Ether should be equal amongst themselves, though equality is more apt for the propagation of the movement. For it is true that inequality will cause a particle by pushing against another larger one to strive to recoil with a part of its movement; but it will thereby merely generate backwards towards the luminous point some partial waves incapable of causing light, and not a wave compounded of many as *CE* was.

Another property of waves of light, and one of the most marvellous, is that when some of them come from different or even from opposing sides, they produce their effect across one another without any hindrance. Whence also it comes about that a number of spectators may view different objects at the same time through the same opening, and that two persons can at the same time see one another's eyes. Now according to the explanation which has been given of the action of light, how the waves do not destroy nor interrupt one another when they cross one another, these effects which I have just mentioned are easily conceived. But in my judgement they are not at all easy to explain according to the views of Mr. Des Cartes, who makes Light to consist in a continuous pressure merely tending to movement. For this pressure not being able to act from two opposite sides at the same time, against bodies which have no inclination to approach one another, it is impossible so to understand what I have been saying about two persons mutually seeing one another's eyes, or how two torches can illuminate one another.

14.3 Study Questions

QUES. 14.1. According to Huygens, how is light like, and how is it unlike, sound in its (i) production, (ii) detection, (iii) speed and (iv) the medium through which it travels?

QUES. 14.2. Can Huygens' theory adequately explain the rectilinear motion of light?

a) Consider Fig. 14.4, which depicts light spreading circularly from a source at point *A*. Where are the barriers, and what arc represents the opening through which it passes?
b) In what region(s) is there illumination? Shadow? Is there a sharp boundary between the two?
c) How does he use his idea that an "infinitude of waves... which issue... from different points of a body unite together... to compose a single wave" to explain the rectilinear path of a wave through an opening?
d) Are there any problems with his explanation? In particular, does his theory really imply that there are no regions within the shadow which are illuminated?

Fig. 14.5 Concentric air density waves spread at the speed
of sound from a vibrating
body

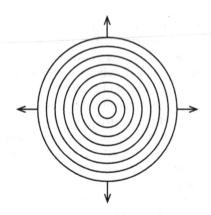

14.4 Exercises

Ex. 14.1 (THE SUPERPOSITION PRINCIPLE AND WAVE INTERFERENCE). According to
the wave theory of sound, a vibrating object (such as a bell when struck) produces
alternating compressions and rarefactions in the adjacent air. These air vibrations,
in turn, propagate away from the object in the form of concentric spherical wave-
fronts.[1] At any particular instant, therefore, the sound source will be surrounded by
a set of concentric spherical shells, so to speak, which represent alternating regions
of high and low air density (see Fig. 14.5). The distance between successive shells
of high (or low) density is called the sound's *wavelength*. The relationship between
the wavelength, λ, speed, v and frequency, f, of the waves is given by

$$v = \lambda f. \tag{14.1}$$

If there are two nearby sources of sound, then a volume of air at a particular location can experience density variations due to two sound waves simultaneously. As
it turns out, if the sound is not too loud then the air density variation at this location is simply the sum of the variations due to each source individually. This is
called the *superposition principle*; Huygens implicitly makes use of it in his *Treatise on Light*.[2] The superposition principle leads to the concept of *constructive and
destructive interference*: if two waves from different sources arrive in phase with
one another at a particular location, their effects add to one another and they are
said to interfere constructively; if they arrive out of phase with one another, their
effects subtract from one another and they are said to interfere destructively.

---

[1] Strictly speaking, the wave-fronts are spherical only if the vibrating body is itself spherically
symmetric.

[2] More generally, the superposition principle finds application in any system which is governed by
a linear wave equation, such as classical electrodynamics in a vacuum, and Schrödinger's quantum
mechanics.

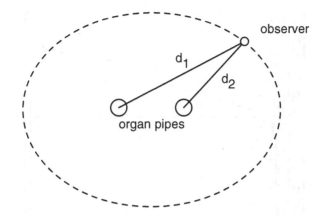

Fig. 14.6 An observer walks in an elliptical path around two organ pipes separated by a distance D

Suppose now that two identical vertical organ pipes, located a distance D apart, are sounded simultaneously. They each emit a single note with wavelength λ. Let us assume that the sound spreads from the top of each pipe in concentric spherical waves. (a) Suppose that $D = \lambda/2$. If a person walks in an elliptical path around the pipes (see Fig. 14.6), how many points of complete constructive interference will he encounter? Where will these be located? What about points of complete destructive interference? (HINT: Where on the ellipse will the distances d_1 and d_2 be equal? How often will they be different by a half-integral number of wavelengths?) (b) Would your answers to the previous questions change if $D = 2\lambda$? If so, how? (c) What about if $D = 2\lambda$ and the two organ pipes were emitting sound waves exactly half a wavelength out of phase with one another?

EX. 14.2 (HUYGENS' PRINCIPLE AND WAVE DIFFRACTION). When sunlight passes through a window into a darkened room it typically forms an illuminated image of the window on the wall (or floor) directly opposite the window. How can this commonplace observation be reconciled with Huygens' wave theory of light, which seems to entail a broad spreading of light waves into the room? In order to address this question, Huygens employed a principle which now bears his name. *Huygens' principle* states that

- each point along a particular wave-front can itself be thought of as a point source of spherical partial waves,
- these partial waves propagate away from their respective source points with a speed which depends on the surrounding medium, and
- the wave-front at any later time can be constructed by drawing a common tangent to the wavefronts of these several partial waves.

In this exercise, we will employ Huygens' principle to find the boundaries between the light and dark regions formed when light passes through a rectangular opening such as a window (see Fig. 14.7). Here, a series of consecutive planar wave-fronts approach from the left, pass through an opening into a room, and illuminate the wall

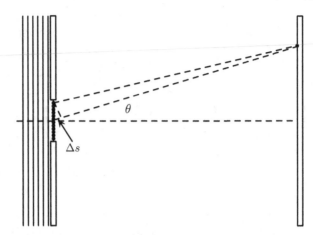

Fig. 14.7 Each point on a planar wave-front incident on an aperture acts as a spherical source of partial waves; these partial waves interfere constructively or destructively, producing bright and dark interference fringes on a screen to the right of the aperture

opposite the opening. According to Huygens' principle, each point across the width of the opening acts as a tiny point-source of spherical partial waves. These partial waves spread into the room and interfere with one another, obeying the superposition principle.[3] Consider now a point on the opposite wall which lies at an angle θ with respect to the initial direction of the incoming waves. If the partial waves emitted from each of the point-sources within the opening arrive in phase at this point on the wall, then they constructively interfere and the point on the wall appears bright. If, on the other hand, the partial waves arrive out of phase, then they destructively interfere and the point on the wall appears dark. By computing how the phase relationship between all the partial waves depends on θ, we can determine the angular width of the illuminated region on the wall opposite the opening.

a) Let us conceptually divide the opening into, say, 12 equally spaced point-sources of partial waves. Consider two of these point-sources (1 and 7) located at the top and at the middle of the opening. Draw two straight lines to the point on the opposite wall, the first from point-source 1 and the second from point-source 7. These lines will be very nearly parallel (provided the aperture is narrow compared to the distance to the wall) and will differ in length by an amount Δs which depends on the opening width, d, and the angle, θ. Prove that $\Delta s = d \sin \theta / 2$.

b) If Δs happens to equal $\lambda/2$, then the partial waves from point-sources 1 and 7 will be out of phase with one another when they arrive at the wall, and will destructively interfere. Now consider the pair of point-sources just below 1 and 7 (2 and 8). Will the waves from these point sources also destructively interfere?

---

[3] See Ex. 14.1, above.

What about waves from 3 and 9? Based on such considerations, argue that the opposite wall will be dark at an angle θ such that[4]

$$\sin \theta = \lambda/d. \qquad \text{(destructive interference condition for single slit)} \quad (14.2)$$

c) Find the angular width of the bright region (twice the angle θ appearing in Eq. 14.2), for 500-nm light waves after passing through a 5 cm wide aperture. Are your results consistent with Huygens' assertion that light travels in a straight line through apertures *despite* its wave-like nature?

d) Similarly, find the angular width for 3 cm sound waves after passing through a 5 cm wide aperture. Are your results consistent with Huygens' assertion that sound spreads significantly when passing through apertures *because of* its wave-like nature?

14.5   Vocabulary

1. Agate
2. Overt
3. Quench
4. Restitution
5. Prodigious
6. Undulation
7. Corpuscle
8. Infinitude

---

[4] This angular position corresponds to the point, C, separating light from shadow in Fig. 14.4 of Huygens' text. Strictly speaking, the region of shadow is neither perfectly dark nor marked by complete destructive interference of light. When light passes through a narrow gap, such as the slit between two facing knife blades, there appear thin (often barely visible) alternating fringes of light and dark within the region of shadow. This was observed by Young, Grimaldi and Newton; see Young's discussion of interference in Chap. 20 of the present volume.

Chapter 15
Reflection of Light Waves

It is evident that one could not demonstrate the equality of the angles of incidence and reflexion by similitude to that which happens to a ball thrown against a wall, of which writers have always made use.

—Christiaan Huygens

15.1 Introduction

What is light? In Chap. I of his *Treatise on Light*, Huygens argued that light consists of successive compressions and rarefactions of a subtle and all-pervasive medium—the æther—which are generated by vibrating particles inside of luminous bodies.[1] This mechanism, he claimed, provides a plausible explanation for the singularly high speed of light, something the particle theory of light could not do. Huygens also attempted to demonstrate that his wave theory can account for the rectilinear propagation of light rays—despite the fact that a wave tends to spread spherically around its source. This he did by invoking what is now known as *Huygens' Principle*.[2] Huygens' principle provides a very powerful recipe for predicting the advance of a wavefront, especially through inhomogeneous media; it is employed extensively nowadays in the fields of optics, acoustics, and even seismology. In what follows, Huygens attempts to use this technique for a more modest aim: to provide a physical explanation of the well-known law of reflection of light. His proof proceeds purely geometrically, assuming only Huygens' principle. To get started, note that the reflective surface in Fig. 15.1, below, is denoted by the line segment AB and that the in-coming wave-front is denoted by the line segment AC. Which line segment denotes the outgoing wave-front? What is the relationship between the trajectories of the in-going and out-going wave-fronts? And most importantly, is Huygen's proof of the law of reflection valid?

---

[1] Huygens' novel theory of light is *not* shared by his famous contemporary, Isaac Newton, who adhered to a corpuscular—or particle—theory of light; see Chaps. 18 and 19 of the present volume.

[2] See Ex. 14.2 in the previous chapter of the present volume.

© Springer International Publishing Switzerland 2016
K. Kuehn, *A Student's Guide Through the Great Physics Texts,*
Undergraduate Lecture Notes in Physics, DOI 10.1007/978-3-319-21816-8_15

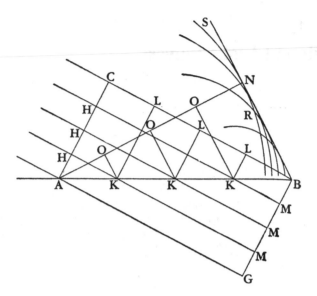

Fig. 15.1 The behavior of light waves incident on a reflecting surface—[*K.K.*]

15.2   Reading: Huygens, *Treatise on Light*

Huygens, C., *Treatise on Light*, Macmillan, London, 1912.

15.2.1   *Chapter II: On Reflexion*

Having explained the effects of waves of light which spread in a homogeneous matter, we will examine next that which happens to them on encountering other bodies. We will first make evident how the Reflexion of light is explained by these same waves, and why it preserves equality of angles.

   Let there be a surface AB (see Fig. 15.1); plane and polished, of some metal, glass, or other body, which at first I will consider as perfectly uniform (reserving to myself to deal at the end of this demonstration with the inequalities from which it cannot be exempt), and let a line AC, inclined to AB, represent a portion of a wave of light, the centre of which is so distant that this portion AC may be considered as a straight line; for I consider all this as in one plane, imagining to myself that the plane in which this figure is, cuts the sphere of the wave through its centre and intersects the plane AB at right angles. This explanation will suffice once for all.

   The piece C of the wave AC, will in a certain space of time advance as far as the plane AB at B, following the straight line CB, which may be supposed to come from the luminous centre, and which in consequence is perpendicular to AC. Now in this same space of time the portion A of the same wave, which has been hindered from communicating its movement beyond the plane AB, or at least partly so, ought to have continued its movement in the matter which is above this plane, and this along a distance equal to CB, making its own partial spherical wave, according to what has been said above. Which wave is here represented by the circumference SNR, the centre of which is A, and its semi-diameter AN equal to CB.

If one considers further the other pieces H of the wave AC, it appears that they will not only have reached the surface AB by straight lines HK parallel to CB, but that in addition they will have generated in the transparent air, from the centres K, K, K, particular spherical waves, represented here by circumferences the semi-diameters of which are equal to KM, that is to say to the continuations of HK as far as the line BG parallel to AC. But all these circumferences have as a common tangent the straight line BN, namely the same which is drawn from B as a tangent to the first of the circles, of which A is the centre, and AN the semi-diameter equal to BC, as is easy to see.

It is then the line BN (comprised between B and the point N where the perpendicular from the point A falls) which is as it were formed by all these circumferences, and which terminates the movement which is made by the reflexion of the wave AC; and it is also the place where the movement occurs in much greater quantity than anywhere else. Wherefore, according to that which has been explained, BN is the propagation of the wave AC at the moment when the piece C of it has arrived at B. For there is no other line which like BN is a common tangent to all the aforesaid circles, except BG below the plane AB; which line BG would be the propagation of the wave if the movement could have spread in a medium homogeneous with that which is above the plane. And if one wishes to see how the wave AC has come successively to BN, one has only to draw in the same figure the straight lines KO parallel to BN, and the straight lines KL parallel to AC. Thus one will see that the straight wave AC has become broken up into all the OKL parts successively, and that it has become straight again at NB.

Now it is apparent here that the angle of reflexion is made equal to the angle of incidence. For the triangles ACB, BNA being rectangular and having the side AB common, and the side CB equal to NA, it follows that the angles opposite to these sides will be equal, and therefore also the angles CBA, NAB. But as CB, perpendicular to CA, marks the direction of the incident ray, so AN, perpendicular to the wave BN, marks the direction of the reflected ray; hence these rays are equally inclined to the plane AB.

But in considering the preceding demonstration, one might aver that it is indeed true that BN is the common tangent of the circular waves in the plane of this figure, but that these waves, being in truth spherical, have still an infinitude of similar tangents, namely all the straight lines which are drawn from the point B in the surface generated by the straight line BN about the axis BA. It remains, therefore, to demonstrate that there is no difficulty herein: and by the same argument one will see why the incident ray and the reflected ray are always in one and the same plane perpendicular to the reflecting plane. I say then that the wave AC, being regarded only as a line, produces no light. For a visible ray of light, however narrow it may be, has always some width, and consequently it is necessary, in representing the

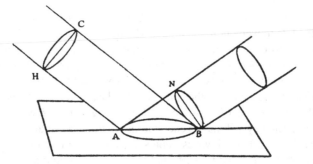

Fig. 15.2 A narrow beam of light reflecting from a surface—[*K.K.*]

wave whose progression constitutes the ray, to put instead of a line AC some plane
figure such as the circle HC in the following figure (see Fig. 15.2), by supposing,
as we have done, the luminous point to be infinitely distant. Now it is easy to see,
following the preceding demonstration, that each small piece of this wave HC having arrived at the plane AB, and there generating each one its particular wave, these
will all have, when C arrives at B, a common plane which will touch them, namely
a circle BN similar to CH; and this will be intersected at its middle and at right
angles by the same plane which likewise intersects the circle CH and the ellipse
AB.

One sees also that the said spheres of the partial waves cannot have any common
tangent plane other than the circle BN; so that it will be this plane where there will
be more reflected movement than anywhere else, and which will therefore carry on
the light in continuance from the wave CH.

I have also stated in the preceding demonstration that the movement of the piece
A of the incident wave is not able to communicate itself beyond the plane AB, or
at least not wholly. Whence it is to be remarked that though the movement of the
ethereal matter might communicate itself partly to that of the reflecting body, this
could in nothing alter the velocity of progression of the waves, on which the angle of
reflexion depends. For a slight percussion ought to generate waves as rapid as strong
percussion in the same matter. This comes about from the property of bodies which
act as springs, of which we have spoken above; namely that whether compressed
little or much they recoil in equal times. Equally so in every reflexion of the light,
against whatever body it may be, the angles of reflexion and incidence ought to be
equal notwithstanding that the body might be of such a nature that it takes away a
portion of the movement made by the incident light. And experiment shows that in
fact there is no polished body the reflexion of which does not follow this rule.

But the thing to be above all remarked in our demonstration is that it does not
require that the reflecting surface should be considered as a uniform plane, as has
been supposed by all those who have tried to explain the effects of reflexion; but only
an evenness such as may be attained by the particles of the matter of the reflecting
body being set near to one another; which particles are larger than those of the ethereal matter, as will appear by what we shall say in treating of the transparency and

opacity of bodies. For the surface consisting thus of particles put together, and the ethereal particles being above, and smaller, it is evident that one could not demonstrate the equality of the angles of incidence and reflexion by similitude to that which happens to a ball thrown against a wall, of which writers have always made use. In our way, on the other hand, the thing is explained without difficulty. For the smallness of the particles of quicksilver, for example, being such that one must conceive millions of them, in the smallest visible surface proposed, arranged like a heap of grains of sand which has been flattened as much as it is capable of being, this surface then becomes for our purpose as even as a polished glass is: and, although it always remains rough with respect to the particles of the Ether it is evident that the centres of all the particular spheres of reflexion, of which we have spoken, are almost in one uniform plane, and that thus the common tangent can fit to them as perfectly as is requisite for the production of light. And this alone is requisite, in our method of demonstration, to cause equality of the said angles without the remainder of the movement reflected from all parts being able to produce any contrary effect.

15.3   Study Questions

QUES. 15.1. Can Huygens's theory of light adequately explain the law of reflection?

a) In Fig. 15.1, which line segment represents the mirror? the incoming wave front? the outgoing wave front?
b) How, exactly, does he construct the outgoing wave front? In particular, what role does the concept of "partial waves" play in his proof? Upon what assumption(s) does his proof rely?
c) What is his conclusion? Is his proof valid? If not, is the problem with one of his assumptions, or with the logic of his subsequent steps?
d) Does the speed of a wave through a medium depend upon the strength of the percussion that generates it? What do experiments indicate?
e) Does Huygens' geometrical proof of the law of reflection apply only to perfectly flat surfaces?
f) Can the corpuscular theory of light account for the law of reflection as well as Huygens' wave theory?

15.4   Exercises

EX. 15.1 (REFLECTION AND REFRACTION OF PARTICLES). According to Newton, light consists of tiny hard particles which propagate through space away from a luminous source.[3] Can this theory account for the observed laws of reflection and

---

[3] Newton defends this claim in his *Opticks*; see Chaps. 18 and 19 of the present volume.

Fig. 15.3 Particle (as opposed to wave) reflection and refraction

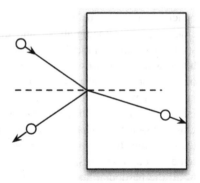

refraction? To address this question, consider Fig. 15.3, which depicts a point particle of mass m which is traveling in a straight line at a speed v toward a stationary flat surface at an incident angle θ_i.

a.) Suppose, first, that the surface is completely hard and thus impenetrable. When the particle strikes the surface what happens to its trajectory? More specifically, at what angle does it recoil from the surface? What laws or principles must you invoke in order to justify your conclusion? Based on your analysis, can the particle theory of light explain the law of reflection?

b.) Suppose now that the surface is penetrable by a particle. At the moment of impact, however, the surface exerts a force which acts perpendicular (and not parallel) to the surface. Consequently, the momentum of the particle in the perpendicular (but not the parallel) direction is changed. If the particle's perpendicular momentum is reduced by an amount Δp, how is its trajectory changed by the action of the surface? What about if its perpendicular momentum is instead increased at the moment of impact? Based on your analysis, can the particle theory of light explain the law of refraction?[4]

15.5   Vocabulary

1. Homogeneous
2. Luminous
3. Tangent
4. Aforesaid

---

[4] For a review of early attempts to reconcile the phenomenon of refraction with the particle theory of light, see Smith, A. M., Descartes's Theory of Light and Refraction: A Discourse on Method, *Transactions of the American Philosophical Society*, 77(3), 1–92, 1987. For a treatment of the law of refraction using the modern photon concept of light, see Drosdoff, D., and A. Widom, Snell's law from an elementary particle viewpoint, *American Journal of Physics*, 73(10), 973–975, 2005.

5. Successive
6. Percussion
7. Ethereal
8. Opacity
9. Quicksilver
10. Requisite

Chapter 16
Opacity, Transparency and Snell's Law

> *Let us pass now to the explanation of the effects of Refraction,*
> *assuming, as we have done, the passage of waves of light*
> *through transparent bodies, and the diminution of velocity*
> *which these same waves suffer in them.*
>
> —Christiaan Huygens

16.1 Introduction

In Chap. 2 of his *Treatise on Light*, Huygens demonstrated geometrically how the law of reflection follows naturally from the wave theory of light. This he did by first considering a wave-front incident at an oblique angle on a reflecting surface. As this wave-front encountered successive points on the surface, they emitted spherical partial waves which propagated away from the surface of the mirror. Then, by drawing a tangent line to these spherical wave-fronts, he was able to reconstruct the outgoing wave, whose direction of propagation was found to obey the law of reflection. With this success, Huygens now turns to the law of refraction. In this case, the incoming wave-front proceeds into a transparent material. This raises the question: why are some some materials transparent, while others reflect light? The discussion turns inevitably to the microscopic structure and composition of matter. Do you find Huygens' ideas plausible? If not, then what do you think might account for the diverse optical properties of materials?

16.2 Reading: Huygens, *Treatise on Light*

Huygens, C., *Treatise on Light*, Macmillan, London, 1912.

16.2.1 Chapter III: On Refraction

In the same way as the effects of Reflexion have been explained by waves of light reflected at the surface of polished bodies, we will explain transparency and the phenomena of refraction by waves which spread within and across diaphanous bodies,

© Springer International Publishing Switzerland 2016
K. Kuehn, *A Student's Guide Through the Great Physics Texts*,
Undergraduate Lecture Notes in Physics, DOI 10.1007/978-3-319-21816-8_16

both solids, such as glass, and liquids, such as water, oils, *etc*. But in order that it may not seem strange to suppose this passage of waves in the interior of these bodies, I will first show that one may conceive it possible in more than one mode.

First, then, if the ethereal matter cannot penetrate transparent bodies at all, their own particles would be able to communicate successively the movement of the waves, the same as do those of the Ether, supposing that, like those, they are of a nature to act as a spring. And this is easy to conceive as regards water and other transparent liquids, they being composed of detached particles. But it may seem more difficult as regards glass and other transparent and hard bodies, because their solidity does not seem to permit them to receive movement except in their whole mass at the same time. This, however, is not necessary because this solidity is not such as it appears to us, it being probable rather that these bodies are composed of particles merely placed close to one another and held together by some pressure from without of some other matter, and by the irregularity of their shapes. For primarily their rarity is shown by the facility with which there passes through them the matter of the vortices of the magnet, and that which causes gravity. Further, one cannot say that these bodies are of a texture similar to that of a sponge or of light bread, because the heat of the fire makes them flow and thereby changes the situation of the particles amongst themselves. It remains then that they are, as has been said, assemblages of particles which touch one another without constituting a continuous solid. This being so, the movement which these particles receive to carry on the waves of light, being merely communicated from some of them to others, without their going for that purpose out of their places or without derangement, it may very well produce its effect without prejudicing in any way the apparent solidity of the compound.

By pressure from without, of which I have spoken, must not be understood that of the air, which would not be sufficient, but that of some other more subtle matter, a pressure which I chanced upon by experiment long ago, namely in the case of water freed from air, which remains suspended in a tube open at its lower end, notwithstanding that the air has been removed from the vessel in which this tube is enclosed.

One can then in this way conceive of transparency in a solid without any necessity that the ethereal matter which serves for light should pass through it, or that it should find pores in which to insinuate itself. But the truth is that this matter not only passes through solids, but does so even with great facility; of which the experiment of Torricelli, above cited, is already a proof. Because on the quicksilver and the water quitting the upper part of the glass tube, it appears that it is immediately filled with ethereal matter, since light passes across it. But here is another argument which proves this ready penetrability, not only in transparent bodies but also in all others.

When light passes across a hollow sphere of glass, closed on all sides, it is certain that it is full of ethereal matter, as much as the spaces outside the sphere. And this ethereal matter, as has been shown above, consists of particles which just touch one another. If then it were enclosed in the sphere in such a way that it could not get out through the pores of the glass, it would be obliged to follow the movement of the sphere when one changes its place: and it would require consequently almost the same force to impress a certain velocity on this sphere, when placed on a horizontal plane, as if it were full of water or perhaps of quicksilver: because every body resists the velocity of the motion which one would give to it, in proportion to the quantity of matter which it contains, and which is obliged to follow this motion. But on the contrary one finds that the sphere resists the impress of movement only in proportion to the quantity of matter of the glass of which it is made. Then it must be that the ethereal matter which is inside is not shut up, but flows through it with very great freedom. We shall demonstrate hereafter that by this process the same penetrability may be inferred also as relating to opaque bodies.

The second mode then of explaining transparency, and one which appears more probably true, is by saying that the waves of light are carried on in the ethereal matter, which continuously occupies the interstices or pores of transparent bodies. For since it passes through them continuously and freely, it follows that they are always full of it. And one may even show that these interstices occupy much more space than the coherent particles which constitute the bodies. For if what we have just said is true: that force is required to impress a certain horizontal velocity on bodies in proportion as they contain coherent matter; and if the proportion of this force follows the law of weights, as is confirmed by experiment, then the quantity of the constituent matter of bodies also follows the proportion of their weights. Now we see that water weighs only one fourteenth part as much as an equal portion of quicksilver: therefore the matter of the water does not occupy the fourteenth part of the space which its mass obtains. It must even occupy much less of it, since quicksilver is less heavy than gold, and the matter of gold is by no means dense, as follows from the fact that the matter of the vortices of the magnet and of that which is the cause of gravity pass very freely through it.

But it may be objected here that if water is a body of so great rarity, and if its particles occupy so small a portion of the space of its apparent bulk, it is very strange how it yet resists Compression so strongly without permitting itself to be condensed by any force which one has hitherto essayed to employ, preserving even its entire liquidity while subjected to this pressure.

This is no small difficulty. It may, however, be resolved by saying that the very violent and rapid motion of the subtle matter which renders water liquid, by agitating the particles of which it is composed, maintains this liquidity in spite of the pressure which hitherto any one has been minded to apply to it.

The rarity of transparent bodies being then such as we have said, one easily conceives that the waves might be carried on in the ethereal matter which fills the interstices of the particles. And, moreover, one may believe that the progression of these waves ought to be a little slower in the interior of bodies, by reason of the small detours which the same particles cause. In which different velocity of light I shall show the cause of refraction to consist.

Before doing so, I will indicate the third and last mode in which transparency may be conceived; which is by supposing that the movement of the waves of light is transmitted indifferently both in the particles of the ethereal matter which occupy the interstices of bodies, and in the particles which compose them, so that the movement passes from one to the other. And it will be seen hereafter that this hypothesis serves excellently to explain the double refraction of certain transparent bodies.

Should it be objected that if the particles of the ether are smaller than those of transparent bodies (since they pass through their intervals), it would follow that they can communicate to them but little of their movement, it may be replied that the particles of these bodies are in turn composed of still smaller particles, and so it will be these secondary particles which will receive the movement from those of the ether.

Furthermore, if the particles of transparent bodies have a recoil a little less prompt than that of the ethereal particles, which nothing hinders us from supposing, it will again follow that the progression of the waves of light will be slower in the interior of such bodies than it is outside in the ethereal matter.

All this I have found as most probable for the mode in which the waves of light pass across transparent bodies. To which it must further be added in what respect these bodies differ from those which are opaque; and the more so since it might seem because of the easy penetration of bodies by the ethereal matter, of which mention has been made, that there would not be any body that was not transparent. For by the same reasoning about the hollow sphere which I have employed to prove the smallness of the density of glass and its easy penetrability by the ethereal matter, one might also prove that the same penetrability obtains for metals and for every other sort of body. For this sphere being for example of silver, it is certain that it contains some of the ethereal matter which serves for light, since this was there as well as in the air when the opening of the sphere was closed. Yet, being closed and placed upon a horizontal plane, it resists the movement which one wishes to give to it, merely according to the quantity of silver of which it is made; so that one must conclude, as above, that the ethereal matter which is enclosed does not follow the movement of the sphere; and that therefore silver, as well as glass, is very easily penetrated by this matter. Some of it is therefore present continuously and in quantities between the particles of silver and of all other opaque bodies: and since it serves for the propagation of light it would seem that these bodies ought also to be transparent, which however is not the case.

Whence then, one will say, does their opacity come? Is it because the particles which compose them are soft; that is to say, these particles being composed of others that are smaller, are they capable of changing their figure on receiving the pressure of the ethereal particles, the motion of which they thereby damp, and so hinder the continuance of the waves of light? That cannot be: for if the particles of the metals are soft, how is it that polished silver and mercury reflect light so strongly? What I find to be most probable herein, is to say that metallic bodies, which are almost the only really opaque ones, have mixed amongst their hard particles some soft ones; so that some serve to cause reflexion and the others to hinder transparency; while, on the other hand, transparent bodies contain only hard particles which have the faculty of recoil, and serve together with those of the ethereal matter for the propagation of the waves of light, as has been said.

Let us pass now to the explanation of the effects of Refraction, assuming, as we have done, the passage of waves of light through transparent bodies, and the diminution of velocity which these same waves suffer in them.

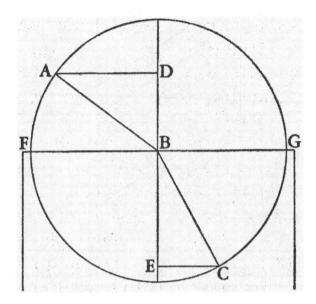

Fig. 16.1 A ray of light bends away from the surface when passing into a medium in which its speed is decreased.—[*K.K.*]

The chief property of Refraction is that a ray of light, such as AB, being in the air, and falling obliquely upon the polished surface of a transparent body, such as FG, is broken at the point of incidence B, in such a way that with the straight line DBE which cuts the surface perpendicularly it makes an angle CBE less than ABD which it made with the same perpendicular when in the air (see Fig. 16.1). And the measure of these angles is found by describing, about the point B, a circle which cuts the radii AB, BC. For the perpendiculars AD, CE, let fall from the points of intersection upon the straight line DE, which are called the Sines of the angles ABD, CBE, have a certain ratio between themselves; which ratio is always the same for all inclinations of the incident ray, at least for a given transparent body. This ratio is, in glass, very nearly as 3 to 2; and in water very nearly as 4 to 3; and is likewise different in other diaphanous bodies.

Another property, similar to this, is that the refractions are reciprocal between the rays entering into a transparent body and those which are leaving it. That is to say that if the ray AB in entering the transparent body is refracted into BC, then likewise CB being taken as a ray in the interior of this body will be refracted, on passing out, into BA.

To explain then the reasons of these phenomena according to our principles, let AB be the straight line which represents a plane surface bounding the transparent substances which lie towards C and towards N (see Fig. 16.2). When I say plane, that does not signify a perfect evenness, but such as has been understood in treating of reflexion, and for the same reason. Let the line AC represent a portion of a wave of light, the centre of which is supposed so distant that this portion may be

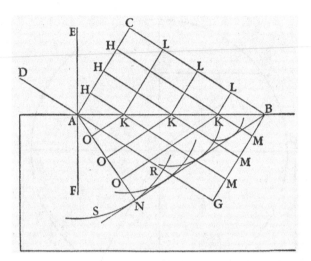

Fig. 16.2 A geometrical construction depicting the behavior of successive light waves as they pass across a surface and into a medium in which their speed is decreased.—[K.K.]

considered as a straight line. The piece C, then, of the wave AC, will in a certain space of time have advanced as far as the plane AB following the straight line CB, which may be imagined as coming from the luminous centre, and which conse- quently will cut AC at right angles. Now in the same time the piece A would have come to G along the straight line AG, equal and parallel to CB; and all the portion of wave AC would be at GB if the matter of the transparent body transmitted the movement of the wave as quickly as the matter of the Ether. But let us suppose that it transmits this movement less quickly, by one-third, for instance. Movement will then be spread from the point A, in the matter of the transparent body through a dis- tance equal to two-thirds of CB, making its own particular spherical wave according to what has been said before. This wave is then represented by the circumference SNR, the centre of which is A, and its semi-diameter equal to two-thirds of CB. Then if one considers in order the other pieces H of the wave AC, it appears that in the same time that the piece C reaches B they will not only have arrived at the surface AB along the straight lines HK parallel to CB, but that, in addition, they will have generated in the diaphanous substance from the centres K, partial waves, represented here by circumferences the semi-diameters of which are equal to two- thirds of the lines KM, that is to say, to two-thirds of the prolongations of HK down to the straight line BG; for these semi-diameters would have been equal to entire lengths of KM if the two transparent substances had been of the same penetrability.

Now all these circumferences have for a common tangent the straight line BN; namely the same line which is drawn as a tangent from the point B to the cir- cumference SNR which we considered first. For it is easy to see that all the other circumferences will touch the same BN, from B up to the point of contact N, which is the same point where AN falls perpendicularly on BN.

It is then BN, which is formed by small arcs of these circumferences, which terminates the movement that the wave AC has communicated within the transparent body, and where this movement occurs in much greater amount than anywhere else. And for that reason this line, in accordance with what has been said more than once, is the propagation of the wave AC at the moment when its piece C has reached B. For there is no other line below the plane AB which is, like BN, a common tangent to all these partial waves. And if one would know how the wave AC has come progressively to BN, it is necessary only to draw in the same figure the straight lines KO parallel to BN, and all the lines KL parallel to AC. Thus one will see that the wave CA, from being a straight line, has become broken in all the positions LKO successively, and that it has again become a straight line at BN. This being evident by what has already been demonstrated, there is no need to explain it further.

Now, in the same figure, if one draws EAF, which cuts the plane AB at right angles at the point A, since AD is perpendicular to the wave AC, it will be DA which will mark the ray of incident light, and AN which was perpendicular to BN, the refracted ray: since the rays are nothing else than the straight lines along which the portions of the waves advance.

Whence it is easy to recognize this chief property of refraction, namely that the Sine of the angle DAE has always the same ratio to the Sine of the angle NAF, whatever be the inclination of the ray DA: and that this ratio is the same as that of the velocity of the waves in the transparent substance which is towards AE to their velocity in the transparent substance towards AF. For, considering AB as the radius of a circle, the Sine of the angle BAC is BC, and the Sine of the angle ABN is AN. But the angle BAC is equal to DAE, since each of them added to CAE makes a right angle. And the angle ABN is equal to NAF, since each of them with BAN makes a right angle. Then also the Sine of the angle DAE is to the Sine of NAF as BC is to AN. But the ratio of BC to AN was the same as that of the velocities of light in the substance which is towards AE and in that which is towards AF; therefore also the Sine of the angle DAE will be to the Sine of the angle NAF the same as the said velocities of light.

To see, consequently, what the refraction will be when the waves of light pass into a substance in which the movement travels more quickly than in that from which they emerge (let us again assume the ratio of 3 to 2), it is only necessary to repeat all the same construction and demonstration which we have just used, merely substituting everywhere ⅔ instead of ⅔. And it will be found by the same reasoning, in this other figure (see Fig. 16.3), that when the piece C of the wave AC shall have reached the surface AB at B, all the portions of the wave AC will have advanced as far as BN, so that BC the perpendicular on AC is to AN the perpendicular on BN as 2 to 3. And there will finally be this same ratio of 2 to 3 between the Sine of the angle BAD and the Sine of the angle FAN.

Hence one sees the reciprocal relation of the refractions of the ray on entering and on leaving one and the same transparent body: namely that if NA falling on the external surface AB is refracted into the direction AD, so the ray AD will be refracted on leaving the transparent body into the direction AN.

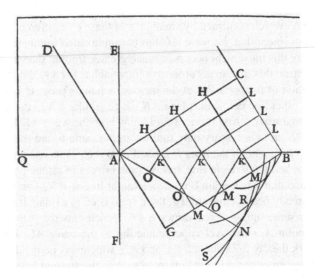

Fig. 16.3 Successive light waves passing across a surface and into a medium in which their speed is increased.—[K.K.]

One sees also the reason for a noteworthy accident which happens in this refraction: which is this, that after a certain obliquity of the incident ray DA, it begins to be quite unable to penetrate into the other transparent substance. For if the angle DAQ or CBA is such that in the triangle ACB, CB is equal to ⅔ of AB, or is greater, then AN cannot form one side of the triangle ANB, since it becomes equal to or greater than AB: so that the portion of wave BN cannot be found anywhere, neither consequently can AN, which ought to be perpendicular to it. And thus the incident ray DA does not then pierce the surface AB.

When the ratio of the velocities of the waves is as two to three, as in our example, which is that which obtains for glass and air, the angle DAQ must be more than $48°$ $11'$ in order that the ray DA may be able to pass by refraction. And when the ratio of the velocities is as 3 to 4, as it is very nearly in water and air, this angle DAQ must exceed $41°$ $24'$. And this accords perfectly with experiment.

But it might here be asked: since the meeting of the wave AC against the surface AB ought to produce movement in the matter which is on the other side, why does no light pass there? To which the reply is easy if one remembers what has been said before. For although it generates an infinitude of partial waves in the matter which is at the other side of AB, these waves never have a common tangent line (either straight or curved) at the same moment; and so there is no line terminating the propagation of the wave AC beyond the plane AB, nor any place where the movement is gathered together in sufficiently great quantity to produce light. And one will easily see the truth of this, namely that CB being larger than ⅔ of AB, the waves excited beyond the plane AB will have no common tangent if about the centres K one then draws circles having radii equal to ½ of the lengths LB to which they correspond. For all these circles will be enclosed in one another and will all pass beyond the point B.

Now it is to be remarked that from the moment when the angle DAQ is smaller than is requisite to permit the refracted ray DA to pass into the other transparent substance, one finds that the interior reflexion which occurs at the surface AB is much augmented in brightness, as is easy to realize by experiment with a triangular prism; and for this our theory can afford this reason. When the angle DAQ is still large enough to enable the ray DA to pass, it is evident that the light from the portion AC of the wave is collected in a minimum space when it reaches BN. It appears also that the wave BN becomes so much the smaller as the angle CBA or DAQ is made less; until when the latter is diminished to the limit indicated a little previously, this wave BN is collected together always at one point. That is to say, that when the piece C of the wave AC has then arrived at B, the wave BN which is the propagation of AC is entirely reduced to the same point B. Similarly when the piece H has reached K, the part AH is entirely reduced to the same point K. This makes it evident that in proportion as the wave CA comes to meet the surface AB, there occurs a great quantity of movement along that surface; which movement ought also to spread within the transparent body and ought to have much re-enforced the partial waves which produce the interior reflexion against the surface AB, according to the laws of reflexion previously explained.

And because a slight diminution of the angle of incidence DAQ causes the wave BN, however great it was, to be reduced to zero, (for this angle being 49° 11′ in the glass, the angle BAN is still 11° 21′, and the same angle being reduced by 1° only the angle BAN is reduced to zero, and so the wave BN reduced to a point) thence it comes about that the interior reflexion from being obscure becomes suddenly bright, so soon as the angle of incidence is such that it no longer gives passage to the refraction.

Now as concerns ordinary external reflexion, that is to say which occurs when the angle of incidence DAQ is still large enough to enable the refracted ray to penetrate beyond the surface AB, this reflexion should occur against the particles of the substance which touches the transparent body on its outside. And it apparently occurs against the particles of the air or others mingled with the ethereal particles and larger than they. So on the other hand the external reflexion of these bodies occurs against the particles which compose them, and which are also larger than those of the ethereal matter, since the latter flows in their interstices. It is true that there remains here some difficulty in those experiments in which this interior reflexion occurs without the particles of air being able to contribute to it, as in vessels or tubes from which the air has been extracted.

Experience, moreover, teaches us that these two reflexions are of nearly equal force, and that in different transparent bodies they are so much the stronger as the refraction of these bodies is the greater. Thus one sees manifestly that the reflexion of glass is stronger than that of water, and that of diamond stronger than that of glass.

I will finish this theory of refraction by demonstrating a remarkable proposition which depends on it; namely, that a ray of light in order to go from one point to another, when these points are in different media, is refracted in such wise at the plane surface which joins these two media that it employs the least possible time: and exactly the same happens in the case of reflexion against a plane surface.

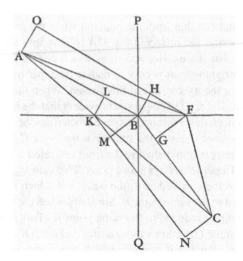

Fig. 16.4 Huygens' explanation of the refraction of a light ray using Fermat's principle of least time.—[*K.K.*]

Mr. Fermat was the first to propound this property of refraction, holding with us, and directly counter to the opinion of Mr. Des Cartes, that light passes more slowly through glass and water than through air. But he assumed besides this a constant ratio of Sines, which we have just proved by these different degrees of velocity alone: or rather, what is equivalent, he assumed not only that the velocities were different but that the light took the least time possible for its passage, and thence deduced the constant ratio of the Sines. His demonstration, which may be seen in his printed works, and in the volume of letters of Mr. Des Cartes, is very long; wherefore I give here another which is simpler and easier.

Let KF be the plane surface (Fig. 16.4); A the point in the medium which the light traverses more easily, as the air; C the point in the other which is more difficult to penetrate, as water. And suppose that a ray has come from A, by B, to C, having been refracted at B according to the law demonstrated a little before; that is to say that, having drawn PBQ, which cuts the plane at right angles, let the sine of the angle ABP have to the sine of the angle CBQ the same ratio as the velocity of light in the medium where A is to the velocity of light in the medium where C is. It is to be shown that the time of passage of light along AB and BC taken together, is the shortest that can be. Let us assume that it may have come by other lines, and, in the first place, along AF, FC, so that the point of refraction F may be further from B than the point A; and let AO be a line perpendicular to AB, and FO parallel to AB; BH perpendicular to FO, and FG to BC.

Since then the angle HBF is equal to PBA, and the angle BFG equal to QBC, it follows that the sine of the angle HBF will also have the same ratio to the sine of BFG, as the velocity of light in the medium A is to its velocity in the medium C. But these sines are the straight lines HF, BG, if we take BF as the semi-diameter of a circle. Then these lines HF, BG, will bear to one another the said ratio of the velocities. And, therefore, the time of the light along HF, supposing that the ray had been OF, would be equal to the time along BG in the interior of the medium C. But the time along AB is equal to the time along OH; therefore the time along OF is equal to the time along AB, BG. Again the time along FC is greater than that along GC; then the time along OFC will be longer than that along ABC. But AF is longer than OF, then the time along AFC will by just so much more exceed the time along ABC.

Now let us assume that the ray has come from A to C along AK, KC; the point of refraction K being nearer to A than the point B is; and let CN be the perpendicular upon BC, KN parallel to BC: BM perpendicular upon KN, and KL upon BA.

Here BL and KM are the sines of angles BKL, KBM; that is to say, of the angles PBA, QBC; and therefore they are to one another as the velocity of light in the medium A is to the velocity in the medium C. Then the time along LB is equal to the time along KM; and since the time along BC is equal to the time along MN, the time along LBC will be equal to the time along KMN. But the time along AK is longer than that along AL: hence the time along AKN is longer than that along ABC. And KC being longer than KN, the time along AKC will exceed, by as much more, the time along ABC. Hence it appears that the time along ABC is the shortest possible; which was to be proven.

16.3   Study Questions

QUES. 16.1. Why are some substances transparent to light while others are opaque?

a) How do waves travel through transparent substances? What three modes of propagation does Huygens suggest?
b) Does Huygens believe that ether can penetrate matter? How does he use the concepts of inertia and drag to justify his views?
c) Do equal volumes of water, quicksilver and gold all have the same amount of empty space within them? How, then, does Huygens explain the incompressibility of water?
d) Is Huygens' theory of matter in contradiction with the fact that some bodies are opaque to light? How does he address this problem? Is his answer convincing? Is it correct?

QUES. 16.2. Can Huygens' theory of light adequately explain the law of refraction?

a) In Fig. 16.2, which line segment represents the surface separating the two media? The incoming wave front? The refracted wave front?
b) Do light waves propagate more slowly, or more quickly, in the interior of bodies? Does his proof of the law of refraction depend upon this? Is his proof valid?
c) How is the situation depicted in Fig. 16.3 different than that depicted in Fig. 16.2. What does this have to do with the speed of light in the medium?
d) Might $\angle BAN$ in Fig. 16.3 ever be zero? Can it be negative? If so, under what conditions? What would this imply?

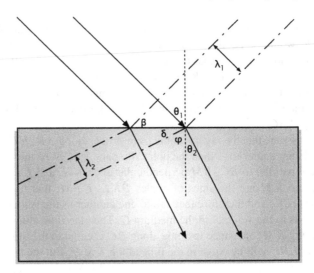

Fig. 16.5 Geometry of the law of refraction

QUES. 16.3. Upon what proposition does Fermat base his derivation of the law of refraction? In what way(s) does Fermat's approach differ from that of Descartes? Of Huygens? Whose approach is superior?

16.4   Exercises

EX. 16.1 (SNELL'S LAW FROM WAVE REFRACTION). In this problem, we will walk through a proof of the law of refraction based on the wave theory of light. You should reproduce this proof and fill in any missing steps. Let us begin with Fig. 16.5, which depicts rays of light incident upon a block of transparent material. The solid lines represent the trajectory of the light as it strikes the interface at an incidence angle θ_1, which is measured with respect to a dotted line constructed perpendicular to the surface of the block. The dashed-dotted lines represent the wavefronts, which lie perpendicular to the trajectory of the light rays, at two successive instants of time. During the intervening time interval, the light which is still outside the block travels one wavelength, λ_1, and the light inside the block travels one wavelength, λ_2.

a) As with sound waves, the speed, frequency, and wavelength of light waves are related by Eq. 14.1. Given that the frequency of light does not change as it crosses the boundary between materials,[1] find a mathematical expression for the ratio v_1/v_2 in terms of λ_1 and λ_2.

---

[1] The frequency of a vibration of wave must be the same on both sides of a material boundary since a particle on the boundary cannot simultaneously be vibrating at two frequencies. In other words, the displacement must be a single-valued function.

b) The *index of refraction* of a particular material is defined by

$$n = c/v, \tag{16.1}$$

where c and v are these speeds of light in empty space and in the material, respectively. What, then, is the ratio n_1/n_2 in terms of λ_1 and λ_2?

c) Using geometry, what is the ratio $\sin(\beta)/\sin(\delta)$ in terms of λ_1 and λ_2? And what is the relationship between β and θ_1? Between δ and θ_2?

d) Put all this together to prove that the incident and refracted waves obey *Snell's law of refraction*:

$$n_1 \sin(\theta_1) = n_2 \sin(\theta_2). \tag{16.2}$$

Ex. 16.2 (SNELL'S LAW FROM FERMAT'S PRINCIPLE). Consider a light source located at position (x_1, y_1) below the flat surface of a uniform medium, such as water or glass. A spectator is located at position (x_2, y_2) above the surface. Demonstrate that a ray of light traveling from the source to the spectator which obeys the law of refraction is the ray of light which requires the minimum transit time. (Hint: One way to approach this problem is to select a point (x, y) on the surface of the medium along the light path, and calculate how changing x changes the transit time.)

16.5 Vocabulary

1. Infinitude
2. Refraction
3. Ether
4. Quicksilver
5. Agate

Chapter 17
Atmospheric Refraction

> *The heights of the Sun and of the Moon, and those of all the*
> *Stars always appear a little greater than they are in reality,*
> *because of these same refractions, as Astronomers know.*
> —Christiaan Huygens

17.1   Introduction

Huygens stated at the outset of his *Treatise on Light* that the aim of natural philosophy is not to merely describe a thing—such as light—but to understand its origin and its cause. So in the subsequent four chapters he presented his wave theory of light and argued that it provides a rational wave-mechanical explanation for all of the known laws of optics. But here is the rub: it requires an all-pervading ethereal medium—comprised of tiny, invisible and extremely hard spheres—to support light waves of such extraordinary speed. While the modern reader may feel uncomfortable with Huygens' constant references to the so-called æther, he or she must keep in mind that all other wave phenomena rely essentially on the existence of some type of medium to support the transmission of waves. After all, what would water waves be if you take away the water? What would sound waves be if you take away the air? And what would "the wave" be in a sports stadium if you take away the cheering fans? In the same way a wave theory of light would be nonsensical without a medium to do the waving. Or so it would seem. In the upcoming readings, we will find that Huygens' famous contemporary, Isaac Newton, offers an alternative to Huygens' wave theory. But before before proceeding to explore Newton's so-called "corpuscular" theory of light, let us remain with Huygens for one more chapter to take a brief look at his wave-mechanical description of how light from distant objects, such as stars or church steeples, bend when passing through Earth's atmosphere.

17.2   Reading: Huygens, *Treatise on Light*

Huygens, C., *Treatise on Light*, Macmillan, London, 1912.

© Springer International Publishing Switzerland 2016   203
K. Kuehn, *A Student's Guide Through the Great Physics Texts,*
Undergraduate Lecture Notes in Physics, DOI 10.1007/978-3-319-21816-8_17

17.2.1   Chapter IV: On the Refraction of the Air

We have shown how the movement which constitutes light spreads by spherical waves in any homogeneous matter. And it is evident that when the matter is not homogeneous, but of such a constitution that the movement is communicated in it more rapidly toward one side than toward another, these waves cannot be spherical: but that they must acquire their figure according to the different distances over which the successive movement passes in equal times.

It is thus that we shall in the first place explain the refractions which occur in the air, which extends from here to the clouds and beyond. The effects of which refractions are very remarkable; for by them we often see objects which the rotundity of the Earth ought otherwise to hide; such as Islands, and the tops of mountains when one is at sea. Because also of them the Sun and the Moon appear as risen before in fact they have, and appear to set later: so that at times the Moon has been seen eclipsed while the Sun appeared still above the horizon. And so also the heights of the Sun and of the Moon, and those of all the Stars always appear a little greater than they are in reality, because of these same refractions, as Astronomers know. But there is one experiment which renders this refraction very evident; which is that of fixing a telescope on some spot so that it views an object, such as a steeple or a house, at a distance of half a league or more. If then you look through it at different hours of the day, leaving it always fixed in the same way, you will see that the same spots of the object will not always appear at the middle of the aperture of the telescope, but that generally in the morning and in the evening, when there are more vapours near the Earth, these objects seem to rise higher, so that the half or more of them will no longer be visible; and so that they seem lower toward mid-day when these vapours are dissipated.

Those who consider refraction to occur only in the surfaces which separate transparent bodies of different nature, would find it difficult to give a reason for all that I have just related; but according to our Theory the thing is quite easy. It is known that the air which surrounds us, besides the particles which are proper to it and which float in the ethereal matter as has been explained, is full also of particles of water which are raised by the action of heat; and it has been ascertained further by some very definite experiments that as one mounts up higher the density of air diminishes in proportion. Now whether the particles of water and those of air take part, by means of the particles of ethereal matter, in the movement which constitutes light, but have a less prompt recoil than these, or whether the encounter and hindrance which these particles of air and water offer to the propagation of movement of the ethereal progress, retard the progression, it follows that both kinds of particles flying amidst the ethereal particles, must render the air, from a great height down to the Earth, gradually less easy for the spreading of the waves of light.

Whence the configuration of the waves ought to become nearly such as this figure represents (see Fig. 17.1): namely, if *A* is a light, or the visible point of a steeple, the waves which start from it ought to spread more widely upwards and less widely downwards, but in other directions more or less as they approximate to these two

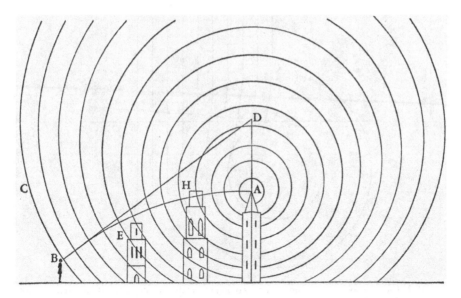

Fig. 17.1 The refraction of light may cause a steeple to appear taller than it is.—[*K.K.*]

extremes. This being so, it necessarily follows that every line intersecting one of these waves at right angles will pass above the point A, always excepting the one line which is perpendicular to the horizon.

Let BC be the wave which brings the light to the spectator who is at B, and let BD be the straight line which intersects this wave at right angles. Now because the ray or straight line by which we judge the spot where the object appears to us is nothing else than the perpendicular to the wave that reaches our eye, as will be understood by what was said above, it is manifest that the point A will be perceived as being in the line BD, and therefore higher than in fact it is.

Similarly if the Earth be AB (see Fig. 17.2), and the top of the Atmosphere CD, which probably is not a well defined spherical surface (since we know that the air becomes rare in proportion as one ascends, for above there is so much less of it to press down upon it), the waves of light from the sun coming, for instance, in such a way that so long as they have not reached the Atmosphere CD the straight line AE intersects them perpendicularly, they ought, when they enter the Atmosphere, to advance more quickly in elevated regions than in regions nearer to the Earth. So that if CA is the wave which brings the light to the spectator at A, its region C will be the furthest advanced; and the straight line AF, which intersects this wave at right angles, and which determines the apparent place of the Sun, will pass above the real Sun, which will be seen along the line AE. And so it may occur that when it ought not to be visible in the absence of vapours, because the line AE encounters the rotundity of the Earth, it will be perceived in the line AF by refraction. But this angle EAF is scarcely ever more than half a degree because the attenuation of the vapours alters the waves of light but little. Furthermore these refractions are not

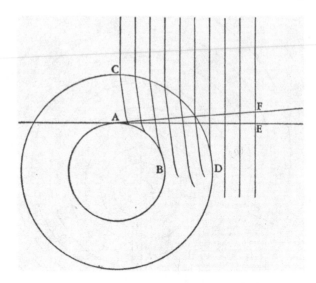

Fig. 17.2 The atmospheric refraction of waves of light from the sun may affect its apparent position in the sky.—[*K.K.*]

altogether constant in all weathers, particularly at small elevations of 2 or 3°; which results from the different quantity of aqueous vapours rising above the Earth.

And this same thing is the cause why at certain times a distant object will be hidden behind another less distant one, and yet may at another time be able to be seen, although the spot from which it is viewed is always the same. But the reason for this effect will be still more evident from what we are going to remark touching the curvature of rays. It appears from the things explained above that the progression or propagation of a small part of a wave of light is properly what one calls a ray. Now these rays, instead of being straight as they are in homogeneous media, ought to be curved in an atmosphere of unequal penetrability. For they necessarily follow from the object to the eye the line which intersects at right angles all the progressions of the waves, as in the first figure (Fig. 17.1) the line *AEB* does, as will be shown hereafter; and it is this line which determines what interposed bodies would or would not hinder us from seeing the object. For although the point of the steeple *A* appears raised to *D*, it would yet not appear to the eye *B* if the tower *H* was between the two, because it crosses the curve *AEB*. But the tower *E*, which is beneath this curve, does not hinder the point *A* from being seen. Now according as the air near the Earth exceeds in density that which is higher, the curvature of the ray *AEB* becomes greater: so that at certain times it passes above the summit *E*, which allows the point *A* to be perceived by the eye at *B*; and at other times it is intercepted by the same tower *E* which hides *A* from this same eye.

But to demonstrate this curvature of the rays conformably to all our preceding Theory, let us imagine that *AB* is a small portion of a wave of light coming from the side *C*, which we may consider as a straight line (see Fig. 17.3). Let us also

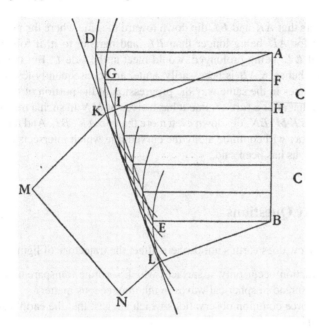

Fig. 17.3 A geometrical demonstration of the curvature of initially horizontal light rays by earth's atmosphere using Huygens' principle.—[*K.K.*]

suppose that it is perpendicular to the Horizon, the portion B being nearer to the Earth than the portion A; and that because the vapours are less hindering at A than at B, the particular wave which comes from the point A spreads through a certain space AD while the particular wave which starts from the point B spreads through a shorter space BE; AD and BE being parallel to the Horizon. Further, supposing the straight lines FG, HI, etc., to be drawn from an infinitude of points in the straight line AB and to terminate on the line DE (which is straight or may be considered as such), let the different penetrabilities at the different heights in the air between A and B be represented by all these lines; so that the particular wave, originating from the point F, will spread across the space FG, and that from the point H across the space HI, while that from the point A spreads across the space AD.

Now if about the centres A, B, one describes the circles DK, EL, which represent the spreading of the waves which originate from these two points, and if one draws the straight line KL which touches these two circles, it is easy to see that this same line will be the common tangent to all the other circles drawn about the centres F, H, etc.; and that all the points of contact will fall within that part of this line which is comprised between the perpendiculars AK, BL. Then it will be the line KL which will terminate the movement of the particular waves originating from the points of the wave AB; and this movement will be stronger between the points KL, than anywhere else at the same instant, since an infinitude of circumferences concur to form this straight line; and consequently KL will be the propagation of the portion of wave AB, as has been said in explaining reflexion and ordinary refraction.

Now it appears that AK and BL dip down toward the side where the air is less easy to penetrate: for AK being longer than BL, and parallel to it, it follows that the lines AB and KL, being prolonged, would meet at the side L. But the angle K is a right angle: hence KAB is necessarily acute, and consequently less than DAB. If one investigates in the same way the progression of the portion of the wave KL, one will find that after a further time it has arrived at MN in such a manner that the perpendiculars KM, LN, dip down even more than do AK, BL. And this suffices to show that the ray will continue along the curved line which intersects all the waves at right angles, as has been said.

17.3   Study Questions

QUES. 17.1. How does earth's atmosphere affect the trajectory of light rays?

a) Does refraction occur only at surfaces which separate transparent bodies? And does light spread in spherical waves in inhomogeneous matter?
b) Describe three common observations which suggest that the earth's atmosphere refracts light.
c) How can one construct the path of a ray of light from its wave-fronts? And how does a spectator infer the position of a distant object based on the path of a light ray from its source?

17.4   Vocabulary

1. League
2. Homogeneous
3. Constitution
4. Rotundity
5. Aperture
6. Ascertain
7. Attenuation
8. Aqueous
9. Interpose
10. Infinitude

Chapter 18
The Particle Theory of Light

Are not the Rays of Light very small Bodies emitted from shining Substances?

—Isaac Newton

18.1 Introduction

According to Huygen's wave theory, light is a compressional (*i.e.* longitudinal) wave propagating through an all-pervading æthereal medium. The propagation of a small part of this wave is what we refer to as a "ray of light." More precisely, according to Huygens a ray of light corresponds to a line drawn such that it intersects—at right angles—any wave-front which it crosses. Thus, when wave-fronts bend and meander as they travel through an inhomogeneous medium, so too do the rays of light follow a curvilinear path through an inhomogeneous medium.[1] So far Huygens. But what, according to Newton, is a ray of light? And why do rays of light obey the laws of optics which he laid out so succinctly in Book I of his *Optics*?[2]

The next two chapters are drawn from Newton's *Queries* included in Book III of his *Opticks*.[3] These *Queries* take the form of rhetorical questions by which Newton articulates his theory of light. He begins slowly, exploring the relationship between light, heat and matter. In particular, in the first 24 *Queries* Newton suggests that when a body vibrates—whether due to heat, friction, percussion, or chemical action—it emits light. Conversely, when light strikes a substance it initiates vibrations within the substance. For instance, when light strikes the eye, vibrations arise in the vitreous humor; these vibrations, in turn, cause the visual perception of color. Likewise, when light strikes a transparent medium—such as water, air or

---

[1] The orthogonality of light rays and wave-fronts is presented in Book IV of Huygens' *Treatise on Light*; see Chap. 17 of the present volume. Notably, Erwin Schrödinger uses Huygens' relationship between light waves and rays to explain wave-particle duality which is an essential features of quantum theory. See Schrödinger's Nobel lecture, included in volume IV.

[2] See Chap. 12 of the present volume.

[3] Originally Newton included only 16 *Queries*; later Latin and English editions included all 31.

© Springer International Publishing Switzerland 2016
K. Kuehn, *A Student's Guide Through the Great Physics Texts*,
Undergraduate Lecture Notes in Physics, DOI 10.1007/978-3-319-21816-8_18

even æther—vibrations arise in the medium; these vibrations, in turn, affect the trajectory of the light within the medium. Indeed, far from rejecting the existence of æther, Newton actually argued that its existence (or rather, inhomogeneities in its density) are responsible for the bending of light rays. Newton, however, carefully distinguishes light itself from vibrations of the æther. This is critical to Newton: vibrations of the æther can certainly *affect* the trajectory of a ray of light, but light *is not* a vibration of the æther. We join the discussion at Query 25, where Newton begins to articulate the specific properties that a ray of light must have in order to explain the curious phenomenon of *birefringence* observed in Iceland crystal.

18.2   Reading: Newton, *Opticks*

Newton, I., *Opticks: or A Treatise of the Reflections, Refractions, Inflections & Colours of Light*, 4th ed., William Innes at the West-End of St. Pauls, London, 1730. Book III, Part 1, Queries 25–30.

Qu. 25. Are there not other original Properties of the Rays of Light, besides those already described? An instance of another original Property we have in the Refraction of Island Crystal, described first by *Erasmus Bartholine,* and afterwards more exactly by *Hugenius* in his Book *De la Lumiere*. This Crystal is a pellucid fissile Stone, clear as Water or Crystal of the Rock, and without Colour; enduring a red Heat without losing its transparency, and in a very strong Heat calcining without Fusion. Steep'd a Day or two in Water, it loses its natural Polish. Being rubb'd on Cloth, it attracts pieces of Straws and other light things, like Ambar or Glass; and with *Aqua fortis* it makes an Ebullition. It seems to be a sort of Talk, and is found in form of an oblique Parallelopiped, with six parallelogram Sides and eight solid Angles (see Fig. 18.1). The obtuse Angles of the Parallelograms are each of them 101° and 52′; the acute ones 78° and 8′. Two of the solid Angles opposite to one another, as *C* and *E*, are compassed each of them with three of these obtuse Angles, and each of the other six with one obtuse and two acute ones. It cleaves easily in planes parallel to any of its Sides, and not in any other Planes. It cleaves with a glossy polite Surface not perfectly plane, but with some little unevenness. It is easily scratch'd, and by reason of its softness it takes a Polish very difficultly. It polishes better upon polish'd Looking-glass than upon Metal, and perhaps better upon Pitch, Leather or Parchment. Afterwards it must be rubb'd with a little Oil or white of an Egg, to fill up its Scratches; whereby it will become very transparent and polite. But for several Experiments, it is not necessary to polish it. If a piece of this crystalline Stone be laid upon a Book, every Letter of the Book seen through it will appear double, by means of a double Refraction. And if any beam of Light falls either perpendicularly, or in any oblique Angle upon any Surface of this Crystal, it becomes divided into two beams by means of the double Refraction. Which beams are of the same Colour with the incident beam of Light, and seem equal to one another in the quantity of their Light, or very nearly equal. One of these Refractions is perform'd

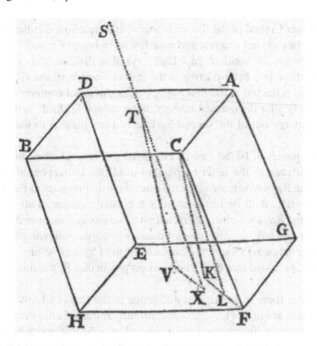

Fig. 18.1 The double refraction of a light ray by an Iceland crystal.—[*K.K.*]

by the usual Rule of Opticks, the Sine of Incidence out of Air into this Crystal being
to the Sine of Refraction, as five to three. The other Refraction, which may be called
the unusual Refraction, is perform'd by the following Rule.

Let $ADBC$ represent the refracting Surface of the Crystal, C the biggest solid
Angle at that Surface, $GEHF$ the opposite Surface, and CK a perpendicular on
that Surface. This perpendicular makes with the edge of the Crystal CF, an Angle
of 19° 3'. Join KF, and in it take KL, so that the Angle KCL be 6° 40', and the
Angle LCF 12° 23'. And if ST represent any beam of Light incident at T in any
Angle upon the refracting Surface $ADBC$, let TV be the refracted beam determin'd
by the given Portion of the Sines 5 to 3, according to the usual Rule of Opticks. Draw
VX parallel and equal to KL. Draw it the same way from V in which L lieth from
K; and joining TX, this line TX shall be the other refracted beam carried from T
to X, by the unusual Refraction.

If therefore the incident beam ST be perpendicular to the refracting Surface, the
two beams TV and TX, into which it shall become divided, shall be parallel to the
lines CK and CL; one of those beams going through the Crystal perpendicularly,
as it ought to do by the usual Laws of Opticks, and the other TX by an unusual
Refraction diverging from the perpendicular, and making with it an Angle VTX of
about 6⅔ °, as is found by Experience. And hence, the Plane VTX, and such like
Planes which are parallel to the Plane CFK, may be called the Planes of perpendicular Refraction. And the Coast towards which the lines KL and VX are drawn,
may be call'd the coast of unusual Refraction.

In like manner Crystal of the Rock has a double Refraction: But the difference of the two Refractions is not so great and manifest as in Island Crystal.

When the beam ST incident on Island Crystal is divided into two beams TV and TX, and these two beams arrive at the farther Surface of the Glass; the beam TV, which was refracted at the first Surface after the usual manner, shall be again refracted entirely after the unusual manner in the second Surface; so that both these beams shall emerge out of the second Surface in lines parallel to the first incident beam ST.

And if two pieces of Island Crystal be placed one after another, in such manner that all the Surfaces of the latter be parallel to all the corresponding Surfaces of the former: The Rays which are refracted after the usual manner in the first Surface of the first Crystal, shall be refracted after the usual manner in all the following Surfaces; and the Rays which are refracted after the unusual manner in the first Surface, shall be refracted after the unusual manner in all the following Surfaces. And the same thing happens, though the Surfaces of the Crystals be any ways inclined to one another, provided that their Planes of perpendicular Refraction be parallel to one another.

And therefore there is an original difference in the Rays of Light, by means of which some Rays are in this Experiment constantly refracted after the usual manner, and others constantly after the unusual manner: For if the difference be not original, but arises from new Modifications impress'd on the Rays at their first Refraction, it would be alter'd by new Modifications in the three following Refractions; whereas it suffers no alteration, but is constant, and has the same effect upon the Rays in all the Refractions. The unusual Refraction is therefore perform'd by an original property of the Rays. And it remains to be enquired, whether the Rays have not more original Properties than are yet discover'd.

Qu. 26. Have not the Rays of Light several sides, endued with several original Properties? For if the Planes of perpendicular Refraction of the second Crystal be at right Angles with the Planes of perpendicular Refraction of the First Crystal, the Rays which are refracted after the usual manner in passing through the first Crystal, will be all of them refracted after the unusual manner in passing through the second Crystal; and the Rays which are Refracted after the unusual manner in passing through the first Crystal, will be all of them refracted after the usual manner in passing through the second Crystal. And therefore there are not two sorts of Rays differing in their nature from one another, one of which is constantly and in all Positions refracted after the usual manner, and the other constantly and in all Positions after the unusual manner. The difference between the two sorts of Rays in the Experiment mention'd in the 25th Question, was only in the Positions of the Sides of the Rays to the Planes of perpendicular Refraction. For one and the same Ray is here refracted sometimes after the usual, and sometimes after the unusual manner, according to the Position which its Sides have to the Crystals. If the Sides of the Ray are posited the same way to both Crystals, it is refracted after the same manner in them both: But if that side of the Ray which looks towards the Coast of the unusual Refraction of the first Crystal, be 90° from that side of the same Ray which looks toward the Coast of the unusual Refraction of the second Crystal, (which may be effected by varying the Position of the second Crystal to the first, and by consequence to the Rays of Light,) the Ray shall be refracted after several manners in the several Crystals. There is nothing more required to determine whether the Rays of Light which fall upon the second Crystal shall be refracted after the usual or after the unusual manner, but to turn about this Crystal, so that the Coast of this Crystal's unusual Refraction may be on this or on that side of the Ray. And therefore every Ray may be consider'd as having four Sides or Quarters, two of which opposite to one another incline the Ray to be refracted after the unusual manner, as often as either of them are turn'd towards the Coast of unusual Refraction; and the other two, whenever either of them are turn'd towards the Coast of unusual Refraction, do not incline it to be otherwise refracted than after the usual manner. The two first may therefore be call'd the Sides of unusual Refraction. And since these Dispositions were in the Rays before their Incidence on the second, third, and fourth Surfaces of the two Crystals, and suffered no alteration (so far as appears,) by the Refraction of the Rays in their passage through those Surfaces, and the Rays were refracted by the same Laws in all the four Surfaces; it appears that those Dispositions were in the Rays originally, and suffer'd no alteration by the first Refraction, and that by means of those Dispositions the Rays were refracted at their Incidence on the first Surface of the first Crystal, some of them after the usual, and some of them after the unusual manner, accordingly as their Sides of unusual Refraction were then turn'd towards the Coast of the unusual Refraction of that Crystal, or sideways from it.

Every Ray of Light has therefore two opposite Sides, originally endued with a Property on which the unusual Refraction depends, and the other two opposite Sides not endued with that Property. And it remains to be enquired, whether there are not more Properties of Light by which the Sides of the Rays differ, and are distinguished from one another.

In explaining the difference of the Sides of the Rays above mention'd, I have supposed that the Rays fall perpendicularly on the first Crystal. But if they fall obliquely on it, the Success is the same. Those Rays which are refracted after the usual manner in the first Crystal, will be refracted after the unusual manner in the second Crystal, supposing the Planes of perpendicular Refraction to be at right Angles with one another, as above; and on the contrary.

If the Planes of the perpendicular Refraction of the two Crystals be neither parallel nor perpendicular to one another, but contain an acute Angle: The two beams of Light which emerge out of the first Crystal, will be each of them divided into two more at their Incidence on the second Crystal. For in this case the Rays in each of two beams will some of them have their Sides of unusual Refraction, and some of them their other Sides turn'd towards the Coast of the unusual Refraction of the second Crystal.

Qu. 27. Are not all Hypotheses erroneous which have hitherto been invented for explaining the Phænomena of Light, by new Modifications of the Rays? For those Phænomena depend not upon new Modifications, as has been supposed, but upon the original and unchangeable Properties of the Rays.

Qu. 28. Are not all Hypotheses erroneous, in which Light is supposed to consist in Pression or Motion, propagated through a fluid Medium? For in all these

Hypotheses the Phænomena of Light have been hitherto explain'd by supposing that they arise from new Modifications of the Rays; which is an erroneous Supposition.

If Light consisted only in Pression propagated without actual Motion, it would not be able to agitate and heat the Bodies which refract and reflect it. If it consisted in Motion propagated to all distances in an instant, it would require an infinite force every moment, in every shining Particle, to generate that Motion. And if it consisted in Pression or Motion, propagated either in an instant or in time, it would bend into the Shadow. For Pression or Motion cannot be propagated in a Fluid in right Lines, beyond an Obstacle which stops part of the Motion, but will bend and spread every way into the quiescent Medium which lies beyond the Obstacle. Gravity tends downwards, but the Pressure of Water arising from Gravity tends every way with equal Force, and is propagated as readily, and with as much force sideways as downwards, and through crooked passages as through strait ones. The Waves on the Surface of stagnating Water, passing by the sides of a broad Obstacle which stops part of them, bend afterwards and dilate themselves gradually into the quiet Water behind the Obstacle. The Waves, Pulses or Vibrations of the Air, wherein Sounds consist, bend manifestly, though not so much as the Waves of Water. For a Bell or a Cannon may be heard beyond a Hill which intercepts the sight of the sounding Body, and Sounds are propagated as readily through crooked Pipes as straight ones. But Light is never known to follow crooked Passages nor to bend into the Shadow. For the fix'd Stars by the Interposition of any of the Planets cease to be seen. And so do the Parts of the Sun by the Interposition of the Moon, *Mercury* or *Venus*. The Rays which pass very near to the edges of any Body, are bent a little by the action of the Body, as we shew'd above; but this bending is not towards but from the Shadow, and is perform'd only in the passage of the Ray by the Body, and at a very small distance from it. So soon as the Ray is past the Body, it goes right on.

To explain the unusual Refraction of Island Crystal by Pression or Motion propagated, has not hitherto been attempted (to my knowledge) except by *Huygens*, who for that end supposed two several vibrating Mediums within that Crystal. But when he tried the Refractions in two successive pieces of that Crystal, and found them such as is mention'd above; he confessed himself at a loss for explaining them. For Pressions or Motions, propagated from a shining Body through an uniform Medium, must be on all sides alike; whereas by those Experiments it appears, that the Rays of Light have different Properties in their different Sides. He suspected that the Pulses of Æther in passing through the first Crystal might receive certain new Modifications, which might determine them to be propagated in this or that Medium within the second Crystal, according to the Position of that Crystal. But what Modifications those might be he could not say, nor think of any thing satisfactory in that Point. And if he had known that the unusual Refraction depends not on new Modifications but on the original and unchangeable Dispositions of the rays, he would have found it as difficult to explain how those Dispositions which he supposed to be impress'd on the Rays by the first Crystal, and in general, how all Rays emitted by shining Bodies, can have those Dispositions in them from the beginning. To me, at least, this seems inexplicable, if Light be nothing else than Pression or Motion propagated through *Æther.*

And it is as difficult to explain by these Hypotheses, how Rays can be alternately in Fits of easy Reflexion and easy Transmission; unless perhaps one might suppose that there are in all Space two Æthereal vibrating Mediums, and that the Vibrations of one of them constitute Light, and the Vibrations of the other are swifter, and as often as they overtake the Vibrations of the first, put them into those Fits. But how two Æthers can be diffused through all Space, one of which acts upon the other, and by consequence is re-acted upon, without retarding, shattering, dispersing and confounding one anothers Motions, is inconceivable. And against filling the Heavens with fluid Mediums, unless they be exceeding rare, a great Objection arises from the regular and very lasting Motions of the Planets and Comets in all manner of Courses through the Heavens. For thence it is manifest, that the Heavens are void of all sensible Resistance, and by consequence of all sensible Matter.

For the resisting Power of fluid Mediums arises partly from the Attrition of the Parts of the Medium, and partly from the *Vis inertiæ* of the Matter. That part of the Resistance of a spherical Body which arises from the Attrition of the Parts of the Medium is very nearly as the Diameter, or, at the most, as the *Factum* of the Diameter, and the Velocity of the spherical Body together.[4] And that part of the Resistance which arises from the *Vis intertiæ* of the Matter, is as the Square of that *Factum*. And by this difference the two sorts of Resistance may be distinguish'd, it will be found that almost all the Resistance of Bodies of a competent Magnitude moving in Air, Water, Quick-silver, and such like Fluids with a competent Velocity, arises from the *Vis intertiæ* of the Parts of the Fluid.

Now that part of the resisting Power of any Medium which arises from the Tenacity, Friction or Attrition of the Parts of the Medium, may be diminish'd by dividing the Matter into smaller Parts, and making the Parts more smooth and slippery: But that part of the Resistance which arises from the *Vis intertiæ*, is proportional to the Density of the Matter, and cannot be diminish'd by dividing the Matter into smaller Parts, nor by any other means than by decreasing the Density of the Medium. And for these Reasons the density of fluid Mediums is very nearly proportional to their Resistance. Liquors which differ not much in Density, as Water, Spirit of Wine, Spirit of Turpentine, hot Oil, differ not much in Resistance. Water is 13 or 14 times lighter than Quick-silver and by consequence 13 or 14 times rarer, and its Resistance is less than that of Quick-silver in the same Proportion, or thereabouts, as I have found by Experiments made with Pendulums. The open Air in which we breathe is 800 or 900 times lighter than Water, and by consequence 800 or 900 times rarer, and accordingly its Resistance is less than that of Water in the same Proportion, or thereabouts; as I have also found by Experiments made with Pendulums. And in thinner Air the Resistance is still less, and at length, by rarifying the Air, becomes insensible. For small Feathers falling in the open Air meet with great Resistance, but in a tall Glass well emptied of Air, they fall as fast as Lead or Gold, as I have seen tried several times. Whence the Resistance seems still to decrease in proportion to the Density of the Fluid. For I do not find by any Experiments that Bodies moving in

---

[4] The factum of the diameter and the velocity is just the product of the two.—[*K.K.*].

Quick-silver, Water or Air, meet with any other sensible Resistance than what arises from the Density and Tenacity of those sensible Fluids, as they would do if the Pores of those Fluids, and all other Spaces, were filled with a dense and subtile Fluid. Now if the Resistance in a Vessel well emptied of Air, was but an hundred times less than in the open Air, it would be about a million of times less than in Quick-silver. But it seems to be much less in such a Vessel, and still much less in the Heavens, at the height of 300 or 400 miles from the Earth, or above. For Mr. *Boyle* has shew'd that Air may be rarified above ten thousand times in Vessels of Glass; and the Heavens are much emptier of Air than any *Vacuum* we can make below. For since the Air is compress'd by the Weight of the incumbent Atmosphere, and the Density of Air is proportional to the Force compressing it, it follows by Computation, that at the height of about seven and a half *English* Miles from the Earth, the Air is four times rarer than at the Surface of the Earth; and at the height of 22½, 30, or 38 miles, it is respectively 64, 256, or 1024 times rarer, or thereabouts; and at the height of 76, 152, 228 miles it is about 1000000, 1000000000000, 1000000000000000000 times rarer; and so on.

Heat promotes Fluidity very much by diminishing the Tenacity of Bodies. It makes many Bodies fluid which are not Fluid in cold, and increases the Fluidity of tenacious Liquids, as of Oil, Balsam, and Honey, and thereby decreases their Resistance. But it decreases not the Resistance of Water considerably, as it would do if any considerable part of the Resistance of Water arose from the Attrition or Tenacity of its Parts. And therefore the Resistance of Water arises principally and almost entirely from the *Vis intertiae* of its Matter; and by consequence, if the Heavens were as dense as Water, they would not have much less Resistance than Water; if as dense as Quick-silver, they would not have much less Resistance than Quick-silver; if absolutely dense, or full of Matter without any *Vacuum*, let the Matter be never so subtil and fluid, they would have a greater Resistance than Quick-silver. A solid Globe in such a Medium would lose above half its Motion in moving three times the length of its Diameter, and a Globe not solid (such as are the Planets,) would be retarded sooner. And therefore to make way for the regular and lasting Motions of the Planets and Comets, it's necessary to empty the Heavens of all Matter, except perhaps some very thin Vapours, Steams, or Effluvia, arising from the Atmospheres of the Earth, Planets, and Comets, and from such an exceedingly rare Æthereal Medium as we described above. A dense Fluid can be of no use for explaining the Phænomena of Nature, the Motions of the Planets and Comets being better explain'd without it. It serves only to disturb and retard the Motions of those great Bodies, and make the Frame of Nature languish: And in the Pores of Bodies, it serves only to stop the vibrating Motions of their Parts, wherein their Heat and Activity consists. And as it is of no use, and hinders the Operations of Nature, and makes her languish, so there is no evidence for its Existence, and therefore it ought to be rejected. And if it be rejected, the Hypotheses that Light consists in Pression or Motion, propagated through such a Medium, are rejected with it.

And for rejecting such a Medium, we have the Authority of those the oldest and most celebrated Philosophers of *Greece* and *Phœnecia,* who made a *Vacuum,* and Atoms, and the Gravity of Atoms, the first Principles of their Philosophy;

tacitly attributing Gravity to some other Cause than dense Matter. Later Philosophers banish the Consideration of such a Cause out of natural Philosophy, feigning Hypotheses, and to deduce Causes from Effects, till we come to the very first Cause, which certainly is not mechanical; and not only to unfold the Mechanism of the World, but chiefly to resolve these and such like Questions. What is there in places almost empty of Matter, and whence is it that the Sun and Planets gravitate towards one another, without dense Matter between them? Whence is it that Nature doth nothing in vain; and whence arises all that Order and Beauty which we see in the World? To what end are Comets, and whence is it that Planets move all one and the same way in Orbs concentrick, while Comets move all manner of ways in Orbs very excentrick; and what hinders the fix'd Stars from falling upon one another? How come the Bodies of Animals to be contrived with so much Art, and for what ends were their several Parts? Was the Eye contrived without Skill in Opticks, and the Ear without Knowledge of Sounds? How do the Motions of the Body follow from the Will, and whence is the Instinct in Animals? Is not the Sensory of Animals that place to which the sensitive Substance is present, and into which the sensible Species of Things are carried through the Nerves and Brain, that there they may be perceived by their immediate presence to that Substance? And these things being rightly dispatch'd, does it not appear from Phænomena that there is a Being incorporeal, living, intelligent, omnipresent, who in infinite Space, as it were in his Sensory, sees the things themselves intimately, and thoroughly perceives them, and comprehends them wholly by their immediate presence to himself: Of which things the Images only carried through the Organs of Sense into our little Sensoriums, are there seen and beheld by that which in us perceives and thinks. And though every true Step made in this Philosophy brings us not immediately to the Knowledge of the first Cause, yet it brings us nearer to it, and on that account is to be highly valued.

Qu. 29. Are not the Rays of Light very small Bodies emitted from shining Substances? For such Bodies will pass through uniform Mediums in right Lines without bending into the Shadow, which is the Nature of the Rays of Light. They will also be capable of several Properties, and be able to conserve their Properties unchanged in passing through several Mediums, which is another Condition of the Rays of Light. Pellucid Substances act upon the Rays of Light at a distance in refracting, reflecting, and inflecting them, and the Rays mutually agitate the Parts of those Substances at a distance for heating them; and this Action and Re-action at a distance very much resembles an attractive Force between Bodies. If Refraction be perform'd by Attraction of the Rays, the Sines of Incidence must be to the Sines of Refraction in a given Proportion, as we shew'd in our Principles of Philosophy: and this Rule is true by Experience. The Rays of Light in going out of Glass into a *Vacuum*, are bent towards the Glass; and if they fall too obliquely on the *Vacuum*, they are bent backwards into the Glass, and totally reflected; and this Reflexion cannot be ascribed to the Resistance of an absolute *Vacuum*, but must be caused by the Power of the Glass attracting the Rays at their going out of it into the *Vacuum*, and bringing them back. For if the farther Surface of the Glass be moisten'd with Water or clear Oil, or liquid and clear Honey, the Rays which would otherwise be reflected will go into the Water, Oil, or Honey; and therefore are not reflected before they arrive at the farther Surface of the

Glass, and begin to go out of it. If they go out of it into the Water, Oil, or Honey, they go on, because the Attraction of the Glass is almost balanced and rendered ineffectual by the contrary Attraction of the Liquor. But if they go out of it into a *Vacuum* which has no Attraction to balance that of the Glass, the Attraction of the Glass either bends and refracts them, or brings them back and reflects them. And this is still more evident by laying together two Prisms of Glass, or two Object-glasses of very long Telescopes, the one plane, the other a little convex, and so compressing them that they do not fully touch, nor are to far asunder. For the Light which falls upon the farther Surface of the First Glass where the Interval between the Glasses is not above the ten hundred thousandth Part of an Inch, will go through that Surface, and through the Air or *Vacuum* between the Glasses, and enter into the second Glass, as was explain'd in the first, fourth, and eighth Observations of the first Part of the second Book. But, if the second Glass be taken away, the Light which goes out of the second Surface of the first Glass into the Air or *Vacuum*, will not go on forwards, but turns back into the first Glass, and is reflected; and therefore it is drawn back by the Power of the first Glass, there being nothing else to turn it back. Nothing more is requisite for producing all the variety of Colours, and degrees of Refrangibility, than that the Rays of Light be Bodies of different Sizes, the least of which may take violet the weakest and darkest of the Colours, and be more easily diverted by refracting Surfaces from the right Course; and the rest as they are bigger and bigger, may make the stronger and more lucid Colours, blue, green, yellow, and red, and be more and more difficultly diverted. Nothing more is requisite for putting the Rays of Light into Fits of easy Reflexion and easy Transmission, than that they be small Bodies which by their attractive Powers, or some other Force, stir up Vibrations in what they act upon, which Vibrations being swifter than the Rays, overtake them successively, and agitate them so as by turns to increase and decrease their Velocities, and thereby put them into those Fits. And lastly, the unusual Refraction of Island-Crystal looks very much as if it were perform'd by some kind of attractive virtue lodged in certain Sides both of the Rays, and of the Particles of the Crystal. For were it not for some kind of Disposition or Virtue lodged in some Sides of the Particles of the Crystal, and not in their other Sides, and which inclines and bends the Rays towards the Coast of unusual Refraction, the Rays which fall perpendicularly on the Crystal, would not be refracted towards that Coast rather than towards any other Coast, both at their Incidence and at their Emergence, so as to emerge perpendicularly by a contrary Situation of the Coast of unusual Refraction at the second Surface; the Crystal acting upon the Rays after they have pass'd through it, and are emerging into the Air; or, if you please, into a *Vacuum*. And since the Crystal by this Disposition in those Sides of the Rays, which answers to and sympathizes with that Virtue or Disposition of the Crystal, as the Poles of two Magnets answer to one another. And as Magnetism may be intended and remitted, and is found only in the Magnet and in Iron: So this Virtue of refracting the perpendicular Rays is greater in Island-Crystal, less in Crystal of the Rock, and is not yet found in other Bodies. I do not say that this Virtue is magnetical: It seems to be of another kind. I only say, that whatever it be, it's difficult to conceive how the Rays of Light, unless they be Bodies, can have a permanent Virtue in two of their Sides which is not in their other

Sides, and this without any regard to their Position to the Space or Medium through which they pass.

What I mean in this Question by a *Vacuum*, and by the Attractions of the Rays of Light towards the Glass or Crystal, may be understood by what was said in the 18th, 19th, and 20th Questions.

Qu. 30. Are not gross Bodies and Light convertible into one another, and may not Bodies receive much of their Activity from the Particles of Light which enter their Composition? For all fix'd Bodies being heated emit Light so long as they continue sufficiently hot, and Light mutually stops in Bodies as often as its Rays strike upon their Parts, as we shew'd above. I know no Body less apt to shine than Water; and yet Water by frequent Distillations changes into fix'd Earth, as Mr. *Boyle* has try'd; and then this Earth being enabled to endure a sufficient Heat, shines by Heat like other Bodies.

The changing of Bodies into Light, and Light into Bodies is very conformable to the Course of Nature, which seems delighted with Transmutations. Water, which is a very fluid tasteless Salt, she changes by Heat into Vapour, which is a sort of Air, and by Cold into Ice, which is a hard, pellucid, brittle fusible Stone; and this Stone returns to Water by Heat, and Vapour returns into Water by Cold. Earth by Heat becomes Fire, and by Cold returns into Earth. Dense Bodies by Fermentation rarify into several sorts of Air, and this Air by Fermentation, and sometimes without it, returns into dense Bodies. Mercury appears sometimes in the form of a fluid Metal, sometimes in the form of a corrosive pellucid Salt call'd Sublimate, sometimes in the form of a tasteless, pellucid, volatile white Earth, call'd *Mercurius Dulcis* or in that of a red opake volatile Earth, call'd Cinnaber; or in that of a red or white Precipitate, or in that of a fluid Salt; and in Distillation it turns into Vapour, and being agitated *in Vacuo*, it shines like Fire. And after all these Changes it returns again into its first form of Mercury. Eggs grow from insensible Magnitudes, and change into Animals; Tadpoles into Frogs; and Worms into Flies. All Birds, Beasts and Fishes, Insects, Trees, and other Vegetables, with their several Parts, grow out of Water and watry Tinctures and Salts, and by Putrefaction return again into watry Substances. And Water standing a few Days in the open Air, yields a Tincture, which (like that of Malt) by standing longer yields a Sediment and a Spirit, but before Putrefaction is fit Nourishment for Animals and Vegetables. And among such various and strange Transmutations, why may not Nature change Bodies into Light, and Light into Bodies?

18.3 Study Questions

QUES. 18.1. What properties of light does Newton infer from the phenomenon of birefringence?

a) What is the appearance, and what are the physical characteristics, of Iceland Crystal?

b) What effect does a single Iceland Crystal have upon a ray of light incident upon its surface? Which lines in Fig. 18.1 represent the incoming beam, the ordinarily refracted beam, and the unusually refracted beam? What happens when the refracted beams strike the opposite side of the crystal?

c) What happens when a ray of light passes through two adjacent Iceland Crystals? Does it matter how the crystals are oriented with respect to one another, and with respect to the incident light ray?

d) What does the double refraction of iceland crystal imply? In particular, does the crystal introduce new properties to incident light rays? Or does it merely act on original and unchanging properties of the incident light rays? And if so, what are these properties?

QUES. 18.2. Why does Newton reject any theory that identifies light with pressure inside a fluid, whether static or in the form of waves?

a) In what way is the behavior of light unlike that exhibited by hydrostatic pressure in fluid media? In particular, is pressure transmitted through tubes and around corners? Is light?

b) In what way is the behavior of light unlike sound waves or water waves passing near an obstacle? In particular, do obstacles bend sound? Do they bend light?

c) According to Newton, can Huygens' wave theory of light account for the birefringence of Iceland crystal? In particular, can longitudinal waves possess different orientations, so to speak?

QUES. 18.3. On what basis does Newton finally reject the existence of æther as a necessary medium for the transmission of light?

a) What type of resistance is attributed to the *attrition*—tenacity or "stickiness"—of a fluid medium? And what type of resistance is attributed to the *inertia* of the fluid medium? How do these two types of resistance depend upon the density of the fluid, or the speed and size of a body moving through them?

b) How does the density of Earth's atmosphere vary with altitude? What is the primary type of resistance (fluid attrition or fluid inertia) offered by air, by water, and by quicksilver? And what effect would such a medium have on the motion of the planets and comets?

c) What does Newton conclude regarding a medium filling space to transmit light waves? Does he appeal to experimental observations? To ancient authorities? Or to both? Which, if any, do you find most convincing?

QUES. 18.4. What is the business of natural philosophy?

a) What types of questions should natural philosophy address? Can the study of philosophy lead man to a knowledge of a first cause?

b) What conclusion(s) does Newton draw from the observed phenomena? Do you agree with his assessment? If not, then what type of conclusions, if any, should be drawn from observed phenomena?

QUES. 18.5. What is Newton's theory of light?

a) Of what is light comprised? How does Newton's theory account for light's (i) linear trajectory, (ii) the heating of bodies, (iii) refraction, (iv) total internal reflection, (v) dispersion and (vi) birefringence?
b) Can bodies turn into light? What examples of transmutations in nature does Newton provide to support his claim? Do you find Newton's analogical reasoning convincing?

18.4 Exercises

Ex. 18.1 (BIREFRINGENCE).   If an incident beam, ST, is perpendicular to the refracting surface of an iceland crystal (as shown in Fig. 18.1), at what angles with respect to ST do the usual and unusual beams emerge after passing through the bottom of the crystal? By how much will they be separated if the crystal is 1 cm tall? Do both the usual and unusual beams obey Snell's law of refraction?

Ex. 18.2 (FELIX BAUMGARTNER AND STRATOSPHERIC DRAG).   On October 14, 2012, Austrian skydiver Felix Baumgartner jumped from a helium balloon at a height of almost 24 miles above Roswell, New Mexico. Calculate his terminal velocity based on the forces of gravity and drag acting on his body.[5] Assume, for the sake of simplicity, that his body is a 90 kg sphere with a density of 1 g/cm^3. According to your calculations, will Baumgartner break the speed of sound? What is the speed of sound at the altitude from which he jumped?

Ex. 18.3 (NATURAL THEOLOGY ESSAY).   According to Newton, to what philosophical—or even theological—conclusion is one inevitably led by the study of natural phenomena? How so? And would you agree?

Ex. 18.4 (BIREFRINGENCE LABORATORY).   Draw a tiny letter A on a sheet of white paper and place a block of iceland spar over it. What do you observe as the iceland spar is rotated? Try to characterize your observations as precisely as possible. What is the correct explanation for this behavior?

---

[5] For the present exercise, you may need to look up the density of air in the stratosphere. The effect of drag on falling bodies is discussed in Ex. 11.3 in the context of Galileo's dialogue on projectile motion, included in Chap. 11 of volume II.

18.5   Vocabulary

1. Fissile
2. Parallelogram
3. Ebullition
4. Manifest
5. Disposition
6. Oblique
7. Contrary
8. Acute
9. Erroneous
10. Hitherto
11. Stagnate
12. Interpose
13. Inexplicable
14. Disperse
15. Thence
16. Tenacity
17. Effluvia
18. Languish
19. Tacit
20. Feign
21. Incorporeal
22. Omnipresent
23. Agitate
24. Ineffectual
25. Convex
26. Asunder
27. Requisite
28. Refrangibility
29. Virtue

Chapter 19
Passive Laws and Active Principles

> *All these things being consider'd, it seems probable to me, that*
> *God in the Beginning form'd Matter in solid, massy, hard,*
> *impenetrable, movable Particles, of such Sizes and Figures, and*
> *with such other Properties, and in such Proportion to Space, as*
> *most conduced to the End for which he form'd them.*
>
> —Isaac Newton

19.1 Introduction

From the first 29 *Queries* in Book III of his *Opticks*, it is apparent that Newton adopts a corpuscular—or particle—theory of light. According to Newton, corpuscles of light are emitted by luminous bodies and fly about like tiny projectiles. But why must these corpuscles obey the laws of reflection and refraction which were laid out so succinctly in Book I of his *Opticks*? What forces must act upon them so as to change their trajectories appropriately? And if these requisite forces have never been directly measured, then can their existence truly be defended? To address questions such as these, in *Qu.* 31 Newton presents a litany of experimental observations which point toward the existence of forces which act between the tiniest constituents of matter.[1] How does the phenomenon of chemical affinity inform his understanding of light? Perhaps most interestingly, toward the end of this *Query* Newton begins to consider whether so-called "passive laws" (such as the principle of inertia), which regulate the behavior of material particles, are able to fully account for the diverse and ordered motions seen in nature. What does Newton conclude? Do you believe that he is justified in asserting the existence of so-called "active principles"? More specifically, is his assertion—that "blind Fate could never make all the Planets move one and the same way in Orbs concentrick"—correct?

---

[1] For the sake of brevity, many of Newton's observations which illustrate chemical affinity have been omitted from the text. The locations of such omissions are indicated by an ellipsis (...).

© Springer International Publishing Switzerland 2016
K. Kuehn, *A Student's Guide Through the Great Physics Texts*,
Undergraduate Lecture Notes in Physics, DOI 10.1007/978-3-319-21816-8_19

19.2 Reading: Newton, *Opticks*

Newton, I., *Opticks: or A Treatise of the Reflections, Refractions, Inflections & Colours of Light*, 4th ed., William Innes at the West-End of St. Pauls, London, 1730. Book III, Part 1.

19.2.1 Excerpts From Qu. 31

Query 31. Have not the small Particles of Bodies certain Powers, Virtues, or Forces, by which they act at a distance, not only upon the Rays of Light for reflecting, refracting, and inflecting them, but also upon one another for producing a great Part of the Phænomena of Nature? For it's well known, that Bodies act one upon another by the Attractions of Gravity, Magnetism, and Electricity; and these Instances shew the Tenor and Course of Nature, and make it not improbable but that there may be more attractive Powers than these. For Nature is very consonant and conformable to her self. How these Attractions may be perform'd, I do not here consider. What I call Attraction may be perform'd by impulse, or by some other means unknown to me. I use that Word here to signify only in general any Force by which Bodies tend towards one another, whatsoever be the Cause. For we must learn from the Phænomena of Nature what Bodies attract one another, and what are the Laws and Properties of the Attraction, before we enquire the Cause by which the Attraction is perform'd. The Attractions of Gravity, Magnetism, and Electricity, reach to very sensible distances, and so have been observed by vulgar Eyes, and there may be others which reach to so small distances as hitherto escape Observation; and perhaps electrical Attraction may reach to such small distances, even without being excited by Friction....

...All Bodies seem to be composed of hard Particles: For otherwise Fluids would not congeal; as Water, Oils, Vinegar, and Spirit or Oil of Vitriol do by freezing; Mercury by Fumes of Lead; Spirit of Nitre and Mercury, by dissolving the Mercury and evaporating the Flegm; Spirit of Wine and Spirit of Urine, by deflegming and mixing them; and Spirit of Urine and Spirit of Salt, by subliming them together to make Sal-armoniac. Even the Rays of Light seem to be hard Bodies; for otherwise they would not retain different Properties in their different Sides. And therefore, Hardness may be reckon'd the Property of all uncompounded Matter. At least, this seems to be as evident as the universal Impenetrability of Matter. For all Bodies, so far as Experience reaches, are either hard, or may be harden'd; and we have no other Evidence of universal Impenetrability, besides a large Experience without an experimental Exception. Now if compound Bodies are so very hard as we find some of them to be, and yet are very porous, and consist of Parts which are only laid together; the simple Particles which are void of Pores, and were never yet divided, must be much harder. For such hard Particles being heaped up together can scarce touch one another in more than a few Points, and therefore must be separable by much less Force than is requisite to break a solid Particle, whose Parts touch in all the Space between them, without any Pores or Interstices to weaken their Cohesion. And how such very hard Particles which are only laid together and touch only in a few Points, can stick together, and that so firmly as they do, without the assistance of something which causes them to be attracted or press'd towards one another, is very difficult to conceive....

...Since Metals dissolved in Acids attract but a small quantity of the Acid, their attractive Force can reach but to a small distance from them. And as in Algebra, where affirmative Quantities vanish and cease, there negative ones begin; so in Mechanicks, where Attraction ceases, there a repulsive Virtue ought to succeed. And that there is such a Virtue, seems to follow from the Reflexions and Inflexions of the Rays of Light. For the Rays are repelled by the Bodies in both these Cases, without the immediate Contact of the reflecting or inflecting Body. It seems also to follow from the Emission of Light; the Ray so soon as it is shaken off from a shining Body by the vibrating Motion of the Parts of the Body, and gets beyond the reach of Attraction, being driven away with exceeding great Velocity. For that Force which is sufficient to turn it back in Reflexion, may be sufficient to emit it. It seems also to follow from the Production of Air and Vapour. The Particles when they are shaken off from Bodies by Heat or Fermentation, so soon as they are beyond the reach of the Attraction of the Body, receding from it, and also from one another with great Strength, and keeping at a distance so sometimes as to take up above a Million of Times more space than they did before in the form of a dense Body. Which vast Contraction and Expansion seems unintelligible, by feigning the Particles of Air to be springy and ramous, or rolled up like Hoops, or by any other means than a repulsive Power. The Particles of Fluids which do not cohere too strongly, and are of such a Smallness as renders them most susceptible of those Agitations which keep Liquors in a Fluor, are most easily separated and rarified into a Vapour, and in the Language of the Chymists, they are volatile, rarifying with an easy Heat, and condensing with Cold. But those which are grosser, and so less susceptible to Agitation, or cohere by a stronger Attraction, are not separated without a stronger Heat, or perhaps not without Fermentation. And these last are the Bodies which Chymists call fix'd, and being rarified by Fermentation, become true permanent Air; those Particles receding from one another with the greatest Force, and being most difficultly brought together, which upon Contact cohere most strongly. And because the Particles of permanent Air are grosser, and arise from denser Substances than those of Vapours, thence it is that true Air is more ponderous than Vapour, and that a moist Atmosphere is lighter than a dry one, quantity for quantity. From the same repelling Power it seems to be that Flies walk upon the Water without wetting their Feet; and that the Object-glasses of long Telescopes lie upon one another without touching; and that dry Powders are difficultly made to touch one another, so as to stick together, unless by melting them, or wetting them with Water, which by exhaling may bring them together; and that two polish'd Marbles, which by immediate Contact stick together, are difficultly brought so close together as to stick.

And thus Nature will be very conformable to her self and very simple, performing all the great Motions of the heavenly Bodies by the Attraction of Gravity which intercedes those Bodies, and almost all the small ones of their Particles by some other attractive and repelling powers which intercede the Particles. The *Vis inertiæ* is a passive Principle by which Bodies persist in their Motion or Rest, receive Motion in proportion to the Force impressing it, and resist as much as they are resisted. By this Principle alone there never could have been any Motion in the World. Some other Principle was necessary for putting Bodies into Motion; and now they are in Motion, some other Principle is necessary for conserving the Motion. For from the various Compositions of two Motions, 'tis very certain that there is not always the same quantity of Motion in the World. For if two Globes joined by a slender Rod, revolve about their common center of Gravity with an uniform Motion, while that Center moves on uniformly in a right Line drawn in the Plane of their circular Motion; the Sum of the two Motions of the Globes, as often as the globes are in the right Line described by their common Center of Gravity, will be bigger than the Sum of their Motions, when they are in a Line perpendicular to that right Line. By this Instance it appears that Motion may be got or lost. But by reason of the Tenacity of Fluids and Attraction of their Parts, and the Weakness of Elasticity in Solids, Motion is much more apt to be lost than got, and is always upon the Decay. For Bodies which are either absolutely hard, or so soft as to be void of Elasticity, will not rebound from one another.[2] Impenetrability makes them only stop. If two equal Bodies meet directly *in vacuo*, they will by the Laws of Motion stop where they meet, and lose all their Motion, and remain in rest, unless they be elastick, and receive new Motion from their Spring. If they have so much Elasticity as suffices to make them re-bound with a quarter or half, or three quarters of the Force with which they come together, they will lose three quarters, or half, or a quarter of their Motion. And this may be try'd, by letting two equal Pendulums fall against one another from equal heights. If the Pendulums be of Lead or soft Clay, they will lose all or almost all their Motions: If of elastic Bodies they will lose all but what they recover from their Elasticity. If it be said, that they can lose no Motion but what they communicate to other Bodies, the consequence is, that *in vacuo* they can lose no Motion, but when they meet they must go on and penetrate one another's Dimensions. If three equal round Vessels be filled, the one with Water, the other with Oil, the third with molten Pitch, and the Liquors be stirred about alike to give them a vortical Motion; the Pitch by its Tenacity will lose its Motion quickly, the Oil being less tenacious will keep it longer, and the Water being less tenacious will keep it longest, but yet will lose it in a short time.[3] Whence it is easy to understand, that if many contiguous Vortices of molten Pitch were each of them as large as those which some suppose to revolve about the Sun and the fix'd Stars, yet these and all

---

[2] The speed of rebound of two colliding bodies depends on the *coefficient of restitution* of each of the bodies involved.—[*K.K.*].

[3] What Newton calls *tenacity* we would today call *viscosity*; see Newton's discussion of viscous (as opposed to inertial) drag in *Qu. 28*, included in Chap. 18 of the present volume.—[*K.K.*].

their Parts would, by their Tenacity and Stiffness, communicate their Motion to one another till they all rested among themselves.[4] Vortices of Oil or Water, or some fluider Matter, might continue longer in Motion; but unless the Matter were void of all Tenacity and Attrition of Parts, and Communication of Motion, (which is not to be supposed), the motion would constantly decay. Seeing therefore the variety of Motion which we find in the World is always decreasing, there is a necessity of conserving and recruiting it by active Principles, such as are the cause of Gravity, by which Planets and Comets keep their Motions in their Orbs, and Bodies acquire great Motion in falling; and the cause of Fermentation, by which the Heart and Blood of Animals are kept in perpetual Motion and Heat; the inward Parts of the Earth are constantly warm'd, and in some places grow very hot; Bodies burn and shine, Mountains take fire, the Caverns of the Earth are blown up, and the Sun continues violently hot and lucid, and warms all things by his Light. For we meet with very little Motion in the World, besides what is owing to these active Principles. And if it were not for these Principles, the Bodies of the Earth, Planets, Comets, Sun and all things in them, would grow cold and freeze, and become inactive Masses; and all Putrefaction, Generation and Life would cease, and the Planets and Comets would not remain in their Orbs.

All these things being consider'd, it seems probable to me, that God in the Beginning form'd Matter in solid, massy, hard, impenetrable, movable Particles, of such Sizes and Figures, and with such other Properties, and in such Proportion to Space, as most conduced to the End for which he form'd them; and that these primitive Particles being Solids, are incomparably harder than any porous Bodies compounded of them; even so very hard, as never to wear or break into pieces; no ordinary Power being able to divide what God himself made one in the first Creation. While the Particles continue entire, they may compose Bodies of one and the same Nature and Texture in all Ages: But should they wear away, or break in pieces, the Nature of Things depending on them, would be changed. Water and Earth, composed of old worn Particles and Fragments of Particles, would not be of the same Nature and Texture now, with Water and Earth composed of entire Particles in the Beginning. And therefore, that Nature may be lasting, the Changes of corporeal Things are to be placed only in the various Separations and new Associations and Motions of these permanent Particles; compound Bodies being apt to break, not in the midst of solid Particles, but where those Particles are laid together, and only touch in a few Points.

It seems to me farther, that these Particles have not only a *Vis inertiæ*, accompanied with such passive Laws of Motion as naturally result from that Force but also that they are moved by certain active Principles, such as is that of Gravity, and that which causes Fermentation, and the Cohesion of Bodies. These Principles I consider, not as occult Qualities, supposed to result from the specifick Forms of Things, but as general Laws of Nature, by which the Things themselves are form'd; Their

---

[4] Newton is probably here criticizing Descartes' theory of vortices, by which he thought the planets to be carried about the sun; see Descartes, R., *The World or Treatise on Light*, Abaris Books, 1979.—[*K.K.*].

Truth appearing to us by Phænomena, though their Causes be not yet discover'd. For these are manifest Qualities, and their Causes only are occult. And the *Aristotelians* gave the Name of occult Qualities, not to manifest Qualities, but to such Qualities only as they supposed to lie hid in Bodies, and to be the unknown Causes of manifest Effects: Such as would be the Causes of Gravity, and of magnetick and electrick Attractions, and of Fermentations, if we should suppose that these Forces or Actions arose from Qualities unknown to us, and uncapable of being discovered and made manifest. Such occult Qualities put a stop to the Improvement of Natural Philosophy, and therefore of late Years have been rejected. To tell us that every Species of Things is endow'd with an occult specifick Quality by which it acts and produced manifest Effects is to tell us nothing: But to derive two or three general Principles of Motion from Phænomena and afterwards to tell us how the Properties and Actions of all corporeal Things follow from those manifest Principles, would be a very great step in Philosophy, though the Causes of those Principles were not yet discover'd: And therefore I scruple not to propose the Principles of Motion above-mention'd, they being of very general Extent, and leave their Causes to be found out.

Now by the help of these Principles, all material Things seems to have been composed of the hard and solid Particles above-mention'd, variously associated in the first Creation by the Counsel of an intelligent Agent. For it became him who created them to get them in order. And if he did so, it's unphilosophical to seek for them any other Origin of the World, or to pretend that it might arise out of a Chaos by the mere Laws of Nature; though being once form'd, it may continue by those Laws for many Ages. For while Comets move in very excentrick Orbs in all manner of Positions, blind Fate could never make all the Planets move one and the same way in Orbs concentrick, some inconsiderable Irregularities excepted, which may have risen from the mutual Actions of Comets and Planets upon one another, and which will be apt to increase, till this System wants a Reformation. Such a wonderful Uniformity in the Planetary System must be allowed the Effect of Choice. And so must the Uniformity in the Bodies of Animals, they having generally a right and a left side shaped alike, and on either side of their Bodies two Legs behind, and either two Arms, two Legs, or two Wings before upon their Shoulders, and between their Shoulders a Neck running down into a Back-bone, and a Head upon it; and in the Head two Ears, two Eyes, a Nose, a Mouth, and a Tongue, alike situated. Also the first Contrivance of those very artificial Parts of Animals, the Eyes, Ears, Brain, Muscles, Heart, Lungs, Midriff, Glands, Larynx, Hands, Wings, swimming Bladders, natural Spectacles, and other Organs of Sense and Motion; and the Instinct of Brutes and Insects, can be the effect of nothing else than the Wisdom and Skill of a powerful ever-living Agent, who being in all Places, is more able by his Will to move the Bodies within his boundless uniform Sensorium, and thereby to form and reform the Parts of the Universe, than we are by our Will to move the Parts of our own Bodies. And yet we are not to consider the World as the Body of God, or the several Parts thereof, as the Parts of God. He is an uniform Being, void of Organs, Members, or Parts, and they are his Creatures subordinate to him, and subservient to his Will; and he is no more the Soul of them, than the Soul of Man is the Soul of the Species of Things carried through the Organs of Sense into the place of its

Sensation, where it perceives them by means of its immediate Presence, without the Intervention of any third thing. The Organs of Sense are not for enabling the Soul to perceive the Species of Things in its Sensorium, but only for conveying them thither; and God has no need of such Organs, he being everywhere present to the Things themselves. And since Space is divisible *in infinitum*, and Matter is not necessarily in all places, it may be also allow'd that God is able to create Particles of Matter of several Sizes and Figures, and in several Proportions to Space, and perhaps of different Densities and Forces, and thereby to vary the Laws of Nature, and make Worlds of several sorts in several Parts of the Universe. At least, I see nothing of Contradiction in all this.

As in Mathematicks, so in Natural Philosophy, the Investigation of difficult Things by the Method of Analysis, ought ever to precede the Method of Composition. This Analysis consists in making Experiments and Observations, and in drawing general Conclusions from them by Induction, and admitting of no Objections against the Conclusions, but such as are taken from Experiments, or other certain Truths. For Hypotheses are not to be regarded in experimental Philosophy. And although the arguing from Experiments and Observations by Induction be no Demonstration of general Conclusions; yet it is the best way of arguing which the Nature of Things admits of, and may be looked upon as so much the stronger, by how much the Induction is more general. And if no Exception occur from Phænomena, the Conclusion may be pronounced generally. But if at any time afterwards any Exception shall occur from Experiments, it may then begin to be pronounced with such Exceptions as occur. By this way of Analysis we may proceed from Compounds to Ingredients, and from Motions to the Forces producing them; and in general from Effects to their Causes, and from particular Causes to more general ones, till the Argument end in the most general. This is the Method of Analysis: And the Synthesis consists in assuming the Causes discover'd, and establish'd as Principles, and by them explaining the Phænomena proceeding from them, and proving the Explanations.

In the first two Books of these Opticks, I proceeded by this Analysis to discover and prove the original Differences of the Rays of Light in respect to Refrangibility, Reflexibility, and Colour, and their alternate Fits of easy Reflexion and easy Transmission, and the Properties of Bodies, both opake and pellucid, on which their Reflexions and Colours depend. And these Discoveries being proved, may be assumed in the Method of Composition for explaining the Phænomena arising from them: An Instance of which Method I gave in the End of the first Book. In this third Book I have only begun the Analysis of what remains to be discover'd about Light and its Effects upon the Frame of Nature, hinting several things about it, and leaving the Hints to be examin'd and improv'd by the farther Experiments and Observations of such as are inquisitive. And if natural Philosophy in all its Parts, by pursuing this Method, shall at length be perfected, the Bounds of Moral Philosophy will be also enlarged. For so far as we can know by natural Philosophy what is the first Cause, what Power he has over us, and what Benefits we receive from him, so far our Duty towards him, as well as towards one another, will appear to us by the Light of Nature. And no doubt, if the Worship of false Gods had not blinded the

Heathen, their moral philosophy would have gone farther than to the four Cardinal Virtues; and instead of teaching the Transmigration of Souls, and to worship the Sun and Moon, and dead Heroes, they would have taught us to worship our true Author and Benefactor, as their Ancestors did under the Government of *Noah* and his Sons before they corrupted themselves.

19.3 Study Questions

QUES. 19.1. How is Newton's theory of light informed by his theory of matter?

a) Of what are bodies composed? Are all bodies divisible?
b) By what forces do bodies act upon one another? Are these forces attractive or repulsive? Can you provide examples of each?
c) Of what is light composed? Why does light obey the laws of reflection and refraction? Do you find Newton's arguments convincing?

QUES. 19.2. Can passive laws of nature explain the cause of motion throughout the world? Or are active principles necessary?

a) What, according to Newton, is the difference between a passive law and an active principle? Does Newton classify his own laws of motion as passive laws or active principles? What about gravity—is this passive law or an active principle?
b) What would happen in the absence of active principles? What examples of such occurrences does Newton provide? Why does Newton reject the vortex theory of planetary motion?
c) What is Newton's view of the origin and nature of matter itself? In particular, what are the essential properties of matter? Why must they have such properties? How can Newton's simple theory of matter account for the diverse properties and changes which matter undergoes?
d) Are active principles, in fact, occult qualities? What is meant by an occult quality? What is the difference between an occult quality and a law of nature?

QUES. 19.3. Can natural philosophy inform moral philosophy?

a) What is natural philosophy? Is the method of natural philosophy the same as that of mathematics? Does it proceed inductively or deductively? What role do exceptions play?
b) Do Newton's views regarding the nature—and the very existence—of God follow rationally from his observations of the World? What is the relationship between the creator and creation?
c) What observational evidence does Newton provide for the "Effect of Choice" in nature? Does Newton believe the laws of nature to be inviolable?

d) What, according to Newton, is the scientific method? In particular, how does one proceed from particular observations to general causes? What does he hold up as an example of the implementation of this method?

e) What is the source of moral law? How can it be apprehended? How can it be hidden or lost? Is Newton correct?

19.4 Exercises

EX. 19.1 (BLIND FATE ESSAY). Is Newton justified in asserting that passive laws alone cannot account for nature? What about his assertion that "blind Fate could never make all the Planets move one and the same way in Orbs concentrick"?

EX. 19.2 (NATURAL THEOLOGY ESSAY). Can the existence of God (or gods) be rationally inferred based on observation of nature? In what way, if any, would such an inference differ from other inferences based on observation of nature? Is Newton right in saying that "If [an intelligent agent created material things, then] it's unphilosophical to seek for any other origin of the world, or to pretend that it might arise out of a chaos by the mere laws of Nature."

EX. 19.3 (NATURAL AND MORAL LAW ESSAY). Do you agree with Newton's conception of moral law? What are the similarities and differences between natural law and moral law? Which is discoverable? Which is reliable? Which can be violated?

EX. 19.4 (ESCAPE VELOCITY OF LIGHT). Suppose, as does Newton, that light is comprised of tiny corpuscles, each with mass m, which are emitted from luminous bodies. According to Newton's universal law of gravitation, when a star of mass M and radius R emits a corpuscle from its surface at the speed of light, c, it is subsequently slowed down by the gravitational force exerted on it by the star itself.[5] In this problem, we will attempt to determine whether such a light corpuscle, considered as a classical mass, can escape the gravitational attraction of its parent star.

a) How does the force of gravity depend upon the distance between the center of the star and the corpuscle?

b) How much work has been done on the corpuscle by the gravitational force by the time the corpuscle has reached a particular distance, r, from the center of the star?

c) What is the initial kinetic energy of the corpuscle? And how does the potential energy of the star-corpuscle system depend on their separation?

d) Under what conditions, if any, is a light corpuscle able to escape the gravitational attraction of its parent star? What does this imply?[6]

---

[5] See the discussion of Newton's universal law of gravitation included in Chap. 27 of volume II.

[6] For an interesting discussion of the relationship between Newton's theory of gravity and the corpuscular theory of light, see McVittie, G., Laplace's alleged "black hole", *The Observatory*, *98*, 272–274, 1978.

19.5 Vocabulary

1. Tenor
2. Consonant
3. Impulse
4. Vulgar
5. Congeal
6. Sublime
7. Impenetrable
8. Requisite
9. Interstices
10. Affirmative
11. Virtue
12. Feign
13. Ramous
14. Cohere
15. Ponderous
16. Conformable
17. Intercede
18. Tenacity
19. Apt
20. Contiguous
21. Lucid
22. Corporeal
23. Occult
24. Manifest
25. Endow
26. Sensorium
27. Subordinate
28. Subservient
29. Induction
30. Heathen
31. Transmigration
32. Asunder

Chapter 20
Measuring Light's Wavelength

The breadth of the undulations constituting the extreme red light must be supposed to be, in the air, about one 36 thousandth of an inch.

—Thomas Young

20.1 Introduction

Thomas Young (1773–1829) was a physician and scientist born to a Quaker family in the village of Milverton in Somerset, England. By the age of 13, he was knowledgeable in Greek, Latin, Hebrew, French and Italian; later he would study near-eastern languages such as Arabic, Persian, Chaldee, Syriac, and Samarian. Young continued his study of languages even while he was practicing medicine. His work on Egyptian hieroglyphics played a significant role in deciphering the demotic script of the Rosetta Stone, which had gone on display in London in 1802 after its recent rediscovery by Napoleon's forces during his campaign in Egypt in 1799.[1] During his teenage years Young had studied Euclid's *Elements* and Newton's *Opticks*. He subsequently published a number of works on the theory of light, sound, colors and the physiology of vision; he is also credited with coining the term "energy" to describe the ability of a system to accomplish work.[2] Young was elected a Fellow the Royal Society in 1794. In 1802 he began delivering lectures at the Royal Institution which covered a wide range of topics, including pure mathematics, motion, friction, architecture, engraving, carpentry, hydraulics, hydrodynamics, sound, heat, weather and the nature of light. In 1807, he published his comprehensive *Course of Lectures on Natural Philosophy and the Mechanical Arts* in two-volumes. The following two reading selections are taken from a lecture in this course entitled *On the nature of light and colours*. Herein, he carefully considers

---

[1] For more information on the advent of modern Egyptology after Napoleon's ill-fated campaign in north Africa, see the biography of the French scientist and mathematician Joseph Fourier in volume IV.

[2] See Hermann von Helmholtz's lecture on the conservation of energy in Chaps. 9–11 of the present volume.

© Springer International Publishing Switzerland 2016
K. Kuehn, *A Student's Guide Through the Great Physics Texts*,
Undergraduate Lecture Notes in Physics, DOI 10.1007/978-3-319-21816-8_20

which of the two prevalent theories of light—the particle theory of Newton or the wave theory of Huygens—might better explain the experimental observations which had been performed by himself and by other scientists. To what conclusion is he led? Do you find his arguments convincing?

20.2   Reading: Young, *On the Nature of Light and Colours*

Young, T., *Course of lectures on natural philosophy and the mechanical arts*, vol. 1, Printed for Joseph Johnson, London, 1807. Lecture 39.

The nature of light is a subject of no material importance to the concerns of life or to the practice of the arts, but it is in many other respects extremely interesting, especially as it tends to assist our views both of the nature of our sensations, and of the constitution of the universe at large. The examination of the production of colours, in a variety of circumstances, is intimately connected with the theory of their essential properties, and their causes; and we shall find that many of these phenomena will afford us considerable assistance in forming our opinion respecting the nature and origin of light in general.

It is allowed on all sides, that light either consists in the emission of very minute particles from luminous substances, which are actually projected, and continue to move, with the velocity commonly attributed to light, or in the excitation of an undulatory motion, analogous to that which constitutes sound, in a highly light and elastic medium pervading the universe; but the judgements of philosophers of all ages have been much divided with respect to the preference of one or the other of these opinions. There are also some circumstances which induce those, who entertain the first hypothesis, either to believe, with Newton, that the emanation of the particles of light is always attended by the undulations of an etherial medium, accompanying it in its passage, or to suppose, with Boscovich, that the minute particles of light themselves receive, at the time of their emission, certain rotatory and vibratory motions, which they retain as long as their projectile motion continues. These additional suppositions, however necessary they may have been thought for explaining some particular phenomena, have never been very generally understood or admitted, although no attempt has been made to accommodate the theory in any other manner to those phenomena.

We shall proceed to examine in detail the manner in which the two principal hypotheses respecting light may be applied to its various properties and affections; and in the first place to the simple propagation of light in right lines through a vacuum, or a very rare homogeneous medium. In this circumstance there is nothing inconsistent with either hypothesis; but it undergoes some modifications, which require to be noticed, when a portion of light is admitted through an aperture, and spreads itself in a slight degree in every direction. In this case it is maintained by Newton that the margin of the aperture possesses an attractive force, which is capable of inflecting the rays: but there is some improbability in supposing that bodies of different forms and of various reflective powers should possess an equal force

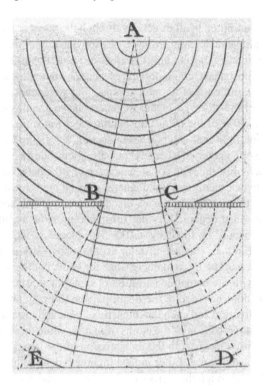

Fig. 20.1 A series of waves, diverging from the centre A, and passing through the aperture BC, extend themselves on each side so as to fill the space $BCDE$, while they affect the parts without this space much less sensibly

of inflection, as they appear to do in the production of these effects; and there is reason to conclude from experiments, that such a force, if it existed, must extend to a very considerable distance from the surfaces concerned, at least a quarter of an inch, perhaps much more, which is a condition not easily reconciled with other phenomena. In the Huygenian system of undulation, this divergence or diffraction is illustrated by a comparison with the motions of waves of water and of sound, both of which diverged when they are admitted into a wide space through an aperture, so much indeed that it has usually been considered as an objection to this opinion, that the rays of light do not diverge in the degree that would be expected if they were analogous to the waves of water. But as it has been remarked by Newton, that the pulses of sound diverge less than the waves of water, so it may fairly be inferred, that in a still more highly elastic medium, the undulations, constituting light, must diverge much less considerably than either (Fig. 20.1).

With respect, however, to the transmission of light through perfectly transparent mediums of considerable density, the system of emanation labours under some difficulties. It is not to be supposed that the particles of light can perforate with freedom the ultimate atoms of matter, which compose a substance of any kind; they must, therefore, be admitted in all directions through the pores or interstices of those atoms: for if we allow such suppositions as Boscovich's, that matter itself is penetrable, that is, immaterial, it is almost useless to argue the question further. It is certain that some substances retain all their properties when they are reduced to the thickness of the 10 millionth of an inch at most, and we cannot therefore suppose that the distances of the atoms of matter in general be so great as the 100 millionth of an inch. Now if 10 ft of the most transparent water transmits, without interruption, one half of the light that enters it, each section or stratum of the thickness of one of these pores of matter must intercept only about one twenty thousand millionth, and so much must the space or area occupied by the particles be smaller than the interstices between them, and the diameter of each atom must be less than the hundred and forty thousandth part of its distance from the neighbouring particles: so that the whole space occupied by the substance must be as little filled, as the whole of England would be filled by a 100 men, placed at the distance of about 30 miles from each other. This astonishing degree of porosity is not indeed absolutely inadmissible, and there are many reasons for believing the statement to agree in some measure with the actual constitution of material substances; but the Huygenian hypothesis does not require the disproportion to be by any means so great, since the general direction and even the intensity of an undulation would be very little affected by the interposition of the atoms of matter, while these atoms may at the same time be supposed to assist in the transmission of the impulse, by propagating it through their own substance. Euler indeed imagined that the undulations of light might be transmitted through the gross substance of material bodies alone, precisely in the same manner as sound is propagated; but this supposition is for many reasons inadmissible.

A very striking circumstance, respecting the propagation of light, is the uniformity of its velocity in the same medium. According to the projectile hypothesis, the force employed in the free emission of light must be about a million million times as great as the force of gravity at the earth's surface; and it must either act with equal intensity on all the particles of light, or must impel some of them through a greater space than others, if its action be less powerful, since the velocity is the same in all cases; for example, if the projectile force is weaker with respect to red light than with respect to violet light, it must continue its action on the red rays to a greater distance than on the violet rays. There is no instance in nature besides of a simple projectile moving with a velocity uniform in all cases, whatever may be its cause, and it is extremely difficult to imagine that so immense a force of repulsion can reside in all substances capable of becoming luminous, so that the light of decaying wood, or of two pebbles rubbed together, may be projected precisely with the same velocity, as the light emitted by iron burning in oxygen gas, or by the reservoir of liquid fire on the surface of the sun. Another cause would also naturally interfere with the uniformity of the velocity of light, if it consisted merely in the motion of projected corpuscles of matter; Mr. Laplace has calculated, that if any of the stars were 250 times as great in diameter as the sun, its attraction would be so strong as to destroy the whole momentum of the corpuscles of light proceeding from it, and to render the star invisible at a great distance; and although there is no reason to imagine that any of the stars are actually of this magnitude, yet some of them are probably many times greater than our sun, and therefore large enough to produce such a retardation in the motion of their light as would materially alter its effects. It is almost unnecessary to observe that the uniformity of the velocity of light, in those spaces which are free from all material substances, is a necessary consequence of the Huygenian hypothesis, since the undulations of every homogeneous elastic medium are always propagated, like those of sound, with the same velocity, as long as the medium remains unaltered.

On either supposition, there is no difficulty in explaining the equality of angles of incidence and reflection; for these angles are equal as well in the collision of common elastic bodies with others incomparably larger, as in the reflections of the waves of water and of the undulations of sound. And it is equally easy to demonstrate, that the sines of the angles of incidence and refraction must be always in the same proportion at the same surface, whether it be supposed to possess an attractive force, capable of acting on the particles of light, or to be the limit of a medium through which the undulations are propagated with a diminished velocity. There are however, some cases of the production of colours, which lead us to suppose that the velocity of light must be smaller in a denser than in a rarer medium; and supposing this fact to be fully established, the existence of such an attractive force could no longer be allowed, nor could the system of emanation be maintained by any one.

The partial reflection from all refracting surfaces is supposed by Newton to arise from certain periodical retardations of the particles of light, caused by undulations, propagated in all cases through an ethereal medium. The mechanism of these supposed undulations is so complicated, and attended by so many difficulties, that the few who have examined them have been in general entirely dissatisfied with them: and the internal vibrations of the particles of light themselves, which Boscovich has imagined, appear scarcely to require a serious discussion. It may, therefore, safely be asserted, that in the projectile hypothesis this separation of the rays of light of the same kind by a partial reflection at every refracting surface, remains wholly unexplained. In the undulatory system, on the contrary, this separation follows as a necessary consequence. It is simplest to consider the ethereal medium which pervades any transparent substance, together with the material atoms of the substance, as constituting together a compound medium denser than the pure ether, but not more elastic; and by comparing the contiguous particles of the rarer and the denser medium with common elastic bodies of different dimensions, we may easily determine not only in what manner but almost in what degree, this reflection must take place in different circumstances. Thus, if one of two equal bodies strikes the other, it communicates to it its whole motion without any reflection; but a smaller body striking a larger one is reflected, with the more force as the difference of their magnitude is greater; and a larger body, striking a smaller one, still proceeds with a diminished velocity; and the remaining motion constituting, in the case of an undulation falling on a rarer medium, a part of a new series of motions which necessarily returns backwards with the appropriate velocity: and we may observe a circumstance nearly similar to this last in a portion of mercury almost in the same manner as from a solid obstacle.

The total reflection of light, falling, with a certain obliquity, on the surface of a rarer medium, becomes, on both suppositions, a particular case of refraction. In the undulatory system, it is convenient to suppose the two mediums to be separated by a short space in which their densities approach by degrees to each other, in order that the undulation may be turned gradually around, so as to be reflected in an equal angle: but this supposition is not absolutely necessary, and the same effects may be expected at the surface of two mediums separated by an abrupt termination.

The chemical process of combustion may easily be imagined either to disengage the particles of light from their various combinations, or to agitate the elastic medium by the intestine motions attending it: but the operations of friction upon substances incapable of undergoing chemical changes, as well as the motions of the electric fluid through imperfect conductors, afford instances of the production of light in which there seems to be no easy way of supposing a decomposition of any kind. The phenomena of solar phosphori appear to resemble greatly the sympathetic sounds of musical instruments, which are agitated by other sounds conveyed to them through the air: it is difficult to understand in what state the corpuscles of light could be retained by these substances so as to be reemitted after a short space or time; and if it is true that diamonds are often found, which exhibit a red light after having received a violet light only, it seems impossible to explain this property, on the supposition of the retention and subsequent emission of the same corpuscles.

The phenomena of the aberration of light agree perfectly well with the system of emanation; and if the ethereal medium, supposed to pervade the earth and its atmosphere, were carried along before it, and partook materially in its motions, these phenomena could not easily be reconciled with the theory of undulation. But there is no kind of necessity for such a supposition: it will not be denied by the advocates of the Newtonian opinion that all material bodies are sufficiently porous to leave a medium pervading them almost absolutely at rest; and if this be granted, the effects of aberration will appear to be precisely the same in either hypothesis.

The unusual refraction of the Iceland spar has been most accurately and satisfactorily described by Huygens, on the simple supposition that this crystal possesses the property of transmitting an impulse more rapidly in one direction than in another; whence he infers that the undulations constituting light must assume a spheroidical instead of a spherical form, and lays down such laws for the direction of its motion, as are incomparably more consistent with experiment than any attempts which have been made to accommodate the phenomena to other principles. It is true that nothing has yet been done to assist us in understanding the effects of a subsequent refraction by a second crystal, unless any person be satisfied with the name of polarity assigned by Newton to a property which he attributes to the particles of light, and which he supposes to direct them in the species of refraction which they are to undergo: but on any hypothesis, until we discover the reason why a part of the light is at first refracted in the usual manner, and another part in the unusual manner, we have no right to expect that we should understand how these dispositions are continued or modified, when the process is repeated.

In order to explain, in the system of emanation, the dispersion of the rays of different colours by means of refraction, it is necessary to suppose that all refractive mediums have an elective attraction, action more powerfully on the violet rays, in proportion to their mass, than on the red. But an elective attraction of this kind is a property foreign to mechanical philosophy, and when we use the term in chemistry, we only confess our incapacity to assign a mechanical cause for the effect, and refer to an analogy with other facts, of which the intimate nature is perfectly unknown to us. It is not indeed very easy to give a demonstrative theory of the dispersion of coloured light upon the supposition of undulatory motion; but we may derive a very satisfactory illustration from the well known effects of waves of different breadths. The simple calculation of the velocity of waves, propagated in a liquid perfectly elastic, or incomprehensible, and free from friction, assigns to them all precisely the same velocity, whatever their breadth may be: the compressibility of the fluids actually existing introduces, however, a necessity for a correction according to the breadth of the wave, and it is very easy to observe, in a river or a pond of considerable depth, that the wider waves proceed much more rapidly than the narrower. We may, therefore, consider the pure ethereal medium as analogous to an infinitely elastic fluid, in which undulations of all kinds move with equal velocity, and material transparent substances, on the contrary, as resembling those fluids, in which we see the large waves advance beyond the smaller; and by supposing the red light to consist of larger or wider undulations and the violet of smaller, we may sufficiently elucidate the greater refrangibility of the red than of the violet light.

It is not, however, merely on the ground of this analogy that we may be induced to suppose the undulations constituting red light to be larger than those of violet light: a very extensive class of phenomena leads us still more directly to the same conclusion; they consist chiefly of the production of colours by means of transparent plates, and by diffraction or inflection, none of which have been explained, upon the supposition of emanation, in a manner sufficiently minute or comprehensive to satisfy the most candid even of the advocates for the projectile system; while on the other hand all of them may be at once understood, from the effect of the interference of double lights, in a manner nearly similar to that which constitutes in sound the sensation of a beat, when two strings, forming an imperfect unison, are heard to vibrate together.

Supposing the light of any given colour to consist of undulations, of a given breadth, or of a given frequency, it follows that these undulations must be liable to those effects which we have already examined in the case of the waves of water, and the pulses of sound. It has been shown that two equal series of waves, proceeding from centres near each other, may be seen to destroy each other's effects at certain points, and other points at to redouble them; and the beating of two sounds has been explained from a similar interference. We are now to apply the same principles to the alternate union and extinction of colours (Fig. 20.2).

In order that the effects of two portions of light may be thus combined, it is necessary that they be derived from the same origin, and that they arrive at the same point by different paths, in directions not much deviating from each other. This deviation may be produced in one or both of the portions by diffraction, by reflection, by refraction, or by any of these effects combined; but the simplest case appears to be, when a beam of homogeneous light falls on a screen in which there

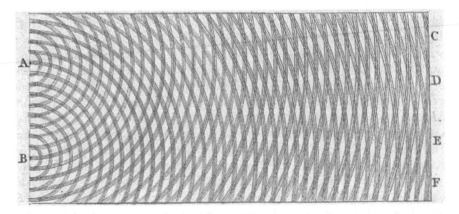

Fig. 20.2 Two equal series of waves, diverging from the centres A and B, and crossing each other in such a manner, that in the lines tending towards C, D, E, and F, they counteract each other's effects and the water remains nearly smooth, while in the intermediate spaces it is agitated

are two very small holes or slits, which may be considered as centres of divergence, from whence the light is diffracted in every direction. In this case, when the two newly formed beams are received on a surface placed so as to intercept them, their light is divided by dark stripes into portions nearly equal, but becoming wider as the surface is more remote from the apertures, so as to subtend very nearly equal angles from the apertures at all distances, and wider also in the same proportion as the apertures are closer to each other. The middle of the two portions is always light, and the bright stripes on each side are at such distances, that the light, coming to them from one of the apertures, must have passed through a longer space than that which comes from the other, by an interval which is equal to the breadth of one, two, three, or more of the supposed undulations, while the intervening dark spaces correspond to a difference of half a supposed undulation, of one and a half, of two and a half, or more.

From a comparison of various experiments, it appears that the breadth of the undulations constituting the extreme red light must be supposed to be, in the air, about one 36 thousandth of an inch, and those of the extreme violet about one 60 thousandth; the mean of the whole spectrum, with respect to the intensity of light, being about one 45 thousandth. From these dimensions it follows, calculating upon the known velocity of light, that almost 500 millions of millions of the slowest of such undulations must enter the eye in a single second. The combination of two portions of white or mixed light, when viewed at a great distance, exhibits a few white and black stripes, corresponding to this interval; although upon closer inspection, the distinct effects of an infinite number of stripes of different breadths appear to be compounded together, so as to produce a beautiful diversity of tints, passing by degrees into each other. The central whiteness is first changed to a yellowish, and then to a tawny colour, succeeded by crimson, and by violet and blue, which together appear, when seen at a distance, as a dark stripe; after this a green light

Fig. 20.3 The manner in which two portions of colored light, admitted through two small apertures, produce light and dark stripes or fringes by their interference, proceeding in the form of hyperbolas; the middle ones are however usually a little dilated, as at *A*

Fig. 20.4 A series of stripes of all colours, of their appropriate breadths, placed side by side in the manner in which they would be separated by refraction, and combined together so as to form the fringes of colours below them, beginning from *white*

appears, and the dark space beyond it has a crimson hue; the subsequent lights are all more or less green, the dark spaces purple and reddish; and the red light appears so far to predominate in all these effects, that the red or purple stripes occupy nearly the same place in the mixed fringes as if their light were received separately.

The comparison of the results of this theory with experiments fully establishes their general coincidence; it indicates, however, a slight correction in some measures, on account of some unknown cause, perhaps connected with the intimate nature of diffraction, which uniformly occasions the portions of light, proceeding in a direction very nearly rectilinear, to be divided into stripes or fringes a little wider than the external stripes, formed by the light which is more bent Figs. 20.3 and 20.4).

When the parallel slits are enlarged, and leave only the intervening substance to cast its shadow, the divergence from its opposite margins still continues to produce the same fringes as before, but they are not easily visible, except within the extent of its shadow, being overpowered in other parts by a stronger light; but if the light thus diffracted be allowed to fall on the eye, either within the shadow, or in its neighbourhood, the stripes will still appear; and in this manner the colours of small fibres are probably formed. Hence if a collection of equal fibres, for example a

Fig. 20.5 A series of coronae, seen round the sun or moon

lock of wool, be held before the eye when we look at a luminous object, the series of stripes belonging to each fibre combine their effects, in such a manner, as to be converted into circular fringes or coronae. This is probably the origin of the coloured circles or coronae sometimes seen round the sun and moon, two or three of them appearing together, nearly at equal distances from each other and from the luminary, the internal ones being, however, like the stripes, a little dilated. It is only necessary that the air should be loaded with globules of moisture, nearly of equal size among themselves, not much exceeding one two thousandth of an inch in diameter, in order that a series of such coronae, at the distance of 2° or 3° from each other, may be exhibited. (Fig. 20.5)

If, on the other hand, we remove the portion of the screen which separates the parallel slits from each other, their external margins will still continue to exhibit the effects of diffracted light in the shadow on each side; and the experiment will assume the form of those which were made by Newton on the light passing between the edges of two knives, brought very nearly into contact; although some of these experiments appear to show the influence of a portion of light reflected by a remoter part of the polished edge of the knives, which indeed must unavoidably constitute a part of the light concerned in the appearance of fringes, wherever their whole breadth exceeds that of the aperture, or of the shadow of the fibre.

The edges of two knives, placed very near each other, may represent the opposite margins of a minute furrow, cut in the surface of a polished substance of any kind, which, when viewed with different degrees of obliquity, present a series of colours nearly resembling those which are exhibited within the shadows of the knives: in this case, however, the paths of the portions of light before their incidence are also to be considered, and the whole difference of these paths will be found to determine the appearance of colour in the usual manner; thus when the surface is so situated, that the image of the luminous point would be seen in it by regular reflection, the difference will vanish, and the light will remain perfectly white, but in other cases various colours will appear, according to the degree of obliquity. These colours may easily be seen, in an irregular form, by looking at any metal, coarsely polished, in the sunshine; but they become more distinct and conspicuous, when a number of

Fig. 20.6 The internal hyper-
bolic fringes of a rectangular
shadow

fine lines of equal strength are drawn parallel to each other, so as to conspire in their
effects.

It sometimes happens that an object, of which a shadow is formed in a beam of
light, admitted through a small aperture, is not terminated by parallel sides; thus the
two portions of light, which are diffracted from two sides of an object, at right angles
with each other, frequently form a short series of curved fringes within the shadow,
situated on each side of the diagonal, which were first observed by Grimaldi, and
which are completely explicable from the general principle, of the interference of
the two portions encroaching perpendicularly on the shadow. (Fig. 20.6)

But the most obvious of all the appearances by this kind is that of the fringes,
which are usually seen beyond the terminations of any shadow, formed in a beam
of light, admitted through a small aperture: in white light three of these fringes are
usually visible, and sometimes four; but in light of one colour only, their number is
greater; and they are always much narrower as they are remoter from the shadow.
Their origin is easily deduced from the interference of the direct light with a portion
of light reflected from the margin of the object which produces them, the obliquity
of its incidence causing a reflection so copious as to exhibit a visible effect, however
narrow that margin may be; the fringes are, however, rendered more obvious as the
quantity of this reflected light is greater.[3] Upon this theory it follows that the distance
of the first dark fringe from the shadow should be half as great as that of the fourth,
the difference of the lengths of the different paths of light being as the squares of
those distances; and the experiment precisely confirms this calculation; with the
same slight correction only as is required in all other cases; the distances of the first
fringes being always a little increased. It may also be observed, that the extent of
the shadow itself is always augmented, and nearly in an equal degree with that of
the fringes: the reason of this circumstance appears to be the gradual loss of light at
the edges of every separate beam, which is so strongly analogous to the phenomena

---

[3] Fresnel disagrees with Young's explanation of single-slit diffraction. See, for example, Fresnel,
A., Diffraction of Light, in *Source Book in Physics*, edited by W. F. Magie, Source books in the
history of science, pp. 318–324, Harvard University Press, Cambridge, Massachusetts, 1963.—
[*K.K.*]

Fig. 20.7 The external fringes seen on each side of the shadow of a hair or wire, which is also divided by its internal fringes. The dotted lines show the natural magnitude of the shadow, independently of diffraction

in waves of water. The same cause may also perhaps have some effect in producing the general modification or correction of the place of the first fringes, although it appears to be scarcely sufficient for explaining the whole of it. (Fig. 20.7)

20.3 Study Questions

QUES. 20.1. How successfully do the wave and particle theories of light account for (a) the propagation of light in a straight line through a vacuum, (b) the spreading of light passing through a narrow aperture, (c) the transmission of light through transparent media, (d) the uniformity of the velocity of light within a particular medium, and from all types of stars, (e) the laws of reflection and refraction, (f) partial reflection, and total internal reflection at the boundary of a medium, (g) the production of light by combustion, friction and phosphorescence, (h) the aberration of starlight, and (i) birefringence and dispersion.

QUES. 20.2. What is the significance of Young's double-slit experiment?

a) What is the relationship between the color of light and its wavelength?
b) Describe Young's double-slit apparatus. How, exactly, can the wavelength of light be measured? Upon what fundamental principle are these measurements based?
c) What is the relationship between the color of light and its frequency? How is this determined?

QUES. 20.3. Can the undulatory theory of light explain (a) the circular fringes appearing when locks of wool are back-lit? (b) Solar and lunar coronæ? (c) The fringes within the shadow of light admitted through a tiny gap between facing knife blades? (d) The colors of light reflected at different angles from coarsely polished surfaces? (e) And the faint fringes appearing in the shadow of an object's edge?

20.4 Exercises

EX. 20.1 (FIBER INTERFERENCE FRINGES). One of Young's early experiments to determine the wavelength of light involved viewing a straight vertically-oriented fiber of hair through an aperture cut into a piece of card-board. The fiber was back-lit

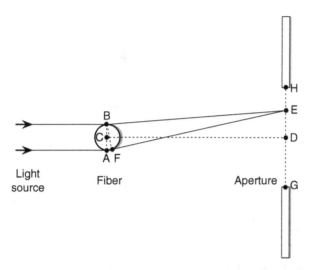

Fig. 20.8 A fiber viewed through an aperture backlit by a candle

with candle-light (see Fig. 20.8). Suppose that the diameter, AB, of the fiber is $1/600$ in. The aperture, HG, has a width of $8/100$ in. The distance, CD, from the fiber to the center of the aperture is $8/10$ in. When peering with one eye through the aperture, vertical fringes of various colors are observed parallel to the fiber. Suppose one can identify two adjacent bright green fringes at locations D and E, where $DE = DH/4$. According to *Huygen's principle*, A and B act as two point-sources of light waves.[4] From this information, you should calculate the wavelength of the green light.[5]

a) First, what is the difference in distance (in wavelengths) traveled by the light waves from the point-sources A and B to point D? Can this explain why there is a bright green fringe at location D?

b) Next, what is the difference in distance (in wavelengths) traveled by the light waves from points-sources A and B to point E? How does your answer rely upon the *superposition principle*?

c) Considering the right triangle ABF, whose hypotenuse is AB, write an equation for the length AF in terms of the distances AB, CD, and DE. (HINT: assume that the distance CD is much larger than DE and construct similar triangles.)

---

[4] Technically speaking, the sides of the fiber act as vertically-oriented line-sources (not point-sources), but this should not distract from the following analysis.

[5] For a review of Huygens' principle, see Chap. 14 of the present volume; this technique was employed previously in Ex. 14.2.

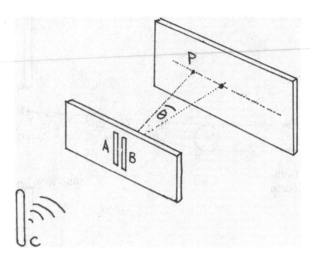

Fig. 20.9 Young's two-slit experiment

d) Combining your answers to the previous questions, write down an equation for the wavelength of green light, λ_g, in terms of AB, CD and DE. From the values provided above, what is λ_g?

e) What would happen to the spacing of the fringes (the distance DE) if the diameter of the fiber were reduced? And what would happen to the spacing if one considered two red fringes, rather than two green fringes?

EX. 20.2 (MEASURING HAIR DIAMETER LABORATORY). In this laboratory experiment, you will attempt to measure the width of one of the hairs of your head using laser light of known wavelength. As preparation, you should work through Ex. 20.1. Once you have done so, you will need little more than a glass microscope slide, a measuring stick, and a laser in order to carry out your measurements.[6] Stretch out one of your hairs on the microscope slide using tape and prop up the slide on a table-top between your laser and a white screen. By carefully measuring the locations of the interference fringes which appear on the white screen when your hair is illuminated with the laser beam, you can calculate the width of your hair. Be careful never to look into the laser beam!

EX. 20.3 (TWO-SLIT INTERFERENCE PATTERN). Consider an opaque barrier containing two thin vertical slits, A and B (see Fig. 20.9). The barrier is illuminated on one side by a monochromatic light source, C.[7] When a (nearly) planar wave-front from the light source strikes the barrier, the slits themselves act as point sources of

---

[6] I have used helium neon lasers from scientific supply companies, such as Sargent Welch, but a low-cost laser pointer works quite well, provided you can keep the "on" button depressed while taking measurements.

[7] A monochromatic light source is one which emits a single wavelength. Sunlight is not monochromatic.

waves.[8] Using *Huygens' principle* and the *superposition principle*, prove that the condition for constructive interference at a point P on the screen is given by

$$d \sin \theta_n = n\lambda, \quad n = \pm 1, \pm 2 \dots \quad \text{(constructive interference condition for two-slits)}$$
$$(20.1)$$

Here, d is the spacing between the two slits, n is called the *order* of the interference maximum, and θ_n is the angular location of the nth order interference maximum (measured with respect to the direction in which the incident wave was initially propagating).

EX. 20.4 (DIFFRACTION GRATINGS). A *diffraction grating* is a very dense series of parallel slits—perhaps a few 100 per inch—cut in an opaque barrier. When a monochromatic light source is viewed through a diffraction grating, a series of very thin parallel bright fringes are observed. Mathematically, these bright fringes—the so-called *principal maxima*—occur at angles θ_n such that

$$d \sin \theta_n = n\lambda, \quad n = 0, \pm 1, \pm 2 \dots \quad \text{(principle maxima for diffraction grating)}$$
$$(20.2)$$

Here, d is the slit spacing and n is (once again) the *order* of the maximum. Notice that the angular positions of the principle maxima of a diffraction grating are the same as the angular positions of the bright fringes produced when monochromatic light passes through two slits (Eq. 20.1). But there is an important difference between the interference patterns produced by a two slits and by a diffraction grating: for a diffraction grating, there appear two distinct types of fringes. *Principle maxima* occur at locations where the light waves from *all of* the slits arrive in phase with one another (Eq. 20.2). *Secondary maxima* occur at locations where the light waves from *not* all of the slits arrive in phase with one another. Thus, secondary maxima appear slightly dimmer than principle maxima. To illustrate this phenomenon, consider the simple case of just three (infinitesimally) narrow slits separated by distances d.

a) Demonstrate that the waves from all three slits are in phase (a principle maximum occurs) when $\theta = 0$, and also when $d \sin \theta = \lambda$.
b) Demonstrate that the waves from all three slits sum to zero (a dark fringe occurs) when $d \sin \theta = \lambda/3$, and also when $d \sin \theta = 2\lambda/3$. This implies that a secondary maximum must occur between these two dark fringes.
c) Argue that, as the number of slits is increased, (i) the dark fringes appearing between principle maxima become more and more numerous, (ii) the secondary maxima thus become more numerous and faint, and (iii) the principle maxima become narrower and brighter.

---

[8] Again technically speaking, the slits act as line-sources.

Ex. 20.5 (Two slits of finite width).   Let us return to the problem of two slits, but this time let us account for the fact that the slits themselves have a finite width. Consider an opaque barrier containing two vertical slits. A monochromatic light source producing light with wavelength $\lambda = 632.8$ nm strikes the slits. Suppose that the slits each have a width of $w = 0.001$ mm, and that they are separated by a distance $d = 0.1$ mm. Ten centimeters behind the opaque barrier is a screen which is used to observe the resulting interference pattern.

a) What is the width of the central bright maximum that arises as a consequence of the finite width of the left slit? Of the right slit? Are these two maxima overlapping? (Hint: look back at Ex. 14.2.)

b) Now consider the locations of the fringes that arise from the fact that there are two slits separated by a distance d. How many of these fringes lie within the central bright fringe(s) calculated in the previous question?

20.5   Vocabulary

1. Constitution
2. Luminous
3. Pervade
4. Induce
5. Emanate
6. Affection
7. Homogeneous
8. Inflect
9. Diverge
10. Analogous
11. Interstices
12. Supposition
13. Porosity
14. Undulate
15. Momentum
16. Polarity
17. Refrangible
18. Diffraction
19. Advocate
20. Interference
21. Redouble
22. Aperture
23. Spectrum
24. Tawny
25. Intervene
26. Coronae
27. Furrow
28. Encroach

Chapter 21
Films, Bubbles and Rainbows

When a film of soapy water is stretched over a wine glass, and placed in a vertical position, its upper edge becomes extremely thin, and appears nearly black, while the parts below are divided by horizontal lines into a series of coloured bands.

—Thomas Young

21.1 Introduction

In the present chapter, we continue our study of Thomas Young's *On the nature of light and colors* from his *Course of Lectures on Natural Philosophy and the Mechanical Arts*. Previously, Young considered the relative merits of the wave and particle theories of light. In particular, he pointed out how the wave theory, unlike the particle theory, is able to explain the interference fringes appearing in shadows cast by the edges of objects. Perhaps most significantly, by measuring the spacing of the interference fringes which appear when light passes through two slits in a barrier, Young was able to measure the wavelength of light. This an amazing experimental feat, since the wavelength of visible light is 100 times smaller than the width of a typical human hair. Perhaps there is a more commonplace phenomenon which supports the wave theory?

21.2 Reading: Young, *On the Nature of Light and Colours*

Young, T., *Course of lectures on natural philosophy and the mechanical arts*, vol. 1, Printed for Joseph Johnson, London, 1807. Lecture 39, continued.

A still more common and convenient method, of exhibiting the effects of the mutual interference of light, is afforded us by the colours of the thin plates of transparent substances. The lights are here derived from the successive partial reflections produced by the upper and under surface of the plate, or when the plate is viewed by transmitted light, from the direct beam which is simply refracted, and that portion of it which is twice reflected within the plate. The appearance in the latter case is much

© Springer International Publishing Switzerland 2016
K. Kuehn, *A Student's Guide Through the Great Physics Texts*,
Undergraduate Lecture Notes in Physics, DOI 10.1007/978-3-319-21816-8_21

less striking than in the former, because the light thus affected is only a small portion of the whole beam, with which it is mixed; while in the former the two reflected portions are nearly of equal intensity, and may be separated from all other light tending to overpower them. In both cases, when the plate is gradually reduced in thickness to an extremely thin edge, the order of colours may be precisely the same as in the stripes and coronae already described; their distance only varying when the surfaces of the plate, instead of being plane, are concave, as it frequently happens in such experiments. The scale of an oxid, which is often formed by the effect of heat on the surface of a metal, in particular of iron, affords us an example of such a series formed in reflected light: this scale is at first inconceivably thin, and destroys none of the light reflected, it soon, however, begins to be of a dull yellow, which changes to red, and then to crimson and blue, after which the effect is destroyed by the opacity which the oxid acquires. Usually however, the series of colours produced in reflected light follows an order somewhat different: the scale of oxid is denser than air, and the iron below than the oxid; but where the mediums above and below the plate are either both rarer or both denser than itself, the different natures of the reflections at its different surfaces appear to produce a modification in the state of the undulations, and the infinitely thin edge of the plate becomes black instead of white, one of the portions of light at once destroying the other, instead of cooperating with it. Thus when a film of soapy water is stretched over a wine glass, and placed in a vertical position, its upper edge becomes extremely thin, and appears nearly black, while the parts below are divided by horizontal lines into a series of coloured bands; and when two glasses, one of which is slightly convex, are pressed together with some force, the plate of air between them exhibits the appearance of coloured rings, beginning from a black spot at the centre, and becoming narrower and narrower, as the curved figure of the glass causes the thickness of the plate of air to increase more and more rapidly. The black is succeeded by a violet, so faint as to be scarcely perceptible; next to this is an orange yellow, and then crimson and blue. When water, or any other fluid, is substituted for the air between the glasses, the rings appear where the thickness is as much less than that of the plate of air, as the refractive density of the fluid is greater; a circumstance which necessarily follows from the proportion of the velocities with which light must, upon the Huygenian hypothesis, be supposed to move in different mediums. It is also a consequence equally necessary in this theory, and equally inconsistent with all others, that when the direction of the light is oblique, the effect of a thicker plate must be the same as that of a thinner plate, when the light falls perpendicularly upon it; the difference of the paths described by the different portions of light precisely corresponding with the observed phenomena (Figs. 21.1, 21.2 and 21.3).

Sir Isaac Newton supposed the colours of natural bodies in general to be similar to these colours of thin plates, and to be governed by the magnitude of their particles. If this opinion were universally true, we might always separate the colours of natural bodies by refraction into a number of different portions, with dark spaces intervening; for every part of a thin plate, which exhibits the appearance of colour, affords such a divided spectrum, when viewed through a prism. There are accordingly many natural colours in which such a separation may be observed; one of the

Fig. 21.1 Analysis of the colours of thin plates seen by reflection, beginning from *black*. A *line* drawn across the curved fringes would show the portions into which the light of any part is divided when viewed through a prism

Fig. 21.2 The coloured stripes of a film of soapy water, covering a wine glass

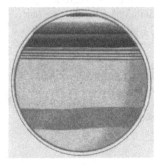

Fig. 21.3 The colours of a thin plate of air or water contained between a convex and a plane glass, as seen by reflection

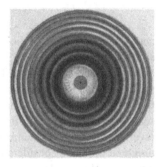

most remarkable of them is that of blue glass, probably coloured with cobalt, which becomes divided into seven distinct portions. It seems, however, impossible to suppose the production of natural colours perfectly identical with those of thin plates, on account of the known minuteness of the particles of colouring bodies, unless the refractive density of these particles be at least 20 or 30 times as great as that of glass or water; which is indeed not at all improbable with respect to the ultimate atoms of bodies, but difficult to believe with respect to any of their arrangements constituting the diversities of material substances.

The colours of mixed plates constitute a distinct variety of the colours of thin plates, which has not been commonly observed. They appear when the interstice between two glasses, nearly in contact, is filled with a great number of minute portions of two different substances, as water and air, oil and air, or oil and water: the light, which passes through one of the mediums, moving with a greater velocity, anticipates the light passing through the other; and their effects on the eye being confounded and combined, their interference produces an appearance of colours

Fig. 21.4 The colours of
a mixed plate; as seen by
partially greasing a lens a
little convex, and a flat glass,
and holding them together
between the eye and the edge
of a dark object. One half of
the series begins from *white*,
the other from *black*, and each
colour is the contrast to that
of the opposite half of the ring

Fig. 21.5 The composition
of the colours of the primary
rainbow, when attended by
supernumerary bows

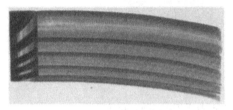

nearly similar to those of the colours of simple thin plates, seen by transmission; but at much greater thicknesses, depending on the difference of the refractive densities of the substances employed. The effect is observed by holding the glasses between the eye and the termination of a bright object, and it is most conspicuous in the portion which is seen on the dark part beyond the object, being produced by the light scattered irregularly from the surfaces of the fluid. Here, however, the effects are inverted, the colours resembling those of the common thin plates, seen by reflection; and the same considerations on the nature of the reflections are applicable to both cases (Fig. 21.4).

The production of the supernumerary rainbows, which are sometimes seen within the primary and without the secondary bow, appears to be intimately connected with that of the colours of the thin plates. We have already seen that the light producing the ordinary rainbow is double, its intensity being only greatest at its termination, where the common bow appears, while the whole light is extended much more widely. The two portions concerned in its production must divide this light into fringes; but unless almost all the drops of a shower happen to be of the same magnitude, the effects of these fringes must be confounded and destroyed: in general, however, they must at least cooperate more or less in producing one dark fringe, which must cut off the common rainbow much more abruptly than it would otherwise have been terminated, and consequently assist the distinctness of its colours. The magnitude of the drops of rain required for producing such of these rainbows as are usually observed, is between the 50th and 100th of an inch: they become gradually narrower as they are more remote from the common rainbows, nearly in the same proportions as the external fringes of a shadow, or the rings seen in a concave plate (Fig. 21.5).

Fig. 21.6 The colours of concave mirrors. The *small circles* in the *middle white* ring represent the aperture by which the light is admitted, and its image; the coloured rings are formed by the light irregularly dissipated before and after reflection

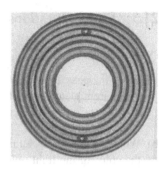

The last species of the colours of double lights, which it will be necessary to notice, constitutes those which have been denominated, from Newton's experiments, the colours of thick plates, but which may be called, with more propriety, the colours of concave mirrors. The anterior surface of a mirror of glass, or any other transparent surface placed before a speculum of metal, dissipates irregularly in every direction two portions of light, one before, and the other after its reflection. When the light falls obliquely on the mirror, being admitted through an aperture near the centre of its curvature, it is easy to show, from the laws of reflection, that the two portions, thus dissipated, will conspire in their effects, throughout the circumference of a circle, passing through the aperture; this circle will consequently be white, and it will be surrounded with circles of colours very nearly at equal distances, resembling the stripes produced by diffraction. The analogy between these colours and those of thin plates is by no means so close as Newton supposed it; since the effect of a plate of any considerable thickness must be absolutely lost in white light, after 10 or 12 alternations of colours at most, while these effects would require the whole process to remain unaltered, or rather to be renewed, after many thousands or millions of changes (Fig. 21.6).

It is presumed, that the accuracy, with which the general law of the interference of light has been shown to be applicable to so great a variety of facts, in circumstances the most dissimilar, will be allowed to establish its validity in the most satisfactory manner. The full confirmation or decided rejection of the theory, by which this law was first suggested, can be expected from time and experience alone; if it be confuted, our prospects will again be confined within their ancient limits, but if it be fully established, we may expect an ample extension of our views with the operations of nature, by means of our acquaintance with a medium, so powerful and so universal, as that to which the propagation of light must be attributed.

21.3   Study Questions

QUES. 21.1. Do the colors of thin plates support the undulatory theory of light?

a) What happens to light which falls perpendicularly upon the surface of a thin transparent plate? Is all of the light transmitted into the medium? Does it all pass through the other side of the plate? What is observed as the thickness of the plate is decreased?
b) What is the color of a thin layer of oxide which has formed atop iron? What is the color of the light reflected from the infinitely thin edge of a layer of soapy water stretched across the rim of a wine glass? What other experiments illustrate similar effects?
c) Can the emanation theory of light account adequately for these effects? What does Young conclude for all these experiments? Is he an advocate of æther? Are his conclusions justified?

21.4   Exercises

EX. 21.1 (SOAP FILM INTERFERENCE). Consider a soap film which is stretched across the rim of a wine glass; a cross-section of the contact point between the glass and film is depicted schematically in Fig. 21.7. Due to the weight of the film, it is thickest at its center, and thinnest near its edges. Suppose that the film makes an angle $\phi = 0.1°$ near the rim of the glass, and that it is illuminated with white light from above. When looking down at the soap film, a circular blue fringe appears at distance AB from the rim, and a circular red fringe appears at distance AC from the rim. The distance BC is 80 microns. First, what color is the film where it touches the rim at A? What does this imply about the phase difference between the waves reflected from the top and bottom surfaces of the film? Finally, can you determine the wavelengths of blue and red light from the information provided?

EX. 21.2 (OPTICAL PATH LENGTH). The speed of light in a particular medium is characterized by the *refractive index* of the medium (Eq. 16.1). Since the speed of light is related to its wavelength and frequency (see Eq. 14.1), one might expect that the frequency or the wavelength (or perhaps both) are different in different media. The frequency of a particular color of light, however, does not change when the light travels from one medium to another; only the wavelength does.[1] As an exercise, suppose that a monochromatic source of red light illuminates a thin sheet of agate, whose refractive index is 1.53 and whose thickness is 0.01 mm. (a) How many wavelengths of light fit inside the agate? (b) How many would fit inside an

---

[1] This must be the case, for otherwise a point located on the boundary between two media would need to be vibrating at two different frequencies as the light passes across the boundary.

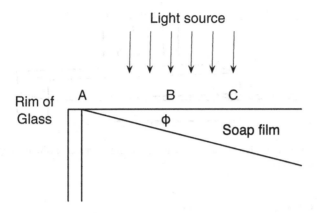

Fig. 21.7 White light incident upon a soap film clinging to the rim of a glass

identical thickness of air? (ANSWER: 14.3) (c) What is the difference in the number of wavelengths, $\Delta N = N_{agate} - N_{air}$? The *optical path length difference* provides a comparison of the effective thicknesses of two different media in terms of the different number of wavelengths of light that fit inside the two media. This can be expressed, for example, using the formula $\lambda_{air} \Delta N$. (d) Now obtain a formula which expresses the optical path length difference of two media in terms of the common physical thickness of the two media, d, and the indices of refraction of the two media.

EX. 21.3 (AIR GAP INTERFERENCE LABORATORY). The wavelength of light may be measured with no more than two rigid rectangular clear glass prisms, some thin paper, rubber bands and a vernier caliper. First, you will need to measure the thickness of a single sheet of paper. To do this, stack several sheets and measure the height of the stack with the caliper. Next, lay the two glass prisms together lengthwise bind them together with a rubber band wrapped around each end; but not before inserting a single sheet of the paper a few millimeters into one end of your "prism sandwich." In this way, you should have establish a thin wedge of air between two glass prisms which are themselves in contact along one edge (see Fig. 21.8). If you have done this properly, you should now be able to observe faint, nearly parallel interference fringes when looking down at the apparatus from where a source of white light illuminates the prisms. By counting the number of fringes of a particular color observed in the span of, say, one centimeter, you should be able to deduce the wavelength of the light. (Hint: Your calculation will involve careful consideration of the interference of light reflecting from the top and bottom of the air gap between the prisms.)

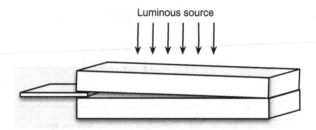

Fig. 21.8 A thin air wedge is formed between two rectangular glass prisms whose edge is propped apart by a sheet of paper

21.5   Vocabulary

1. Fringe
2. Oxid
3. Prism
4. Concave
5. Anterior
6. Oblique
7. Emanation
8. Corpuscle
9. Refraction
10. Stratum
11. Dispersion
12. Explicable
13. Confute
14. Contiguous
15. Dilate
16. Spheroid

Chapter 22
Producing and Detecting Polarization

> *These experiments and reasoning, if only thoroughly studied and understood, will form a solid groundwork for the analysis of the splendid optical phenomena next to be considered.*
>
> —John Tyndall

22.1 Introduction

John Tyndall (1820–1893) was born into a protestant family in Leighlinbridge in southern Ireland. As a young man, he worked as a civil servant in the Irish Ordinance Survey and then the English Survey in Preston. He was dismissed in 1843 for his outspoken criticism of government corruption, so he went to work at a private surveyor's office. In 1847 he took a position as a mathematics teacher at Queenwood School in Hampshire. Drawn to natural philosophy, he moved to Germany and earned his doctoral degree at Marburg University under the guidance of Robert Bunsen. He then returned to England in 1851 to seek a university appointment, and in 1853 was finally awarded a Professorship in Natural Philosophy at the Royal Institution of London. Years later, he would succeed Michael Faraday as Superintendent of the Royal Institution. During his career, he published dozens of scientific articles and several books; he was well-known for his pioneering work on the scattering of light by atmospheric particles such as air pollutants. But Tyndall was perhaps best known for his popular scientific lectures which drew large audiences both in England and abroad. As a member of a group of intellectuals called the "X-Club"— along with biologist T.H. Huxley and philosopher Herbert Spencer—he argued for a philosophy of strict scientific naturalism subsidized by generous public funding. He forcefully opposed the natural theology of leading scientists such as James Clerk Maxwell and even Isaac Newton. In his famous 1874 Belfast Address before the British Association for the Advancement of Science, he exalted the work of Charles Darwin and argued that science "shall wrest from theology, the entire domain of cosmological theory."[1] Tyndall travelled to the United States in the 1870's where

---

[1] This controversial and influential lecture can be found in Tyndall, J., *Address Delivered Before The British Association Assembled At Belfast with Additions*, Longmans, Green, and Co., 1874.

© Springer International Publishing Switzerland 2016
K. Kuehn, *A Student's Guide Through the Great Physics Texts*,
Undergraduate Lecture Notes in Physics, DOI 10.1007/978-3-319-21816-8_22

he delivered popular lectures to packed houses in Boston, Philadelphia, Baltimore, Washington and New York. The readings included in the following chapters are drawn from a published version of these lectures in which Tyndall "sought to raise the Wave-theory of Light to adequate clearness in the reader's mind, and to show its power as an organizer of optical phenomena." His first lecture covers geometrical optics and philosophical topics such as the "Sterility of the Middle Ages." The second lecture covers wave optics and contains a critical review of Newton's particle theory of light. The third lecture begins with a discussion of the relationship between the architecture of crystals and how their structure might influence the polarization of light passing through them. We start our reading of Tyndall at this point (§4) in his third lecture (Fig. 22.1).

22.2   Reading: Tyndall, *Six Lectures on Light*

Tyndall, J., *Six lectures on light delivered in America in 1872–1878*, D. Appleton and Company, New York, 1886. Lecture III.

22.2.1   *Double Refraction of Light Explained by the Undulatory Theory*

The two elements of rapidity of propagation, both of sound and light, in any substance whatever, are *elasticity* and *density*, the speed increasing with the former and diminishing with the latter. The enormous velocity of light in stellar space is attainable because the ether is at the same time of infinitesimal density and of enormous elasticity. Now the ether surrounds the atoms of all bodies, but it is not independent of them. In ponderable matter it acts as if its density were increased without a proportionate increase of elasticity; and this accounts for the diminished velocity of light in refracting bodies. We here reach a point of cardinal importance. In virtue of the crystalline architecture that we have been considering, the ether in many crystals possesses different densities, and hence different elasticities, in two different directions; and the consequence is, that some of these media transmit light with two different velocities. But as refraction depends wholly upon the change of velocity on entering the refracting medium, and is greatest where the change of velocity is greatest, we have in many crystals two different refractions. By such crystals a beam of light is divided into two. This effect is called *double refraction.*

In ordinary water, for example, there is nothing in the grouping of the molecules to interfere with the perfect homogeneity of the ether; but, when water crystallizes to ice, the case is different. In a plate of ice the elasticity of the ether in a direction perpendicular to the surface of freezing is different from what it is parallel to the surface of freezing; ice is, therefore, a double refracting substance. Double refraction is displayed in a particularly impressive manner by Iceland spar, which is crystallized carbonate of lime. The difference of ethereal density in two directions in this crystal is very great, the separation of the beam into the two halves being, therefore, particularly striking.

I am unwilling to quit this subject before raising it to unmistakable clearness in your minds. The vibrations of light being transversal, the elasticity concerned in the propagation of any ray is the elasticity at right angles to the direction of propagation. In Iceland spar there is one direction round which the crystalline molecules are symmetrically built. This direction is called the axis of the crystal. In consequence of this symmetry the elasticity is the same in all directions perpendicular to the axis, and hence a ray transmitted along the axis suffers no double refraction. But the elasticity along the axis is greater than the elasticity at right angles to it. Consider, then, a system of waves crossing the crystal in a direction perpendicular to the axis. Two directions of vibration are open to such waves: the ether particles can vibrate parallel to the axis or perpendicular to it. *They do both*, and hence immediately divide themselves into two systems propagated with different velocities. Double refraction is the necessary consequence.

By means of Iceland spar cut in the proper direction, double refraction is capable of easy illustration. Causing the beam which builds the image of our carbon-points to pass through the spar, the single image is instantly divided into two. Projecting (by the lens E, Fig. 22.2) an image of the aperture (L) through which the light issues from the electric lamp, and introducing the spar (P), two luminous disks (E O) appear immediately upon the screen instead of one.

The two beams into which the spar divides the single incident-beam have been subjected to the closest examination. They do not behave alike. One of them obeys the ordinary law of refraction discovered by Snell, and is, therefore, called the *ordinary ray*: its index of refraction is 1.654. The other does not obey this law. Its index of refraction, for example, is not constant, but varies from a maximum of 1.654 to a minimum of 1.483; nor in this case do the incident and refracted rays always lie in the same plane. It is, therefore, called the *extraordinary ray*. In calc-spar, as just stated, the ordinary ray is the most refracted. One consequence of this merits a passing notice. Pour water and bisulphide of carbon into two cups of the same depth; the cup that contains the more strongly refracting liquid will appear shallower than

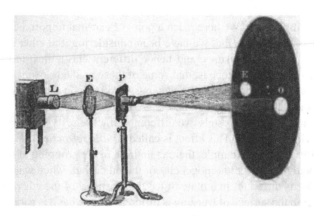

Fig. 22.2 Light from an aperture (L) focused by a lens (E) onto Iceland spar (P) is divided into an ordinary (O) and an extraordinary (E) beam.—[*K.K.*]

the other. Place a piece of Iceland spar over a dot of ink; two dots are seen, the one appearing nearer than the other to the eye. The nearest dot belongs to the most strongly-refracted ray, exactly as the nearest cup-bottom belongs to the most highly refracting liquid. When you turn the spar round, the extraordinary image of the dot rotates round the ordinary one, which remains fixed. This is also the deportment of our two disks upon the screen.

22.2.2   Polarization of Light Explained by the Undulatory Theory

The double refraction of Iceland spar was first treated in a work published by Erasmus Bartholinus, in 1669. The celebrated Huyghens sought to account for this phenomenon on the principles of the wave theory, and he succeeded in doing so. He, moreover, made highly important observations on the distinctive character of the two beams transmitted by the spar, admitting, with resigned candour, that he had not solved them, and leaving that solution to future times. Newton, reflecting on the observations of Huyghens, came to the conclusion that each of the beams transmitted by Iceland spar had two sides; and from the analogy of this *two-sidedness* with the *two-endedness* of a magnet, wherein consists its polarity, the two beams came subsequently to be described as *polarized.*

We may begin the study of the polarization of light, with ease and profit, by means of a crystal of tourmaline. But we must start with a clear conception of an ordinary beam of light. It has been already explained that the vibrations of the individual ether-particles are executed *across* the line of propagation. In the case of ordinary light we are to figure the ether-particles as vibrating in all directions, or azimuths, as it is sometimes expressed, across this line.

Now, in the case of a plate of tourmaline cut parallel to the axis of the crystal, a beam of light incident upon the plate is divided into two, the one vibrating parallel

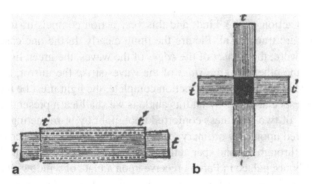

Fig. 22.3 Looking through stacked plates of tourmaline whose crystal axes are (**a**) parallel and (**b**) crossed.—[*K.K.*]

to the axis of the crystal, the other at right angles to the axis. The grouping of the molecules, and of the ether associated with the molecules, reduces all the vibrations incident upon the crystal to these two directions. One of these beams, namely, that whose vibrations are perpendicular to the axis, is quenched with exceeding rapidity by the tourmaline. To such vibrations many specimens of the crystal are highly opaque; so that, after having passed through a very small thickness of the tourmaline, the light emerges with all its vibrations reduced to a single plane. In this condition it is what we call *plane polarized light.*

A moment's reflection will show that, if what is here stated be correct, on placing a second plate of tourmaline with its axis parallel to the first, the light will pass through both; but that, if the axes be crossed, the light that passes through the one plate will be quenched by the other, a total interception of the light being the consequence. Let us test this conclusion by experiment. The image of a plate of tourmaline (*tt*, Fig. 22.3a) is now before you. I place parallel to it another plate (*t't'*): the green of the crystal is a little deepened, nothing more; this agrees with our conclusion. By means of an endless screw, I now turn one of tile crystals gradually round, and you observe that as long as the two plates are oblique to each other, a certain portion of light gets through; but that when they are at right angles to with each other, the space common to both is a space of darkness (Fig. 22.3b). Our conclusion, arrived at prior to experiment, is thus verified.

Let us now return to a single plate; and here let me say that it is on the green light transmitted by the tourmaline that you are to fix your attention. We have to illustrate the two-sidedness of that green light, in contrast to the all-sidedness of ordinary light. The light surrounding the green image, being ordinary light, is reflected by a plane glass mirror in all directions; the green light, on the contrary, is not so reflected. The image of the tourmaline is now horizontal: reflected upwards, it is still green; reflected sideways, the image is reduced to blackness, because of the incompetency of the green light to be reflected in this direction. Making the plate of tourmaline vertical, and reflecting it as before, it is in the upper image that the light is quenched; in the side image you have now the green. This is a result of the greatest significance. If the vibrations of light were longitudinal, like those of sound, you could have no action of this kind; and this very action compels us to assume that the vibrations are transversal. Picture the thing clearly. In the one case the mirror receives, as it were, the impact of the *edges* of the waves, the green light being then quenched. In the other case the *sides* of the waves strike the mirror, and the green light is reflected. To render the extinction complete, the light must be received upon the mirror at a special angle. What this angle is we shall learn presently.

The quality of two-sidedness conferred upon light by bi-refracting crystals may also be conferred upon it by ordinary reflection. Malus made this discovery in 1808, while looking through Iceland spar at the light of the sun reflected from the windows of the Luxembourg palace in Paris. I receive upon a plate of window-glass the beam from our lamp; a great portion of the light reflected from the glass is polarized. The vibrations of this reflected beam are executed, for the most part, parallel to the surface of the glass, and when the glass is held so that the beam shall make an angle of 58° with the perpendicular to the glass, the *whole* of the reflected beam is polarized. It was at this angle that the image of the tourmaline was completely quenched in our former experiment. It is called *the polarizing angle*.

Sir David Brewster proved the angle of polarization of a medium to be that particular angle at which the refracted and reflected rays inclose a right angle.[2] The polarizing angle augments with the index of refraction. For water it is $52\frac{1}{2}°$; for glass, as already stated, 58°; while for diamond it is 68°.

And now let us try to make substantially the experiment of Malus. The beam from the lamp is received at the proper angle upon a plate of glass and reflected through the spar. Instead of two images, you see but one. So that the light, when polarized, as it now is by reflection, can only get through the spar in one direction, and consequently produce but one image. Why is this? In the Iceland spar, as in the tourmaline, all the vibrations of the ordinary light are reduced to two planes at right angles to each other; but, unlike the tourmaline, both beams are transmitted

---

[2] This beautiful law is usually thus expressed: *The index of refraction of any substance is the tangent of its polarizing angle.* With the aid of this law and an apparatus similar to that [shown in 4 of Tyndall's first lecture, reproduced here— *K.K.*],

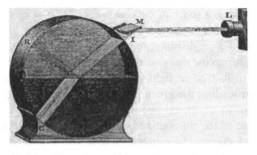

we can readily determine the index of refracting any liquid. The refracted and reflected beams being visible, they can readily be caused to enclose a right angle. The polarizing angle of the liquid may be thus found with the sharpest precision. It is then only necessary to seek out its natural tangent to obtain the index of refraction.

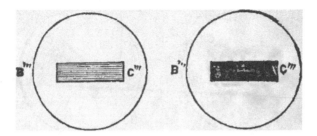

Fig. 22.4 When plates of tourmaline, with their axes parallel, are placed before each of the beams emerging from Iceland spar, one of the beams is transmitted, the other is blocked.—[*K.K.*]

with equal facility by the spar. The two beams, in short, emergent from the spar, are polarized, their directions of vibration being at right angles to each other. It is important to remember this. When, therefore, the light was polarized by reflection, the direction of vibration in the spar which coincided with the direction of vibration of the polarized beam transmitted it, and that direction only. Only one image, therefore, was possible under the conditions.

You will now observe that such logic as connects our experiments is simply a transcript of the logic of Nature. On the screen before you are two disks of light produced by the double refraction of Iceland spar. They are, as you know, two images of the aperture through which the light issues from the camera. Placing the tourmaline in front of the aperture, two images of the crystal will also be obtained; but now let us reason out beforehand what is to be expected from this experiment. The light emergent from the tourmaline is polarized. Placing the crystal with its axis horizontal, the vibration of its transmitted light will be horizontal. Now the spar, as already stated, has two directions of vibration, one of which at the present moment is vertical, the other horizontal. What are we to conclude? That the green light will be transmitted along the latter, which is parallel to the axis of the tourmaline, and not along the former, which is perpendicular to that axis. Hence we may infer that one image of the tourmaline will show the ordinary green light of the crystal, while the other image will be black. Tested by experiment, our reasoning is verified to the letter (Fig. 22.4).

Let us push our test still further. By means of an endless screw, the crystal can be turned 90° round. The black image, as I turn, becomes gradually brighter, and the bright one gradually darker; at an angle of 45° both images are equally bright (Fig. 22.5); while, when 90° have been obtained, the axis of tho crystal being then vertical, the bright and black images have changed places exactly as reasoning would have led us to suppose (Fig. 22.6).

Given the two beams transmitted through Iceland spar, it is perfectly manifest that we have it in our power to determine instantly, by means of a plate of tourmaline, the directions in which the ether-particles vibrate in the two beams. The double refracting spar might be placed in any position whatever. A minute's trial with the tourmaline would enable you to determine the position which yields a black and a bright image, and from this you would at once infer the directions of vibration.

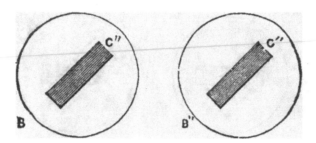

Fig. 22.5 As in Fig. 22.4, but with the axes of the tourmaline rotated by 45° with respect to the Iceland spar axes.—[*K.K.*]

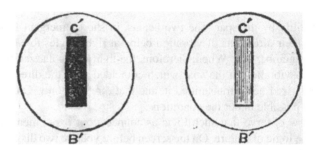

Fig. 22.6 As in Fig. 22.4, but with the axes of the tourmaline rotated by 90° with respect to the Iceland spar axes.—[*K.K.*]

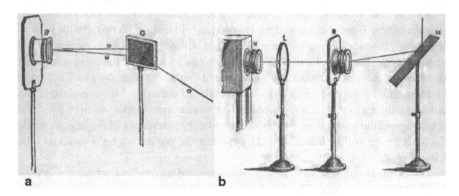

Fig. 22.7 (*B* is the bi-refracting spar, dividing the incident light into the two beams *o* and *e*, *G* is the mirror.) The beam is here **a** reflected *laterally* (**a**). When the reflection is *upwards*, the other beam is reflected as shown in (**b**)

Let us reason still further together. The two beams from the spar being thus polarized, it is plain that if they be suitably received upon a plate of glass at the polarizing angle, one of them will be reflected, the other not. This is a simple inference from our previous knowledge; but you observe that the inference is justified by experiment. (Figs. 22.7a and b.)

I have said that the whole of the beam reflected from glass at the polarizing angle is polarized; a word must now be added regarding the far larger portion of the light which is *transmitted* by the glass. The transmitted beam contains a quantity of polarized light equal to the reflected beam: but this is only a fraction of the whole transmitted light. By taking two plates of glass instead of one, we augment the quantity of the transmitted polarized light; and by taking *a bundle* of plates, we so increase the quantity as to render the transmitted beam, for all practical purposes, *perfectly* polarized. Indeed, bundles of glass plates are often employed as a means of furnishing polarized light. Interposing such a bundle at the proper angle into the paths of the two beams emergent from Iceland spar, that which, in the last experiment, failed to be reflected, is here transmitted. The plane of vibration of this transmitted light is at right angles to that of the reflected light.

One word more. When the tourmalines are crossed, the space where they cross each other is black. But we have seen that the least obliquity on the part of the crystals permits light to get through both. Now suppose, when the two plates are crossed, that we interpose a third plate of tourmaline between them, with its axis oblique to both. A portion of the light transmitted by the first plate will get through this intermediate one. But, after it has got through, *its plane of vibration is changed*: it is no longer perpendicular to the axis of the crystal in front. Hence it will get through that crystal. Thus, by pure reasoning, we infer that the interposition of a third plate of tourmaline will in part abolish the darkness produced by the perpendicular crossing of the other two plates. I have not a third plate of tourmaline; but the talc or mica which you employ in your stoves is a more convenient substance, which acts in the same way. Between the crossed tourmalines, I introduce a film of this crystal with its axis oblique to theirs. You see the edge of the film slowly descending, and as it descends, light takes the place of darkness. The darkness, in fact, seems scraped away, as if it were something material. This effect bas been called, naturally but improperly, *depolarization*. Its proper meaning will be disclosed in our next lecture.

These experiments and reasoning, if only thoroughly studied and understood, will form a solid groundwork for the analysis of the splendid optical phenomena next to be considered.

22.3 Study Questions

QUES. 22.1. Why do some types of substances exhibit double refraction while others do not?

a) What two properties dictate the speed of sound waves through a substance? What does this imply about the nature of the ether filling both interstellar space and the interior of ponderable bodies?
b) What effect does crystalline structure have on the speed of light in a refracting medium? Does water exhibit double refraction? Why or why not? What about ice?

c) Why does Iceland Spar exhibit double refraction? Does it always? What is its index of refraction? Does the extraordinary beam from Iceland spar obey the law of refraction? And which beam is more strongly refracted, the ordinary or extraordinary beam?

d) How can one determine the polarization of the ordinary and extraordinary beams of light passing through an Iceland crystal?

QUES. 22.2. What does it mean for a ray of light to be plane polarized?

a) Where does the term "polarization" come from? How did Newton explain the double refraction of Iceland spar?

b) According to the wave theory, what is the relationship between the motion of ether particles and the polarization of a light beam? Is an ordinary beam of light polarized?

c) What effect does a crystal of tourmaline have on a beam of light? What effect do two crystals with crossed axes have on a beam of light?

d) What happens when a third crystal of tourmaline is inserted obliquely between two crystals of tourmaline whose axes are crossed? Is tourmaline the only substance which exhibits this effect?

QUES. 22.3. What effect does reflection have on a beam of light?

a) What did Malus observe when viewing the light reflected from the windows of the Luxembourg palace?

b) What is the polarizing angle of plate glass? Of water? Of diamond? What is the relationship between the polarizing angle of a substance and its index of refraction?

c) Under what conditions can a beam of light transmitted through a plate of glass be perfectly polarized?

22.4   Exercises

EX. 22.1 (BREWSTER'S ANGLE AND REFRACTIVE INDEX). Suppose that an unpolarized beam of light, when reflected from a transparent material, is entirely polarized when the angle of incidence is 60°. (a) In which orientation (with respect to the plane of incidence) are the reflected and transmitted beams polarized? (b) What is the angle of refraction of the transmitted beam? (c) Starting from the law of refraction, prove mathematically that the index of refraction of any substance is the tangent of its polarizing angle. (d) What is the index of refraction of this particular material?

EX. 22.2 (BREWSTER'S ANGLE LABORATORY). Using an available light source and a polarizing filter, determine the index of refraction of a plate of glass (or other transparent material) by measuring the polarization angle.[3] Compare your results to either (i) a published value of the refractive index of this material or (ii) another laboratory measurement of the refractive index of this material (such as the pin-method used for Ex. 12.9).

22.5 Vocabulary

1. Rapidity
2. Infinitesimal
3. Ponderable
4. Homogeneity
5. Candour
6. Polarized
7. Tourmaline
8. Azimuth
9. Transverse
10. Quench
11. Confer
12. Augment

---

[3] Brewster, D., On the laws which regulate the polarisation of light by reflexion from transparent bodies., *Philosophical Transactions of the Royal Society of London (1776-1886)*, *105*, 125–159, 1815.

Chapter 23
Crystal Symmetry and Light Rotation

> *The crystal thus cut possesses the extraordinary power of*
> *twisting the plane of vibration of a polarized ray to an extent*
> *dependent on the thickness of the crystal.*
>
> —John Tyndall

23.1   Introduction

In the third of his *Six lectures on light*, John Tyndall described how a light beam is
actually a transverse wave whose polarization may be determined in one of two
ways. The first method, discovered by Malus and perfected by Brewster, is by
reflecting the beam from a transparent substance at a special *polarizing angle*. In
this unique case, if the incident beam is polarized in the plane of incidence, then
the beam is entirely transmitted by refraction; if the beam is polarized perpendicular to the plane of incidence, then the beam is entirely reflected. Notably, if the
incident beam is initially unpolarized, then the polarizing angle of a particular substance is such that the reflected and transmitted beams are perfectly orthogonal to
one another. The second method of determining a beam's polarization is by passing the beam through a transparent tourmaline crystal. If its plane of polarization is
perpendicular to the crystal axis of the tourmaline, then the beam will be blocked;
if its plane is parallel, then it will pass. Moreover, when a thin crystal of mica is
inserted at an oblique angle between two crystals of tourmaline whose axes are
themselves crossed, light is—perhaps surprisingly—transmitted through all three.
It is this curious phenomenon which Tyndall now endeavors to explain in his fourth
lecture.

23.2   Reading: Tyndall, *Six Lectures on Light*

Tyndall, J., *Six lectures on light delivered in America in 1872–1878*, D. Appleton
and Company, New York, 1886. Lecture IV.

© Springer International Publishing Switzerland 2016   269
K. Kuehn, *A Student's Guide Through the Great Physics Texts,*
Undergraduate Lecture Notes in Physics, DOI 10.1007/978-3-319-21816-8_23

23.2.1 Action of Crystals on Polarized Light: The Nicol Prism

WE have this evening to examine and illustrate the chromatic phenomena produced by the action of crystals, and double-refracting bodies generally, upon polarized light, and to apply the Undulatory Theory to their elucidation. For a long time investigators were compelled to employ plates of tourmaline for this purpose, and the progress they made with so defective a means of inquiry is astonishing. But these men had their hearts in their work, and were on this account enabled to extract great results from small instrumental appliances. But for our present purpose we need far larger apparatus; and, happily, in these later times this need has been to a great extent satisfied. We have seen and examined the two beams emergent from Iceland spar, and have proved them to be polarized. If, at the sacrifice of half the light, we could abolish one of these, the other would place at our disposal a beam of polarized light, incomparably stronger than any attainable from tourmaline.

The beams, as you know, are refracted differently, and from this, as made plain in Lecture I, we are able to infer that the one may be totally reflected, when the other is not. An able optician, named Nicol, cut a crystal of Iceland spar in two halves in a certain direction. He polished the severed surfaces, and reunited them by Canada balsam, the surface of union being so inclined to the beam traversing the spar that the ordinary ray, which is the most highly refracted, was totally reflected by the balsam, while the extraordinary ray was permitted to pass on.

Let bx, cy (Fig. 23.1) represent the section of an elongated rhomb of Iceland spar cloven from the crystal. Let this rhomb be cut along the plane bc; and the two severed surfaces, after having been polished, reunited by Canada balsam. We learned, in our first lecture, that total reflection only takes place when a ray seeks to escape from a more refracting to a less refracting medium, and that it always, under these circumstances, takes place when the obliquity is sufficient. Now the refractive index of Iceland spar is, for the extraordinary ray less, and for the ordinary greater, than for Canada balsam. Hence, in passing from the spar to the balsam, the extraordinary ray passes from a less refracting to a more refracting medium, where total reflection cannot occur; while the ordinary ray passes from a more refracting to a less refracting medium, where total reflection can occur. The requisite obliquity is secured by making the rhomb of such a length that the plane of which bc is the section shall be perpendicular, or nearly so, to the two end surfaces of the rhomb bx, cy.

The invention of the Nicol prism was a great step in practical optics, and quite recently such prisms have been constructed of a size and purity which enable audiences like the present to witness the chromatic phenomena of polarized light to a degree altogether unattainable a short time ago. The two prisms here before you belong to my excellent friend Mr. William Spottiswoode, and they were manufactured by Mr. Ladd. I have with me another pair of very noble prisms, still larger

Fig. 23.1 An ordinary beam
of light is reflected from the
joined surfaces of Iceland
spar within a Nicol's prism.
—[K.K.]

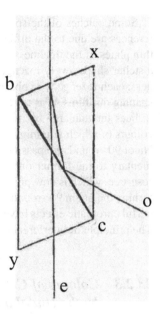

than these, manufactured for me by Mr. Browning, who has gained so high and
well-merited a reputation in the construction of spectroscopes.[1]

23.2.2   Colours of Films of Selenite in Polarized Light

These two Nicol prisms play the same part as the two plates of tourmaline. Placed
with their directions of vibration parallel, the light passes through both; while
when these directions are crossed the light is quenched. Introducing a film of mica
between the prisms, the light, as in the case of the tourmaline, is restored. But notice,
when the film of mica is *thin* you have sometimes not only light, but *coloured* light.
Our work for some time to come will consist of the examination of such colours.
With this view, I will take a representative crystal, one easily dealt with, because
it cleaves with great facility—the crystal gypsum, or selenite, which is crystallized
sulphate of lime. Between the crossed Nicols I place a thick plate of this crystal;
like the mica, it restores the light, but it produces no colour. With my penknife I
take a thin splinter from the crystal and place it between the prisms; the image of
the splinter glows with the richest colours. Turning the prism in front, these colours
gradually fade and disappear, but, by continuing the rotation until the vibrating sections of the prisms are parallel to each other, vivid colours again arise, but these
colours are complementary to the former ones.

---

[1] The largest and purest prism hitherto made has been recently constructed for Mr. Spottiswoode
by Messrs. Tisley & Spiller.

Some patches of the splinter appear of one colour, some of another. These differences are due to the different thicknesses of the film. As in the case of Hooke's thin plates, if the thickness be uniform, the colour is uniform. Here, for instance, is a stellar shape, every lozenge of the star being a film of gypsum of uniform thickness: each lozenge, you observe, shows a brilliant and uniform colour. It is easy, by shaping our films so as to represent flowers or other objects, to exhibit such objects in hues unattainable by art. Here, for example, is a specimen of heart's-ease, the colours of which you might safely defy the artist to reproduce. By turning the front Nicol 90° round, we pass through a colourless phase to a series of colours complementary to the former ones. This change is still more strikingly represented by a rose-tree, which is now presented in its natural hues—a red flower and green leaves; turning the prism 90° round, we obtain a green flower and red leaves. All these wonderful chromatic effects have definite mechanical causes in the motions of the ether. The principle of interference duly applied and interpreted explains them all.

23.2.3 Colours of Crystals in Polarized Light Explained by the Undulatory Theory

By this time you have learned that the word 'light' may be used in two different senses; it may mean the impression made upon consciousness, or it may mean the physical agent which makes the impression. It is with the agent that we have to occupy ourselves at present. That agent is a substance which fills all space, and surrounds the atoms and molecules of bodies. To this interstellar and interatomic medium definite mechanical properties are ascribed, and we deal with it in our reasonings and calculations as a body possessed of these properties. In mechanics we have the composition and resolution of forces and of motions, extending to the composition and resolution of vibrations. We treat the luminiferous ether on mechanical principles, and, from the composition, resolution, and interference of its vibrations we deduce all the phenomena displayed by crystals in polarized light.

Let us take, as an example, the crystal of tourmaline, with which we are now so familiar. Let a vibration cross this crystal oblique to its axis. Experiment has assured us that a portion of the light will pass through. The quantity which passes we determine in this way. Let AB (Fig. 23.2) be the axis of the tourmaline, and let ab represent the amplitude of the ethereal vibration before it reaches AB. From a and b let the two perpendiculars ac and bd be drawn upon the axis: then cd will be the amplitude of the transmitted vibration.

I shall immediately ask you to follow me while I endeavour to explain the effects observed when a film of gypsum is placed between the two Nicol's prisms. But, prior to this, it will be desirable to establish still further the analogy between the action of the prisms and that of the two plates of tourmaline. The magnified images of these plates, with their axes at right-angles to each other, are now before you. Introducing between them a film of selenite, you observe that by turning the film round it may be placed in a position where it has no power to abolish the darkness of the superposed

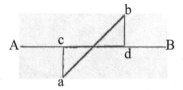

Fig. 23.2 Only the horizontal component of an oblique light vibration is transmitted through a tourmaline crystal whose axis is horizontal.—[*K.K.*]

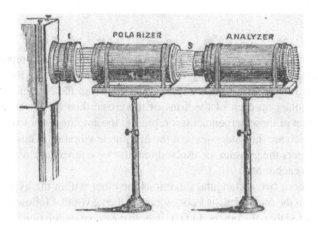

Fig. 23.3 Light from an electric lamp illuminates a thin film of transparent gypsum inserted between two Nicol prisms; the first is the *polarizer*, the second is the *analyzer*.—[*K.K.*]

portions of the tourmalines. Why is this? The answer is, that in the gypsum there are two directions, at right angles to each other, in which alone vibrations can take place, and that in our present experiment one of these directions is parallel to one of the axes of the tourmaline, and the other parallel to the other axis. When this is the case, the film exercises no sensible action upon the light. But now I turn the film so as to render its directions of vibration *oblique* to the two tourmaline axes; then, you see it exercises the power, demonstrated in the last lecture, of restoring the light.

Let us now mount our Nicol's prisms, and cross them as we crossed the tourmalines. Introducing our film of gypsum between them, you notice that in one particular position the film has no power whatever over the field of view. But, when the film is turned a little way round, the light passes. We have now to understand the mechanism by which this is effected.

Firstly, then, we have a prism which receives the light from the electric lamp, and which is called the *polarizer*. Then we have the plate of gypsum (supposed to be placed at *S*, Fig. 23.3), and then the prism in front, which is called the *analyzer*. On its emergence from the first prism, the light is polarized; and, in the particular case now before us, its vibrations are executed in a horizontal plane. We have to examine what occurs when the two directions of vibration in the gypsum are oblique

Fig. 23.4 Horizontally polarized light vibrations from
the first Nicol depicted in
Fig. 23.3 are resolved into perpendicular components along
the fast and slow axes of an
obliquely oriented transparent
gypsum crystal.—[*K.K.*]

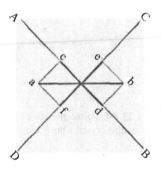

to the horizon. Draw a rectangular cross (*AB*, *CD*, Fig. 23.4) to represent these
two directions. Draw a line (*ab*) to represent the amplitude of the vibration on the
emergence of the light from the first Nicol. Let fall from the two ends of this line
two perpendiculars on each of the arms of the cross; then the distances (*cd*, *ef*)
between the feet of these perpendiculars represent the amplitudes of two rectangular
vibrations, which are the *components* of the first single vibration. Thus the polarized
ray, when it enters the gypsum, is resolved into its two equivalents, which vibrate at
right angles to each other.

In one of these two rectangular directions the ether within the gypsum is more
sluggish than in the other; and, as a consequence, the waves that follow this direction
are more retarded than the others. In fact, in both cases the undulations are shortened
when they enter the gypsum, but in the one case they are more shortened than in
the other. You can readily imagine that in this way the one system of waves may
get half a wave-length, or indeed any number of half-wave lengths, in advance of
the other. The possibility of interference here at once flashes upon the mind. A
little consideration, however, will render it evident that, as long as the vibrations are
executed at right angles to each other, they cannot quench each other, no matter what
the retardation may be. This brings us at once to the part played by the analyzer. Its
sole function is to recompound the two vibrations emergent from the gypsum. It
reduces them to a single plane, where, if one of them be retarded by the proper
amount, extinction will occur.

But here, as in the case of thin films, the different lengths of the waves of light
come into play. Red will require a greater thickness to produce the retardation necessary for extinction than blue; consequently, when the longer waves have been
withdrawn by interference, the shorter ones remain, the film of gypsum shining
with the colours which they confer. Conversely, when the shorter waves have been
withdrawn, the thickness is such that the longer waves remain. An elementary consideration suffices to show that, when the directions of vibration of the prisms and
the gypsum enclose an angle of 45°, the colours are at their maximum brilliancy.
When the film is turned from this direction, the colours gradually fade, until, at the
point where the directions of vibration are parallel, they disappear altogether.

The best way of obtaining a knowledge of these phenomena is to construct a
model of thin wood or pasteboard, representing the plate of gypsum, its planes of

vibration, and also those of the polarizer and analyzer. Two parallel pieces of the board are to be separated by an interval which shall represent the thickness of the film of gypsum. Between them, two other pieces, intersecting each other at a right angle, are to represent the planes of vibration within the film; while attached to the two parallel surfaces outside are two other pieces of board to represent the planes of vibration of the polarizer and analyzer. On the two intersecting planes the waves are to be drawn, showing the resolution of the first polarized beam into two others, and then the subsequent reduction of the two systems of vibrations to a common plane by the analyzer. Following out rigidly the interaction of the two systems of waves, we are taught by such a model that all the phenomena of colour obtained by the combination of the waves when the planes of vibration of the two Nicols are parallel are displaced by the *complementary* phenomena when the planes of vibration are perpendicular to each other.

In considering the next point, we will operate, for the sake of simplicity, with monochromatic light—with red light, for example, which is easily obtained pure by red glass. Supposing a certain thickness of the gypsum produces a retardation of half a wave-length, twice this thickness will produce a retardation of two half wavelengths, three times this thickness a retardation of three half-wave lengths, and so on. Now, when the Nicols are parallel, the retardation of half a wave-length, or of any *odd* number of half wave-lengths, produces extinction; at all thicknesses, on the other hand, which correspond to a retardation of an *even* number of half wave-lengths, the two beams support each other, when they are brought to a common plane by the analyzer. Supposing, then, that we take a plate of a wedge-form, which grows gradually thicker from edge to back, we ought to expect in red light a series of recurrent bands of light and darkness; the dark bands occurring at thicknesses which produce retardations of one, three, five, *etc.*, half wave-lengths, while the bright bands occur between the dark ones. Experiment proves the wedge-shaped film to show these bands. They are also beautifully shown by a circular film, so worked as to be thinnest at the centre, and gradually increasing in thickness from the centre outwards. A splendid series of rings of light and darkness is thus produced.

When, instead of employing red light, we employ blue, the rings are also seen: but, as they occur at thinner portions of the film, they are smaller than the rings obtained with the red light. The consequence of employing white light may be now inferred; inasmuch as the red and the blue fall in different places, we have *iris-coloured* rings produced by the white light.

Some of the chromatic effects of irregular crystallization are beautiful in the extreme. Could I introduce between our Nicols a pane of glass covered by those frost-ferns which the cold weather renders now so frequent, rich colours would be the result. The beautiful effects of the irregular crystallization of tartaric acid and other substances on glass plates, now presented to you, illustrate what you might expect from the frosted window-pane. And not only do crystalline bodies act thus upon light, but almost all bodies that possess a definite structure do the same. As a general rule, organic bodies act thus upon light; for their architecture implies an arrangement of the molecules, and of the ether, which involves double refraction. A film of horn, or the section of a shell, for example, yields very beautiful colours in polarized light. In a tree, the ether certainly possesses different degrees of elasticity along and across the fibre; and, were wood transparent, this peculiarity of molecular structure would infallibly reveal itself by chromatic phenomena like those that you have seen.

23.2.4   Colours Produced by Strain and Pressure

But not only do natural bodies behave in this way, but it is possible, as shown by Brewster, to confer, by artificial strain or pressure, a temporary double-refracting structure upon non-crystalline bodies, such as common glass. This is a point worthy of illustration. When I place a bar of wood across my knee and seek to break it, what is the mechanical condition of the bar? It bends, and its convex surface is *strained* longitudinally; its concave surface, that next my knee, is longitudinally *pressed*. Both in the strained portion and in the pressed portion the ether is thrown into a condition which would render the wood, were it transparent, double-refracting. For, in cases like the present, the drawing of the molecules asunder longitudinally is always accompanied by their approach to each other laterally; while the longitudinal squeezing is accompanied by lateral retreat. Each half of the bar exhibits this antithesis, and is therefore double-refracting.

Let us now repeat this experiment with a bar of glass. Between the crossed Nicols I introduce such a bar. By the dim residue of light lingering upon the screen, you see the image of the glass, but it has no effect upon the light. I simply bend the glass bar with my finger and thumb, keeping its length oblique to the directions of vibration in the Nicols. Instantly light flashes out upon the screen. The two sides of the bar are illuminated, the edges most, for here the strain and pressure are greatest. In passing from longitudinal strain to longitudinal pressure, we cross a portion of the glass where neither is exerted. This is the so-called neutral axis of the bar of glass, and along it you see a dark band, indicating that the glass along this axis exercises no action upon the light. By employing the force of a press, instead of the force of my finger and thumb, the brilliancy of the light is greatly augmented.

Again, I have here a square of glass which can be inserted into a press of another kind. Introducing the uncompressed square between the prisms, its neutrality is declared; but it can hardly be held sufficiently loosely in the press to prevent its action from manifesting itself. Already, though the pressure is infinitesimal, you see spots of light at the points where the press is in contact with the glass. On turning a screw the image of the square of glass flashes out upon the screen. Luminous spaces are seen separated from each other by dark bands.

Every two adjacent luminous spaces are in opposite mechanical conditions. On one side of the dark band we have strain, on the other side pressure; while the dark band marks the neutral axis between both. I now tighten the vice, and you see colour; tighten still more, and the colours appear as rich as those presented by crystals. Releasing the vice, the colours suddenly vanish; tightening suddenly, they reappear. From the colours of a soap-bubble Newton was able to infer the thickness of the

Fig. 23.5 The proximity
of the *horizontal* lines in
these two consecutive images
represent the relative density
at various positions along a
vibrating rectangular glass
rod; the mid-point undergoes
alternating compressions and
rarefactions, while the ends
remain unaffected.—[*K.K.*]

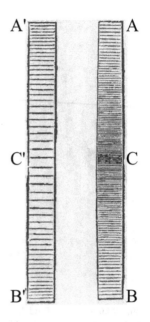

bubble, thus uniting by the bond of thought apparently incongruous things. From
the colours here presented to you, the magnitude of the pressure employed might be
inferred. Indeed, the late M. Wertheim, of Paris, invented an instrument for the determination of strains and pressures, by the colours of polarized light, which exceeded
in accuracy all previous instruments of the kind.

And now we have to push these considerations to a final illustration. Polarized
light may be turned to account in various ways as an analyzer of molecular condition. It may, for instance, be applied to reveal the condition of a solid body when
it becomes sonorous. A strip of glass 6 ft long, 2 in. wide, and a quarter of an
inch thick, is held at the centre between the finger and thumb. On sweeping a wet
woollen rag over one of its halves, you hear an acute sound due to the vibrations
of the glass. What is the condition of the glass while the sound is heard? This: its
two halves lengthen and shorten in quick succession. Its two ends, therefore, are
in a state of quick vibration; but at the centre the pulses from the two ends alternately meet and retreat from each other. Between their opposing actions, the glass
at the centre is kept motionless; but, on the other hand, it is alternately strained and
compressed. The state of the glass may be illustrated by a row of spots of light,
as the propagation of a sonorous pulse was illustrated in our second lecture. By a
simple mechanical contrivance the spots are made to vibrate to and fro: the terminal
dots have the largest amplitude of vibration, while those at the centre are alternately
crowded together and drawn asunder, the centre one not moving at all. (In Fig. 23.5,
AB may be taken to represent the glass rectangle with its centre condensed; while
A′B′ represents the same rectangle with its centre rarefied. The ends of the strip
suffer neither condensation nor rarefaction.)

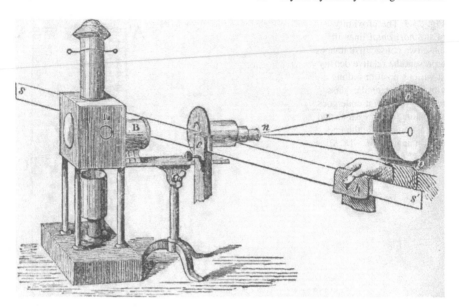

Fig. 23.6 Biot and Tyndall's method of observing sound vibrations within a rubbed transparent rod using polarized light.—[*K.K.*]

If we introduce the strip of glass (*ss'*, Fig. 23.6) between the crossed Nicols, taking care to keep it oblique to the directions of vibration of the Nicols, and sweep our wet rubber over the glass, this may be expected to occur: At every moment of compression the light will flash through; at every moment of strain the light will also flash through; and these states of strain and pressure will follow each other so rapidly that we may expect a permanent luminous impression to be made upon the eye. By pure reasoning, therefore, we reach the conclusion that the light will be revived whenever the glass is sounded. That it is so, experiment testifies: at every sweep of the rubber, a fine luminous disk (*o*) flashes out upon the screen. The experiment may be varied in this way: Placing in front of the polarizer a plate of unannealed glass, you have a series of beautifully coloured rings, intersected by a black cross. Every sweep of the rubber not only abolishes the rings, but introduces complementary ones, the black cross being, for the moment, supplanted by a white one. This is a modification of a beautiful experiment which we owe to Biot. His apparatus, however, confined the observation of it to a single person at a time.

23.2.5 Colours of Unannealed Glass

Bodies are usually expanded by heat and contracted by cold. If the heat be applied with perfect uniformity, no local strains or pressures come into play; but, if one portion of a solid be heated and other portions not, the expansion of the heated

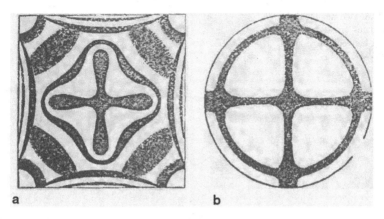

a   b

Fig. 23.7 Unannealed glass plates of different shapes inserted between crossed Nicols.—[*K.K.*]

portion introduces strains and pressures which reveal themselves under the scrutiny of polarized light. When a square of common window-glass is placed between the Nicols, you see its dim outline, but it exerts no action on the polarized light. Held for a moment over the flame of a spirit-lamp, on reintroducing it between the Nicols, light flashes out upon the screen. Here, as in the case of mechanical action, you have luminous spaces of strain divided by dark neutral axes from spaces of pressure.

Let us apply the heat more symmetrically. A small square of glass is perforated at the centre, and into the orifice a bit of copper wire is introduced. Placing the square between the prisms, and heating the wire, the heat passes by conduction to the glass, through which it spreads from the centre outwards. You immediately see, bounding four luminous quadrants, a dim cross, which becomes gradually blacker by comparison with the adjacent brightness. And as, in the case of pressure, we produced colours, so here also, by the proper application of heat, gorgeous chromatic effects may be produced. The condition necessary to the production of these colours may be rendered permanent by first heating the glass sufficiently, and then cooling it, so that the chilled mass shall remain in a state of permanent strain and pressure. Two or three examples will illustrate this point. Figures 23.7a and 23.7b represent the figures obtained with two pieces of glass thus prepared. Two rectangular pieces of unannealed glass, crossed and placed between the polarizer and analyzer, exhibit the beautiful iris fringes represented in Fig. 23.8.

23.2.6 Circular Polarization

But we have to follow the ether still further into its hiding-places. Suspended before you is a pendulum, which, when drawn aside and liberated, oscillates to and fro. If, when the pendulum is passing the middle point of its excursion, I impart a shock to it tending to drive it at right angles to its present course, what occurs? The two

Fig. 23.8 Two unannealed glass plates crossed and stacked between crossed Nicols.—[*K.K.*]

impulses compound themselves to a vibration oblique in direction to the former one, but the pendulum still oscillates in a *plane*. But, if the rectangular shock be imparted to the pendulum when it is at the limit of its swing, then the compounding of the two impulses causes the suspended ball to describe not a straight line, but an ellipse; and, if the shock be competent of itself to produce a vibration of the same amplitude as the first one, the ellipse becomes a circle.

Why do I dwell upon these things? Simply to make known to you the resemblance of these gross mechanical vibrations to the vibrations of light. I hold in my hand a plate of quartz cut from the crystal perpendicular to its axis. The crystal thus cut possesses the extraordinary power of twisting the plane of vibration of a polarized ray to an extent dependent on the thickness of the crystal. And the more refrangible the light the greater is the amount of twisting; so that, when white light is employed, its constituent colours are thus drawn asunder. Placing the quartz between the polarizer and analyzer, you see this vivid red, and, turning the analyzer in front, from right to left, the other colours of the spectrum appear in succession. Specimens of quartz have been found which require the analyzer to be turned from left to right to obtain the same succession of colours. Crystals of the first class are therefore called right-handed, and of the second class, left-handed crystals.

With profound sagacity, Fresnel, to whose genius we mainly owe the expansion and final triumph of the undulatory theory of light, reproduced mentally the mechanism of these crystals, and showed their action to be due to the circumstance that, in them, the waves of ether so act upon each other as to produce the condition represented by our rotating pendulum. Instead of being plane polarized, the light in rock crystal is *circularly polarized*. Two such rays, transmitted along the axis of the crystal, and rotating in opposite directions, when brought to interference by the analyzer, are demonstrably competent to produce all the observed phenomena.

23.2.7   Complementary Colours of Bi-refracting Spar in Circularly Polarized Light. Proof that Yellow and Blue Are Complementary

I now remove the analyzer, and put in its place the piece of Iceland spar with which we have already illustrated double refraction. The two images of the carbon- points are now before you, produced, as you know, by two beams vibrating at right angles to each other. Introducing a plate of quartz between the polarizer and the spar, the two images glow with complementary colours. Employing the image of an aperture instead of that of the carbon-points, we have two coloured circles. As the analyzer is caused to rotate, the colours pass through various changes; but they are always complementary. When the one is red, the other is green; when the one is yellow, the other is blue. Here we have it in our power to demonstrate afresh a statement made in our first lecture, that, although the mixture of blue and yellow pigments produces green, the mixture of blue and yellow lights produces white. By enlarging our aperture, the two images produced by the spar are caused to approach each other, and finally to overlap. The one is now a vivid yellow, the other a vivid blue, and you notice that where the colours are superposed we have a pure white. (See Fig. 23.9, where N is the end of the polarizer, Q the quartz plate, L a lens, and B the bi-refracting spar. The two images overlap at O, and produce white by their mixture.)

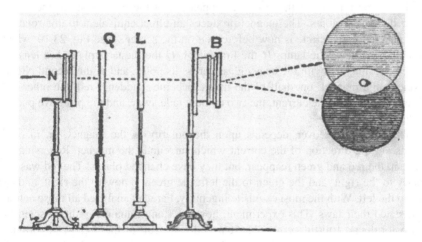

Fig. 23.9 Pure *white* appears in the region where *blue* and *yellow* light overlap.—[*K.K.*]

23.2.8   The Magnetization of Light

This brings us to a point of our inquiries which, though rarely illustrated in lectures, is nevertheless so likely to affect profoundly the future course of scientific thought that I am unwilling to pass it over without reference. I refer to the experiment which Faraday, its discoverer, called the 'magnetization of light.' The arrangement for this celebrated experiment is now before you. We have first our electric lamp, then a Nicol prism, to polarize the beam emergent from the lamp; then an electro-magnet, then a second Nicol, and finally our screen. At the present moment the prisms are crossed, and the screen is dark. I place from pole to pole of the electro-magnet a cylinder of a peculiar kind of glass, first made by Faraday, and called Faraday's heavy glass. Through this glass the beam from the polarizer now passes, being intercepted by the Nicol in front. On exciting the magnet light instantly appears upon the screen. By the action of the magnet upon the ether contained within the heavy glass, the plane of vibration is caused to rotate, the light being thus enabled to get through the analyzer.

The two classes into which quartz-crystals are divided have been already mentioned. In my hand I hold a compound plate, one half of it taken from a right-handed, and the other from a left-handed crystal. Placing the plate in front of the polarizer, I turn one of the Nicols until the two halves of the plate show a common puce colour. This yields an exceedingly sensitive means of rendering visible the action of a magnet upon light. By turning either the polarizer or the analyzer through the smallest angle, the uniformity of the colour disappears, and the two halves of the quartz show different colours. The magnet produces an effect equivalent to this rotation. The puce-coloured circle is now before you on the screen. (See Fig. 23.10, where N is the nozzle of the lamp, H the first Nicol, Q the biquartz plate, L a lens, M the electro-magnet, with the heavy glass across its poles, and P the second Nicol.) Exciting the magnet, one half of the image becomes suddenly red, the other half green. Interrupting the current, the two colours fade away, and the primitive puce is restored.

The action, moreover, depends upon the polarity of the magnet, or, in other words, on the direction of the current which surrounds the magnet. Reversing the current, the red and green reappear, but they have changed places. The red was formerly to the right, and the green to the left; the green is now to the right, and the red to the left. With the most exquisite ingenuity, Faraday analyzed all those actions and stated their laws. This experiment, however, long remained rather a scientific curiosity than a fruitful germ. That it would bear fruit of the highest importance, Faraday felt profoundly convinced, and recent researches are on the way to verify his conviction.

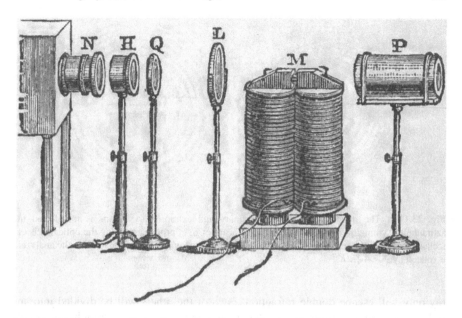

Fig. 23.10 Faraday's apparatus for magnetically rotating the plane of polarization of a beam of light.—[*K.K.*]

23.2.9 Iris-Rings Surrounding the Axes of Crystals

A few words more are necessary to complete our knowledge of the wonderful interaction between ponderable molecules and the ether interfused among them. Symmetry of molecular arrangement implies symmetry on the part of the ether; atomic dissymmetry, on the other hand, involves the dissymmetry of the ether, and, as a consequence, double refraction. In a certain class of crystals the structure is homogeneous, and such crystals produce no double refraction. In certain other crystals the molecules are ranged symmetrically round a certain line, and not around others. Along the former, therefore, the ray is undivided, while along all the others we have double refraction. Ice is a familiar example: its molecules are built with perfect symmetry around the perpendiculars to the planes of freezing, and a ray sent through ice in this direction is not doubly refracted; whereas, in all other directions, it is. Iceland spar is another example of the same kind: its molecules are built symmetrically round the line uniting the two blunt angles of the rhomb. In this direction a ray suffers no double refraction, in all others it does. This direction of no double refraction is called the *optic axis* of the crystal.

Hence, if a plate be cut from a crystal of Iceland spar perpendicular to the axis, all rays sent across this plate in the direction of the axis will produce but one image. But, the moment we deviate from the parallelism with the axis, double refraction sets in. If, therefore, a beam that has been rendered *conical* by a converging lens be sent through the spar so that the central ray of the cone passes along the axis, this

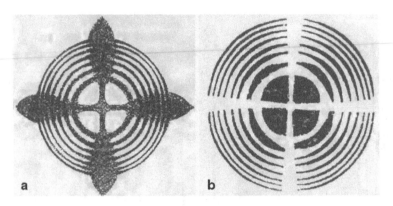

Fig. 23.11 a The interference between ordinary and extraordinary beams is manifested in azimuthally symmetric fringes when a conical beam of light propagates along the optical axis of Iceland spar inserted between crossed Nicols. **b** Complementary colors appear when the analyzer is rotated by 90°.—[*K.K.*]

ray only will escape double refraction. Each of the others will be divided into an ordinary and an extraordinary ray, the one moving more slowly through the crystal than the other; the one, therefore, retarded with reference to the other. Here, then, we have the conditions for interference, when the waves are reduced by the analyzer to a common plane.

Placing the plate of Iceland spar between the crossed Nicol's prisms, and employing the conical beam, we have upon the screen a beautiful system of iris-rings surrounding the end of the optic axis, the circular bands of colour being intersected by a black cross (Fig. 23.11a). The arms of this cross are parallel to the two directions of vibration in the polarizer and analyzer. It is easy to see that those rays whose planes of vibration within the spar coincide with the plane of vibration of *either* prism, cannot get through *both*. This complete interception produces the arms of the cross. With monochromatic light the rings would be simply bright and black—the bright rings occurring at those thicknesses of the spar which cause the rays to conspire; the black rings at those thicknesses which cause them to quench each other. Turning the analyzer 90° round, we obtain the complementary phenomena. The black cross gives place to a bright one, and every dark ring is supplanted also by a bright one (Fig. 23.11b). Here, as elsewhere, the different lengths of the light-waves give rise to iris-colours when white light is employed.

Besides the *regular* crystals which produce double refraction in no direction, and the *uniaxal* crystals which produce it in all directions but one, Brewster discovered that in a large class of crystals there are *two* directions in which double refraction does not take place. These are called *biaxal* crystals. When plates of these crystals, suitably cut, are placed between the polarizer and analyzer, the axes (*AA′*, Fig. 23.12) are seen surrounded, not by circles, but by curves of another order and of a perfectly definite mathematical character. Each band, as proved experimentally

Fig. 23.12 Lemniscata
formed when a plate of biaxial crystal is placed between
a polarizer and analyzer and
illuminated.—[*K.K.*]

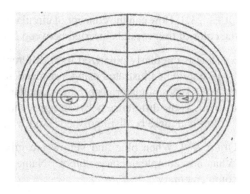

by Herschel, forms a *lemniscata*; but the experimental proof was here, as in num-
berless other cases, preceded by the deduction which showed that, according to the
undulatory theory, the bands must possess this special character.

23.3 Study Questions

QUES. 23.1. What is the cause of the diverse colors observed when a thin film of
selenite or mica is placed between crossed Nicol prisms?

a) What is a Nicol prism and how does it work? Why is a Nicol prism superior to a
tourmaline crystal for producing polarized light?
b) What happens when a film of mica is placed between crossed Nicols? What
is notable when the film is thin? What about when a thin film of selenite
(transparent gypsum) is rotated between two crossed Nicols?
c) In what sense is a light beam like a mechanical vibration? What happens if these
vibrations happen to be oblique to the axis of a tourmaline crystal or a Nicol?
Will the beam pass through?
d) In Fig. 23.4, which line segment represents the amplitude of vibration of the
light upon passing through the polarizer?Which line segments represent the
amplitudes of vibration of the light passing through the selenite?
e) Do perpendicular vibrations travel with the same speed through the selenite? Do
they interfere with one another while in the selenite?
f) What effect does the analyzer have on the perpendicular vibrations leaving the
selenite crystal? Under what condition(s) will a particular color of light be
blocked by the analyzer? Transmitted?
g) What happens if a crystal of non-uniform thickness is inserted between the
polarizer and analyzer?

QUES. 23.2. How can polarized light be used to detect strain within materials caused
by (i) pressure, (ii) sound, and (iii) heat?Why are strained bodies double refracting?
And what is meant by the *neutral axis* of a strained bar?

QUES. 23.3. How can the concept of circular polarization be employed to understand the color of thin crystals inserted between Nicols?

a) How is circularly polarized light different than plane polarized light? How can its nature be illustrated using a suspended pendulum bob?
b) What types of materials produce circularly polarized light when illuminated?
c) What is the difference between right- and left-handed circular polarization? And how can these be distinguished? And what types of materials produce each?

QUES. 23.4. When blue and yellow are mixed, do they produce green or white? What are complementary colors? Name a few. How do you know they are complementary?

QUES. 23.5. In what way can a magnet affect light? Why do you suppose Faraday thought this to be significant? Does this suggest that light itself is magnetic?

QUES. 23.6. Why are some crystals birefringent, while others are not?

a) What is the essential feature necessary for birefringence?
b) What is the structure of ice, and what are the implications for light propagation through ice? What about Iceland spar? In which direction is its optical axis?
c) How do *biaxial* crystals differ from *regular* and *uniaxial* crystals?

23.4   Exercises

EX. 23.1 (BIREFRINGENCE, QUARTER-WAVE PLATES AND CIRCULAR POLARIZA-TION). For a beam of red light ($\lambda = 633$ nm) a certain birefringent crystal has a refractive index of 1.544 along its fast axis and 1.553 along its slow axis. A thin plate of this crystal is inserted between crossed Nicol prisms so that its fast and slow axes are at 45° angles with respect to the polarizer and analyzer (see Figs. 23.3 and 23.4).

a) Is the fast or slow axis of the crystal associated with the ordinary beam? Which travels faster, the ordinary or the extraordinary beam? Which has a longer wavelength?
b) How thick must the plate be in order to generate a one-half wavelength phase difference between the ordinary and extraordinary beams? This is called a *half-wave plate*.
c) After exiting the half-wave plate (but before striking the analyzer), what is the state of polarization of the ordinary and extraordinary beams? What about the sum of the two beams? Is it plane (linearly) polarized? If so, in which direction?
d) What happens when the ordinary and extraordinary beams which exit the crystal strike the analyzer—do they pass through or are they blocked? What if the analyzer is rotated 90° so that it is parallel to the polarizer?
e) How thick must the plate be in order to generate a one-quarter wavelength phase difference between the ordinary and extraordinary beams? After exiting this

quarter-wave plate (but before striking the analyzer) the light is now *circularly polarized*: the polarization plane rotates around the direction of travel—like a ribbon of paper twisted about its long axis.

EX. 23.2 (LIGHT ROTATION LABORATORY). Study how commercial polarizers transmit (or block) light when two of them are oriented in the same, and in opposite, directions.[2] What happens when a third polarizer is placed between two crossed polarizers? Is the result the same if the third polarizer is placed on the outside of the other two? Once you understand how the polarizers work, explore how each of the following materials act on light when inserted between a set of polarizing filters. Carefully explain your observations and their implications. (a) a thin piece of mica (b) a piece of stressed plexiglass (c) frost ferns on a window in cold weather (d) cellophane from product wrappers (such as DVDs) (e) tempered glass windows (such as automobile or shower glass) (f) window glass heated with a small flame (g) several overlapping pieces of cellophane tape on a sheet of acetate[3] (try to measure the thickness of a single sheet of cellophane tape using this method) (h) optically active solutions;[4] (are these substances right-handed or left handed?)

23.5 Vocabulary

1. Chromatic
2. Elucidation
3. Tourmaline
4. Balsam
5. Oblique
6. Requisite
7. Spectroscope
8. Gypsum
9. Selenite
10. Mica
11. Complementary
12. Ascribe
13. Composition
14. Resolution
15. Superpose
16. Component

---

[2] A Polarizing Filter Demo Kit, which includes much of the equipment used in this laboratory experiment, is available from Educational Innovations, Inc., Bethel, CT.

[3] *Not* Scotch Magic tape. For this experiment you might try to start with a simple pattern (*e.g.* one or two pieces of tape) before advancing to more complex ones.

[4] An optically active solution may be obtained by mixing one part water and one part Karo syrup, fructose, turpentine or honey.

17. Retard
18. Antithesis
19. Augment
20. Manifest
21. Infinitesimal
22. Infer
23. Sonorous
24. Asunder
25. Demonstrably
26. Profound
27. Biaxial
28. Lemniscata

Chapter 24
Light Scattering

> *How, if the air be blue, can the light of sunrise and sunset,*
> *which travels through vast distances of air, be yellow, orange, or*
> *even red?*
>
> —John Tyndall

24.1 Introduction

In the first nine sections of his fourth lecture on light, John Tyndall explained how
crystals rotate the plane of polarization of light passing through them: it is a consequence of the different refractive indices along different crystal axes. Thus, Tyndall
identifies structural asymmetry—and its effect on the transverse vibrations of interatomic ether—as the cause of birefringence in crystals. In the final sections of his
fourth lecture on light, included in the present chapter, we will see how particles
suspended in a transparent medium also affect light propagation. But before moving on to this fascinating topic, Tyndall pauses to emphasizes how the undulatory
theory of light provides a powerful organizing principle for understanding all of the
surprising and beautiful phenomena thus far observed. Does Tyndall believe that
the undulatory theory of light is true, or merely useful? How would you describe his
philosophy of science?

24.2 Reading: Tyndall, *Six Lectures on Light*

Tyndall, J., *Six lectures on light delivered in America in 1872–1878*, D. Appleton
and Company, New York, 1886. Lecture IV, beginning at §10.

24.2.1 *Power of the Undulatory Theory*

I have taken this somewhat wide range over polarization itself, and over the phenomena exhibited by crystals in polarized light, in order to give you some notion

© Springer International Publishing Switzerland 2016
K. Kuehn, *A Student's Guide Through the Great Physics Texts,*
Undergraduate Lecture Notes in Physics, DOI 10.1007/978-3-319-21816-8_24

of the firmness and completeness of the theory which grasps them all. Starting from the single assumption of transverse undulations, we first of all determine the wave-lengths, and find all the phenomena of colour dependent on this element. The wave-lengths may be determined in many independent ways. Newton virtually determined them when he measured the periods of his Fits: the length of a fit, in fact, is that of a quarter of an undulation. The wave-lengths may be determined by diffraction at the edges of a slit (as in the Appendix to these Lectures); they may be deduced from the interference fringes produced by reflection; from the fringes produced by refraction; also by lines drawn with a diamond upon glass at measured distances asunder. And when the lengths determined by these independent methods are compared together, the strictest agreement is found to exist between them.

With the wave-lengths at our disposal, we follow the ether into the most complicated cases of interaction between it and ordinary matter, 'the theory is equal to them all. It makes not a single new physical hypothesis; but out of its original stock of principles it educes the counterparts of all that observation shows. It accounts for, explains, simplifies the most entangled cases; corrects known laws and facts; predicts and discloses unknown ones; becomes the guide of its former teacher Observation; and, enlightened by mechanical conceptions, acquires an insight which pierces through shape and colour to force and cause.'[1]

But, while I have thus endeavoured to illustrate before you the power of the undulatory theory as a solver of all the difficulties of optics, do I therefore wish you to close your eyes to any evidence that may arise against it? By no means. You may urge, and justly urge, that a 100 years ago another theory was held by the most eminent men, and that, as the theory then held had to yield, the undulatory theory may have to yield also. This seems reasonable; but let us understand the precise value of the argument. In similar language a person in the time of Newton, or even in our time, might reason thus: Hipparchus and Ptolemy, and numbers of great men after them, believed that the earth was the centre of the solar system. But this deep-set theoretic notion had to give way, and the theory of gravitation may, in its turn, have to give way also. This is just as reasonable as the first argument. Wherein consists the strength of the theory of gravitation? Solely in its competence to account for all the phenomena of the solar system. Wherein consists the strength of the theory of undulation ? Solely in its competence to disentangle and explain phenomena a 100-fold more complex than those of the solar system. Accept if you will the scepticism of Mr. Mill[2] regarding the undulatory theory; but if your scepticism be philosophical, it will wrap the theory of gravitation in the same or greater doubt.[3]

---

[1] Whewell.

[2] Removed from us since these words were written.

[3] The only essay known to me on the Undulatory Theory, from the pen of an American writer, is an excellent one by President Barnard, published in the Smithsonian Report for 1862.

24.2.2   The Blue of the Sky

I am unwilling to quit these chromatic phenomena without referring to a source of colour which has often come before me of late in the blue of your skies at noon, and the deep crimson of your horizon after the set of sun. I will here summarise and extend what I have already said upon this subject in another place. Proofs of the most cogent description could be adduced to show that the blue light of the firmament is reflected light. That light comes to us across the direction of the solar rays, and even against the direction of the solar rays; and this lateral and opposing rush of wave-motion can only be due to the rebound of the waves from the air itself, or from something suspended in the air. The solar light, moreover, is not reflected by the sky in the proportions which produce white. The sky is blue, which indicates an excess of the smaller waves. The blueness of the air has been given as a reason for the blueness of the sky; but then the question arises. How, if the air be blue, can the light of sunrise and sunset, which travels through vast distances of air, be yellow, orange, or even red? The passage of the white solar light through a blue medium could by no possibility redden the light; the hypothesis of a blue air is therefore untenable. In fact the agent, whatever it be, which sends us the light of the sky, exercises in so doing a dichroitic action. The light reflected is blue, the light transmitted is orange or red. A marked distinction is thus exhibited between reflection from the sky and that from an ordinary cloud, which exercises no such dichroitic action.

The cloud, in fact, takes no note of size on the part of the waves of ether, but reflects them all alike. Now the cause of this may be that the cloud particles are so large in comparison with the size of the waves of ether as to scatter them all indifferently. A broad cliff reflects an Atlantic roller as easily as a ripple produced by a sea-bird's wing; and in the presence of large reflecting surfaces, the existing differences of magnitude among the waves of ether may also disappear. But supposing the reflecting particles, instead of being very large, to be very small, in comparison with the size of the waves. Then, instead of the whole wave being fronted and in great part thrown back, a small portion only is shivered off by the obstacle. Suppose, then, such minute foreign particles to be diffused in our atmosphere. Waves of all sizes impinge upon them, and at every collision a portion of the impinging wave is struck off. All the waves of the spectrum, from the extreme red to the extreme violet, are thus acted upon; but in what proportions will they be scattered? Largeness is a thing of relation; and the smaller the wave, the greater is the relative size of any particle on which the wave impinges, and the greater also the relative reflection.

A small pebble placed in the way of the ring-ripples produced by heavy rain-drops on a tranquil pond will throw back a large fraction of each ripple incident upon it, while the fractional part of a larger wave thrown back by the same pebble might be infinitesimal. Now to preserve the solar light white, its constituent proportions must not be altered; but in the scattering of the light by these very small particles we see that the proportions *are* altered. The smaller waves are in excess, and, as a consequence, in the scattered light blue will be the predominant colour. The other colours of the spectrum must, to some extent, be associated with the blue: they are not absent, but deficient. We ought, in fact, to have them all, but in diminishing proportions, from the violet to the red.

We have thus reasoned our way to the conclusion, that were particles, small in comparison to the size of the ether waves, sown in our atmosphere, the light scattered by those particles would be exactly such as we observe in our azure skies. And, indeed, when this light is analysed, all the colours of the spectrum are found in the proportions indicated by our conclusion.

By its successive collisions with the particles the white light is more and more robbed of its shorter waves; it therefore loses more and more of its due proportion of blue. The result may be anticipated. The transmitted light, where short distances are involved, will appear yellowish. But as the sun sinks towards the horizon the atmospheric distance increases, and consequently the number of the scattering particles. They weaken in succession the violet, the indigo, the blue, and even disturb the proportions of green. The transmitted light under such circumstances must pass from yellow through orange to red. This also is exactly what we find in nature. Thus, while the reflected light gives us, at noon, the deep azure of the Alpine skies, the transmitted light gives us, at sunset, the warm crimson of the Alpine snows.

But can small particles be really proved to act in the manner indicated? No doubt of it. Each one of you can submit the question to an experimental test. Water will not dissolve resin, but spirit will; and when spirit which holds resin in solution is dropped into water, the resin immediately separates in solid particles, which render the water milky. The coarseness of this precipitate depends on the quantity of the dissolved resin. Professor Brücke has given us the proportions which produce particles particularly suited to our present purpose. One gramme of clean mastic is dissolved in 87 g of absolute alcohol, and the transparent solution is allowed to drop into a beaker containing clear water briskly stirred. An exceedingly fine precipitate is thus formed, which declares its presence by its action upon light. Placing a dark surface behind the beaker, and permitting the light to fall into it from the top or front, the medium is seen to be of a very fair sky-blue. A trace of soap in water gives a tint of blue. London milk makes an approximation to the same colour through the operation of the same cause: and Helmholtz has irreverently disclosed the fact that a blue eye is simply a turbid medium.

24.2.3 Artificial Sky

But we have it in our power to imitate far more closely the natural conditions of this problem. We can generate in air artificial skies, and prove their perfect identity with the natural one, as regards the exhibition of a number of wholly unexpected phenomena. It has been recently shown in a great number of instances that waves of ether issuing from a strong source, such as the sun or the electric light, are competent to shake asunder the atoms of gaseous molecules. The apparatus used to illustrate this consists of a glass tube about a yard in length, and from $2\frac{1}{2}$ to 3 in. internal diameter. The gas or vapour to be examined is introduced into this tube, and upon it the condensed beam of the electric lamp is permitted to act. The vapour is so chosen that one, at least, of its products of decomposition, as soon as it is formed, shall be *precipitated* to a kind of cloud. By graduating the quantity of the vapour, this precipitation may be rendered of any degree of fineness, forming particles distinguishable by the naked eye, or particles which are probably far beyond the reach of our highest microscopic powers. I have no reason to doubt that particles may be thus obtained whose diameters constitute but a very small fraction of the length of a wave of violet light.

Now, in all such cases when suitable vapours are employed in a sufficiently attenuated state, no matter what the vapour may be, the visible action commences with the formation of a *blue cloud*. Let me guard myself at the outset against all misconception as to the use of this term. The blue cloud here referred to is totally invisible in ordinary daylight. To be seen, it requires to be surrounded by darkness, *it only* being illuminated by a powerful beam of light. This cloud differs in many important particulars from the finest ordinary clouds, and might justly have assigned to it an intermediate position between these clouds and true cloudless vapour.

It is possible to make the particles of this *actinic cloud* grow from an infinitesimal and altogether ultramicroscopic size to particles of sensible magnitude; and by means of these, in a certain stage of their growth, we produce a blue which rivals, if it does not transcend, that of the deepest and purest Italian sky. Introducing into our tube a quantity of mixed air and nitrite of butyl vapour sufficient to depress the mercurial column of an air-pump one-twentieth of an inch, adding a quantity of air and hydrochloric acid sufficient to depress the mercury half an inch further, and sending through this compound and highly attenuated atmosphere, the beam of the electric light; gradually within the tube arises a splendid azure, which strengthens for a time, reaches a maximum of depth and purity, and then, as the particles grow larger, passes into whitish blue. This experiment is representative, and it illustrates a general principle. Various other colourless substances of the most diverse properties, optical and chemical, might be employed for this experiment. The *incipient cloud*, in every case, would exhibit this superb blue; thus proving to demonstration that particles of infinitesimal size, without any colour of their own, and irrespective of those optical properties exhibited by the substance in a massive state, are competent to produce the blue colour of the sky.

24.2.4   *Polarization of Sky-Light*

But there is another subject connected with our firmament, of a more subtle and recondite character than even its colour. I mean that 'mysterious and beautiful phenomenon,' the polarization of the light of the sky. Looking at various points of the blue firmament through a Nicol's prism, and turning the prism round its axis, we soon notice variations of brightness. In certain positions of the prism, and from certain points of the firmament, the light appears to be wholly transmitted, while it is only necessary to turn the prism round its axis through an angle of 90° to materially diminish the intensity of the light. Experiments of this kind prove that the blue light sent to us by the firmament is polarized, and on close scrutiny it is also found that the direction of most perfect polarization is perpendicular to the solar rays. Were the heavenly azure like the ordinary light of the sun, the turning of the prism would have no effect upon it; it would be transmitted equally during the entire rotation of the prism. The light of the sky is in great part quenched, because it is in great part polarized.

The same phenomenon is exhibited in perfection by our actinic clouds, the only condition necessary to its production being the smallness of the particles. In all cases, and with all substances, the cloud formed at the commencement, when the precipitated particles are sufficiently fine, is *blue*. In all cases, moreover, this fine blue cloud polarizes *perfectly* the beam which illuminates it, the direction of polarization enclosing an angle of 90° with the axis of the illuminating beam.

It is exceedingly interesting to observe both the growth and the decay of this polarization. For 10 or 15 min after its first appearance the light from a vividly illuminated incipient cloud, looked at horizontally, is absolutely quenched by a Nicol's prism with its longer diagonal vertical. But as the sky-blue is gradually rendered impure by the introduction of particles of too large a size, in other words, as real clouds begin to be formed, the polarization begins to deteriorate, a portion of the light passing through the prism in all its positions, as it does in the case of skylight. It is worthy of note that for some time after the cessation of perfect polarization the *residual* light which passes, when the Nicol is in its position of minimum transmission, is of a gorgeous blue, the whiter light of the cloud being extinguished. When the cloud texture has become sufficiently coarse to approximate to that of ordinary clouds, the rotation of the Nicol ceases to have any sensible effect on the quantity of the light discharged at right angles to the beam.

The perfection of the polarization in a direction perpendicular to the illuminating beam was also illustrated by the following experiment executed with many vapours. A Nicol's prism large enough to embrace the entire beam of the electric lamp was placed between the lamp and the experimental tube. Sending the beam polarized by the Nicol through the tube, I placed myself in front of it, the eyes being on a level with its axis, my assistant occupying a similar position behind the tube. The short diagonal of the large Nicol was in the first instance vertical, the plane of vibration of the emergent beam being therefore also vertical. As the light continued to act, a superb blue cloud visible to both my assistant and myself was slowly formed. But this cloud, so deep and rich when looked at from the positions mentioned, utterly disappeared when looked at vertically downwards, or vertically upwards. Reflection from the cloud was not possible in these directions. When the large Nicol was slowly turned round its axis, the eye of the observer being on the level of the beam, and the line of vision perpendicular to it, entire extinction of the light emitted horizontally occurred when the longer diagonal of the large Nicol was vertical. But a vivid blue cloud was seen when looked at downwards or upwards. This truly fine experiment, which I should certainly have made without suggestion, was, as a matter of fact, first definitely suggested by a remark addressed to me in a letter by Professor Stokes.

All the phenomena of colour and of polarization observable in the case of sky-light are manifested by those actinic clouds; and they exhibit additional phenomena which it would be neither convenient to pursue, nor perhaps possible to detect, in the actual firmament. They enable us, for example, to follow the polarization from its first appearance on the barely visible blue to its final extinction in the coarser cloud. These changes, as far as it is now necessary to refer to them, may be thus summed up:— (1) The actinic cloud, as long as it continues blue, discharges polarized light in all directions, but the direction of maximum polarization, like that of skylight, is at right angles to the direction of the illuminating beam. (2) As long as the cloud remains distinctly blue the light discharged from it at right angles to the illuminat-ing beam is *perfectly* polarized. It may be utterly quenched by a Nicol's prism, the cloud from which it issues being caused to disappear. Any deviation from the per-pendicular enables a portion of the light to get through the prism. (3) The direction of vibration of the polarized light is at right angles to the illuminating beam. Hence a. plate of tourmaline, with its axis parallel to the beam, stops the light, and with the axis perpendicular to the beam transmits the light. (4) A plate of selenite placed between the Nicol and the actinic cloud shows the colours of polarized light; in fact, the cloud itself plays the part of a polarizing Nicol. (5) The particles of the blue cloud are immeasurably small, but they increase gradually in size, and at a certain period of their growth cease to discharge perfectly polarized light. For some time afterwards the light that reaches the eye through the Nicol is of a magnificent blue, far exceeding in depth and purity that of the purest sky; thus the waves that first feel the influence of size, at both limits of the polarization, are the shortest waves of the spectrum. These are the first to accept polarization, and they are the first to escape from it.

24.3 Study Questions

QUES. 24.1. What are some different ways in which the wavelength of light may be measured? Are these measurements consistent with one another?

QUES. 24.2. What, according to Tyndall, is the primary strength of the wave theory, or for that matter *any* scientific theory?

QUES. 24.3. Why is the sky blue? And why are some eyes blue?

a) If interstellar space is dark, then why doesn't the sky directly overhead look black? And if sunlight is really reflected from the sky overhead, then why doesn't it look white?

b) What does it mean that the sky is *dichroic*? Are clouds dichroic, too? Which color of light is most strongly scattered by the atmosphere? Which color is most weakly scattered?

c) How can the theory that small particles scatter different colors differently be put to the test in the laboratory? What other common turbid media display a blue color?

d) What is an *actinic cloud*? What is it made of and under what condition(s) does it appear blue?

QUES. 24.4. In what sense is sky-light polarized? How can this be tested out-of-doors, or in the laboratory using an actinic cloud? Do these observations depend on the size of the scattering particles?

24.4 Exercises

EX. 24.1 (BLOOD-MOON). Why does the moon typically look red during a lunar eclipse? For that matter, why is the moon visible *at all* during a lunar eclipse?

EX. 24.2 (TYNDALL SCATTERING LABORATORY). Shine an unpolarized laser beam through a transparent cuvette full of water.[4] Can you see the beam in the water? Now add a bit of skim milk to the water. Can you see the beam in the cloudy water? Using a polarizing filter, observe the scattered light coming from the cuvette from many different angles: from directly above the cuvette, along the direction of the beam (be careful not to look at the laser beam!), and at several angles around the sides of the cuvette. For each angle of observation, determine if the beam is polarized, and if so, in which orientation. Be as precise as possible, making sketches in your lab book. If you are unsure of the polarization orientation of your filter, you may calibrate it by observing the light reflected at the polarizing angle from a glass surface (see the discussion of Brewster's angle in Exs. 22.1 and 22.2). Your description should be unambiguous. Why does the scattered light behave like this?

EX. 24.3 (RAYLEIGH SCATTERING LABORATORY). Observe the blue sky through a polarizing filter. Can you discern the polarization of the scattered light from the sky? Is the light from the clouds similarly polarized?[5]

24.5 Vocabulary

1. Transverse
2. Educe
3. Cogent
4. Adduce
5. Firmament
6. Endeavor

---

[4] I have used helium neon lasers from scientific supply companies such as Sargent Welch, but a low-cost laser pointer is sufficient.

[5] For a comprehensive and mathematically sophisticated treatment of light scattering, see Sects. 9.6 and 9.7 of Jackson, J. D., *Classical Electrodynamics*, second ed., John Wiley & Sons, New York, 1975. For a gentler treatment, see Sect. 32.5 in Volume I of Feynman, R.P., Leighton, R.B., and Sands, M.L., *The Feynman Lectures on Physics*, Commemorative ed., Addison-Wesley Publishing Co., 1989.

7. Dichroitic
8. Impinge
9. Constituent
10. Spectrum
11. Resin
12. Turbid
13. Precipitate
14. Attenuate
15. Azure
16. Recondite
17. Incipient
18. Manifest

Chapter 25
Induction of Electrical Currents

> *These considerations, with their consequence, the hope of obtaining electricity from ordinary magnetism, have stimulated me at various times to investigate experimentally the inductive effect of electric currents.*
>
> —Michael Faraday

25.1 Introduction

Michael Faraday (1791–1867) was born in Newington Butts, which is now part of London.[1] He had little formal education beyond day school as a young boy. Yet from a position as apprentice to a bookbinder at the age of 14, Faraday would become the Superintendent and Director of the laboratory at the Royal Institution, where his careful experimental work on electro-magnetism laid the foundation for Maxwell's electromagnetic theory of light.[2] Faraday's scientific career began in 1813, after he presented Sir Humphry Davy with a carefully prepared transcript of notes which he had taken after attending Davy's chemistry lectures in 1812. Impressed, Davy asked Faraday to fill a position as his laboratory assistant at the Royal Institution, which Faraday did. Shortly thereafter, he was invited to serve as Davy's secretary on a tour of Europe, where he met and talked with many leading scientific figures, including Ampère in Paris and Volta in Milan. Returning to England, Faraday's laboratory work initially focused on chemistry and electrolysis Fig. 25.1, and he invented many of the terms now used in electrochemistry, such as *electrolyte*, *electrode*, *anode* and *cathode*. Faraday liquefied chlorine for the first time in 1823, and 2 years later he isolated and identified the organic compound benzene, which he called "bicarburet of hydrogen." Faraday was elected a Fellow of the Royal Society in 1824, and in 1826 he initiated a series of Christmas lectures for children, which have been held at the Royal Institute ever since.[3] Inspired by the work of Oersted, Ampère and Arago,

---

[1] For biographical notes on Faraday, see, for instance, Bence, J., *The life and letters of Faraday*, Longmans, Green and Co., London, 1870.

[2] Maxwell's electromagnetic theory of light will be developed in Chap. 31 of the present volume.

[3] Faraday's final series of Christmas lectures are wonderfully clear and enjoyable to read. See Faraday, M., *The Chemical History of a Candle: to which is added a Lecture on Platinum*, Harper

© Springer International Publishing Switzerland 2016
K. Kuehn, *A Student's Guide Through the Great Physics Texts*,
Undergraduate Lecture Notes in Physics, DOI 10.1007/978-3-319-21816-8_25

Fig. 25.1 Michael Faraday's Laboratory at the Royal Institution. From Jones, *The life and letters of Faraday*, 1870

Faraday's attention turned increasingly to electrical and magnetic phenomena. A comprehensive description of his work may be found in a series of articles which was published in the *Philosophical Transactions of the Royal Society of London* under the title *Experimental Researches in Electricity*. Among the seminal discoveries included herein are the rotation of polarized light by an electro-magnet[4] and the phenomenon of diamagnetism.[5] But Faraday's most significant discovery, and the topic of the reading selection below, was presented initially in his address to the Royal Society on November 24, 1831. Herein, he describes his attempts to obtain "electricity from ordinary magnetism." The lecture begins with a brief reminder of how bodies held in electrical "tension"—we would say they have an electrical potential, or voltage, difference—induce an electrical polarization in nearby materials.[6] Before proceeding, he explains how his work was inspired, at least in part,

---

& Brothers, New York, 1861 and Faraday, M., *A Course of Six Lectures on the Forces of Matter and Their Relations to Each Other*, Richard Griffin and Company, London and Glasgow, 1860.

[4] The rotation of light by a magnetic field is now called *Faraday rotation*. This is described by John Tyndall in his lectures on light; see Chap. 23 of the present volume.

[5] See Chap. 29 for a discussion of diamagnetic materials, and how they differ from ordinary magnetic materials.

[6] The use of the term "electric tension" for electrical potential difference survives today: overhead high-voltage power lines are sometimes still referred to as "high tension" lines.

by Arago's recently observed magnetic phenomena[7] and by Ampère's "beautiful theory" of magnets.[8]

25.2   Reading: Faraday, *Experimental Researches in Electricity*

Faraday, M., *Experimental Researches in Electricity*, vol. 1, Taylor and Francis, London, 1839. First series, beginning at §1. Read November 24, 1831.

1. The power which electricity of tension possesses of causing an opposite electrical state in its vicinity has been expressed by the general term Induction; which, as it has been received into scientific language, may also, with propriety, be used in the same general sense to express the power which electrical currents may possess of inducing any particular state upon matter in their immediate neighbourhood, otherwise indifferent. It is with this meaning that I purpose using it in the present paper.

2. Certain effects of the induction of electrical currents have already been recognised and described: as those of magnetization; Ampère's experiments of bringing a copper disk near to a flat spiral; his repetition with electromagnets of Arago's extraordinary experiments, and perhaps a few others. Still it appeared unlikely that these could be all the effects which induction by currents could produce; especially as, upon dispensing with iron, almost the whole of them disappear, whilst yet an infinity of bodies, exhibiting definite phenomena of induction with electricity of tension, still remain to be acted upon by the induction of electricity in motion.

3. Further: Whether Ampère's beautiful theory were adopted, or any other, or whatever reservation were mentally made, still it appeared very extraordinary, that as every electric current was accompanied by a corresponding intensity of magnetic action at right angles to the current, good conductors of electricity, when placed within the sphere of this action, should not have any current induced through them, or some sensible effect produced equivalent in force to such a current.

4. These considerations, with their consequence, the hope of obtaining electricity from ordinary magnetism, have stimulated me at various times to investigate experimentally the inductive effect of electric currents. I lately arrived at positive results; and not only had my hopes fulfilled, but obtained a key which appeared to me to open out a full explanation of Arago's magnetic phenomena, and also to discover a new state, which may probably have great influence in some of the most important effects of electric currents.

5. These results I purpose describing, not as they were obtained, but in such a manner as to give the most concise view of the whole.

---

[7] See Faraday's discussion of Arago's wheel in Chap. 26 of the present volume.

[8] Ampère discusses how permanent magnets may be understood in terms of circulating electrical currents at the end of the reading included in Chap. 8 of the present volume.

25.2.1 On the Induction of Electric Currents

6. About 26 ft of copper wire one twentieth of an inch in diameter were wound round a cylinder of wood as a helix, the different spires of which were prevented from touching by a thin interposed twine. This helix was covered with calico, and then a second wire applied in the same manner. In this way 12 helices were superposed, each containing an averaged length of wire of 27 ft, and all in the same direction. The first, third, fifth, seventh, ninth, and eleventh of these helices were connected at their extremities end to end, so as to form one helix; the others were connected in a similar manner; and thus two principal helices were produced, closely interposed, having the same direction, not touching anywhere, and each containing 155 ft in length of wire.

7. One of these helices was connected with a galvanometer, the other with a voltaic battery of ten pairs of plates 4 in.^{2}, with double coppers and well charged; yet not the slightest sensible deflection of the galvanometer needle could be observed.

8. A similar compound helix, consisting of six lengths of copper and six of soft iron wire, was constructed. The resulting iron helix contained 214 ft of wire, the resulting copper helix 208 ft; but wether the current from the trough was passed through the copper or the iron helix, no effect upon the other could be perceived at the galvanometer.

9. In these and many other similar experiments no difference in action of any kind appeared between iron and other metals.

10. Two hundred and three feet of copper wire in one length were passed round a large block of wood; other 203 ft of similar wire were interposed as a spiral between the turns of the first, and metallic contact everywhere prevented by twine. One of these helices was connected with a galvanometer and the other with a battery of 100 pairs of plates 4 in.^{2}, with double copper, and well charged. When the contact was made there was a sudden and very slight effect at the galvanometer, and there was a similar slight effect when the contact with the battery was broken. But whilst the voltaic current was continuing to pass through the one helix, no galvanometrical appearances of any effect like induction upon the other helix could be perceived, although the active power of the battery was proved to be great, by its heating the whole of its own helix, and by the brilliancy of the discharge when made through charcoal.

11. Repetition of the experiments with a battery of 120 pairs of plates produced no other effects; but it was ascertained, both at this and the former time, that the slight deflection of the needle occurring at the moment of completing the connexion, was always in one direction, and that the equally slight deflection produced when the contact was broken, was in the other direction; and also, that these effects occurred when the first helices were used (6. 8).

12. The results which I had by this time obtained with magnets led me to believe that the battery current through one wire, did, in reality, induce a similar current through the other wire, but that it continued for an instant only, and partook

more of the nature of the electrical wave passed through from the shock of a common Leyden jar than of that from a voltaic battery, and therefore might magnetise a steel needle, although it scarcely affected the galvanometer.

13. This expectation was confirmed; for on substituting a small hollow helix, formed round a glass tube, for the galvanometer, introducing a steel needle, making contact as before between the battery and the inducing wire (7. 10), and then removing the needle before the battery contact was broken, it was found magnetised.

14. When the battery contact was first made, then an unmagnetised needle introduced into the small indicating helix, and lastly the battery contact broken, the needle was found magnetised to an equal degree apparently with the first; but the poles were of the contrary kind.

15. The same effects took place on using the large compound helices first described (6. 8).

16. When the unmagnetised needle was put into the indicating helix, before contact of the inducing wire with the battery, and remained there until the contact was broken, it exhibited little or no magnetism; the first effect having been nearly neutralised by the second (13. 14). The force of the induced current upon making contact was found always to exceed that of the induced current at breaking of contact; and if therefore the contact was made and broken many times in succession, whilst the needle remained in the indicating helix, it as last came out not unmagnetised, but a needle magnetised as if the induced current upon making contact had acted alone on it. This effect may be due to the accumulation (as it is called) at the poles of the unconnected pile, rendering the current upon first making contact more powerful than what it is afterwards, at the moment of breaking contact.

17. If the circuit between the helix or wire under induction and the galvanometer or indicating spiral was not rendered complete *before* the connexion between the battery and the inducing wire was completed or broken, then no effects were perceived at the galvanometer. Thus, if the battery communications were first made, and then the wire under induction connected with the indicating helix, no magnetising power was there exhibited. But still retaining the latter communication, when those with the battery were broken, a magnet was formed in the helix, but of the second kind, *i.e.* with poles indicating a current in the same direction to that belonging to the battery current, or to that always induced by that current in the first instance.

18. In the preceding experiments the wires were placed near to each other, and the contact of the inducing one with the battery made when the inductive effect was required; but as some particular action might be supposed to be exerted at the moments of making and breaking contact, the induction was produced in another way. Several feet of copper wire were stretched in wide zigzag forms, representing the letter W, on one surface of a broad board; a second wire was stretched in precisely similar forms on a second board, so that when brought near the first, the wires should everywhere touch, except that a sheet of thick paper was interposed. One of these wires was connected with the galvanometer,

and the other with a voltaic battery. The first wire was then moved towards the second, and as it approached, the needle was deflected. Being then removed, the needle was deflected in the opposite direction. By first making the wires approach and then recede, simultaneously with the vibrations of the needle, the latter soon became very extensive; but when the wires ceased to move from or towards each other, the galvanometer needle soon came to its usual position.

19. As the wires approximated, the induced current was in the *contrary* direction to the inducing current. As the wires receded, the induced current was in the *same* direction as the inducing current. When the wires remained stationary, there was no induced current (54).

20. When a small voltaic arrangement was introduced into the circuit between the galvanometer (10) and its helix or wire, so as to cause a permanent deflection of 30° or 40°, and then the battery of 100 pairs of plates connected with the inducing wire, there was an instantaneous action as before (11); but the galvanometer-needle immediately resumed and retained its place unaltered, notwithstanding the continued contact of the inducing wire with the trough: such was the case in whichever way the contacts were made (33).

21. Hence it would appear that collateral currents, either in the same or in opposite directions, exert no permanent inducing power on each other, affecting their quantity or tension.

22. I could obtain no evidence by the tongue, by spark, or by heating fine wire or charcoal, of the electricity passing through the wire under induction; neither could I obtain any chemical effects, though the contacts with metallic and other solutions were made and broken alternately with those of the battery, so that the second effect of induction should not oppose or neutralize the first (13. 16).

23. This deficiency of effect is not because the induced current of electricity cannot pass fluids, but probably because of its brief duration and feeble intensity; for on introducing two large copper plates into the circuit on the induced side (20), the plates being immersed in brine, but prevented from touching each other by an interposed cloth, the effect at the indicating galvanometer, or helix, occurred as before. The induced electricity could also pass through the trough (20). When, however, the quantity of fluid was reduced to a drop, the galvanometer gave no indication.

24. Attempts to obtain similar effects to these by the use of wires conveying ordinary electricity were doubtful in the results. A compound helix similar to that already described (6), and containing eight elementary helices was used. Four of the helices had their similar ends bound together by wire, and the two general terminations thus produced connected with the small magnetising helix contained an unmagnetised needle (13). The other four helices were similarly arranged, but their ends connected with a Leyden jar. On passing the discharge, the needle was found to be a magnet; but it appeared probable that a part of the electricity of the jar had passed off to the small helix, and so magnetised the needle. There was indeed no reason to expect that the electricity of a jar possessing as it does great tension, would not diffuse itself through all the metallic matter interposed between the coatings.

25. Still it does not follow that the discharge of ordinary electricity through a wire does not produce analogous phenomena to those arising from voltaic electricity; but as it appears impossible to separate the effects produced at the moment when the discharge begins to pass, from the equal and contrary effects produced when it ceases to pass (16), inasmuch as with ordinary electricity these periods are simultaneous, so there can be scarcely any hope that in this form of the experiment they can be perceived.

26. Hence it is evident that currents of voltaic electricity present phenomena of induction somewhat analogous to those produced by electricity of tension, although, as will be seen hereafter, many differences exist between them. The result is the production of other currents, (but which are only momentary,) parallel, or tending to parallelism, with the inducing current. By reference to the poles of the needle formed in the indicating helix (13. 14) and to the deflections of the galvanometer-needle (11), it was found in all cases that the induced current, produced by the first action of the inducing current, was in the contrary direction to the latter, but that the current produced by the cessation of the inducing current was in the same direction. For the purpose of avoiding periphrasis, I propose to call this action of the current from the voltaic battery, *volta-electric induction*. The properties of the wire after induction has developed the first current, and whilst the electricity from the battery continues to flow through its inducing neighbor (10. 18), constitute a peculiar electric condition, the consideration of which will be resumed hereafter.

25.2.2   Evolution of Electricity from Magnetism

27. A welded ring was made of soft round bar-iron, the metal being seven eighths of an inch in thickness, and the ring 6 in. in external diameter. Three helices were put round one part of this ring, each containing about 24 ft of copper wire one twentieth of an inch thick; they were insulated from the iron and each other, and superposed in the manner before described (6.), occupying about 9 in. in length upon the ring. They could be used separately or arranged together; the group may be distinguished by the mark A (Fig. 25.2). On the other part of the ring about 60 ft of similar copper wire in two pieces were applied in the same manner, forming a helix B, which had the same common direction with the helices of A, but being separated from it at each extremity by about half an inch of the uncovered iron.

28. The helix B was connected by copper wires with a galvanometer 3 ft from the ring. The wires of A were connected end to end so as to form one long helix, the extremities were connected with a battery of ten pairs of plates 4 in.^{2}. The galvanometer was immediately affected, and to a degree far beyond what has been described, when with a battery of tenfold power helices without iron were used (10); but though the contact was continued, the effect was not permanent, for the needle soon came to rest in its natural position, as if quite indifferent to

Fig. 25.2 Two copper wires
wrapped around an iron ring.
—[*K.K.*]

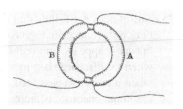

the attached electro-magnetic arrangement. Upon breaking the contact with the
battery, the needle was again powerfully deflected, but in the contrary direction
to that induced in the first instance.

29. Upon arranging the apparatus so that *B* should be out of use, the galvanometer
    be connected with one of the three wires of *A*, and the other two made into a
    helix through which the current from the trough (28) was passed; similar but
    rather more powerful effects were produced.

30. When the battery contact was made in one direction, the galvanometer needle
    was deflected on the one side; if made in the other direction, the deflection was
    on the other side. The deflection on making a battery contact always indicated an
    induced current in the opposite direction to that from the battery; but on break-
    ing the contact the deflection indicated an induced current in the same direction
    as that of the battery. No making or breaking of the contact at *B* side, or in any
    part of the galvanometer circuit, produced any effect at the galvanometer. No
    continuance of the battery current caused any deflection of the galvanometer-
    needle. As the above results are common to all these experiments, and to similar
    ones with ordinary magnets to be hereafter detailed, they need not be again
    particularly described.

31. Upon using the power of 100 pairs of plates (10) with this ring, the impulse
    at the galvanometer, when contact was completed or broken, was so great as
    to make the needle spin round rapidly four or five times before the air and
    terrestrial magnetism could reduce its motion to mere oscillation.

32. By using charcoal at the ends of the *B* helix, a minute spark could be perceived
    when the contact of the battery with *A* was completed. This spark could not
    be due to any diversion of a part of the current of the battery through the iron
    to the helix *B*; for when the battery contact was continued, the galvanometer
    still resumed its perfectly indifferent state (28). The spark was rarely seen on
    breaking contact. A small platina wire could not be ignited by this induced
    current; but there seems every reason to believe that the effect would be obtained
    by using a stronger original current or a more powerful arrangement of helices.

33. A feeble voltaic current was sent through the helix *B* and the galvanometer, so
    as to deflect the needle of the latter 30° or 40°, and then the battery of 100 pairs
    of plates connected with *A*; but after the first effect was over, the galvanometer
    needle resumed exactly the position due to the feeble current transmitted by its
    own wire. This took place in whichever way the battery contacts were made,
    and shows that here again (20) no permanent influence of the currents upon
    each other, as to their quantity and tension, exists.

Fig. 25.3 A galvanometer measures the electrical current in a copper wire wrapped around an iron cylinder placed between the poles of two bar magnets. —[*K.K.*]

34. Another arrangement was then used connecting the former experiments on volta-electric induction with the present. A combination of helices like that already described (6) was constructed upon a hollow cylinder of pasteboard: there were eight lengths of copper wire, containing altogether 220 ft; four of these helices were connected end to end, and then with the galvanometer (7); the other intervening four were also connected end to end, and the battery of 100 pairs discharged through them. In this form the effect on the galvanometer was hardly sensible (11), but magnets could be made by the induced current (13.). But when a soft iron cylinder seven eighths of an inch thick, and 12 in. long, was introduced into the pasteboard tube, surrounded by the helices, then the induced current affected the galvanometer powerfully, and with all the phenomena just described (30). It possessed also the power of making magnets with more energy, apparently, than when no iron cylinder was present.

35. When the iron cylinder was replaced by an equal cylinder of copper, no effect beyond that of the helices alone was produced. The iron cylinder arrangement was not so powerful as the ring arrangement already described (27).

36. Similar effects were then produced by *ordinary magnets*: thus the hollow helix just described (34) had all its elementary helices connected with the galvanometer by two copper wires, each 5 ft in length; the soft iron cylinder was introduced into its axis; a couple of bar magnets, each 24 in. long, were arranged with their opposite poles at one end in contact, so as to resemble a horse-shoe magnet, and then contact made between the other poles and the ends of the iron cylinder, so as to convert it for the time into a magnet (Fig. 25.3): by breaking the magnetic contacts, or reversing them, the magnetism of the iron cylinder could be destroyed or reversed at pleasure.

37. Upon making magnetic contact, the needle was deflected; continuing the contact, the needle became indifferent, and resumed its first position; on breaking the contact, it was again deflected, but in the opposite direction to the first effect, and then it again became indifferent. When the magnetic contacts were reversed, the deflections were reversed.

38. When the magnetic contact was made, the deflection was such as to indicate an induced current of electricity in the opposite direction to that fitted to form a magnet having the same polarity as that really produced by contact with the bar magnets. Thus when the marked and unmarked poles were placed as in Fig. 25.4, the current in the helix was in the direction represented, *P* being supposed to be the end of the wire going to the positive pole of the battery, or that end towards which the zinc plates face, and *N* the negative wire. Such a current would have converted the cylinder into a magnet of the opposite kind

Fig. 25.4 The direction of
the current induced in a
wire looped around an iron
rod when touched across
the poles of a horseshoe
magnet.
—[*K.K.*]

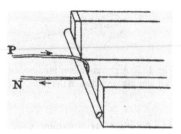

to that formed by contact with the poles *A* and *B*; and such a current moves in
the opposite direction to the currents which in M. Ampère's beautiful theory are
considered as constituting a magnet in the position figured.[9]

39. But as it might be supposed that in all the preceding experiments of this section it was by some peculiar effect taking place during the formation of the
magnet, and not by its mere virtual approximation, that the momentary induced
current was excited, the following experiment was made. All the similar ends of
the compound hollow helix (34) were bound together by copper wire, forming
two general terminations, and these were connected with the galvanometer. The
soft iron cylinder (34) was removed, and a cylindrical magnet, three quarters
of an inch in diameter and 8 in. and a half in length, used instead. One end
of this magnet was introduced into the axis of the helix (Fig. 25.5), and then,
the galvanometer-needle being stationary, the magnet was suddenly thrust in;
immediately the needle was deflected in the same direction as if the magnet had
been formed by either of the two preceding processes (34. 36). Being left in, the
needle resumed its first position, and then the magnet being withdrawn the needle was deflected in the opposite direction. These effects were not great; but by
introducing and withdrawing the magnet, so that the impulse each time should
be added to those previously communicated to the needle, the latter could be
made to vibrate through an arc of 180° or more.

40. In this experiment the magnet must not be passed entirely through the helix,
for then a second action occurs. When the magnet is introduced, the needle at
the galvanometer is deflected in a certain direction; but being in, whether it be

---

[9] The relative position of an electric current and a magnet is by most persons found very difficult
to remember, and three or four helps to the memory have been devised by M. Ampère and others.
I venture to suggest the following as a very simple and effectual assistance in these and similar
latitudes. Let the experimenter think he is looking down upon a dipping needle, or upon the pole
of the earth, and then let him think upon the direction of the motion of the hands of a watch, or of
a screw moving direct; currents in that direction round a needle would make it into such a magnet
as the dipping needle, or would themselves constitute an electro-magnet of similar qualities; or if
brought near a magnet would tend to make it take that direction; or would themselves be moved
into that position by a magnet so placed; or in M. Ampère's theory are considered as moving in
that direction in the magnet. These two points of the position of the dipping-needle and the motion
of the watch-hands being remembered, any other relation of the current and magnet can be at once
deduced from it.

Fig. 25.5 A magnet introduced into the axis of a helical
coil of wire. —[*K.K.*]

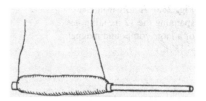

pushed quite through or withdrawn, the needle is deflected in a direction the
reverse of that previously produced. When the magnet is passed in and through
at one continuous motion, the needle moves one way, is then suddenly stopped,
and finally moves the other way.

41. If such a hollow helix as that described (34) be laid east and west (or in any
other constant position), and a magnet be retained east and west, its marked
pole always being one way; then whichever end of the helix the magnet goes in
at, and consequently whichever pole of the magnet enters first, still the needle
is deflected the same way: on the other hand, whichever direction is followed in
withdrawing the magnet, the deflection is constant, but contrary to that due to
its entrance.

42. These effects are simple consequences of the law hereafter described (114).

43. When the eight elementary helices were made of one long helix, the effect was
not so great as in the arrangement described. When only one of the eight helices
was used, the effect was also much diminished. All care was taken to guard
against any direct action of the inducing magnet upon the galvanometer, and it
was found that by moving the magnet in the same direction, and to the same
degree on the outside of the helix, no effect on the needle was produced.

44. The Royal Society are in possession of a large compound magnet formerly
belonging to Dr. Gowin Knight, which, by permission of the President and
Council, I was allowed to use in the prosecution of these experiments: it is
at present in the charge of Mr. Christie, at his house at Woolwich, where, by
Mr. Christie's kindness, I was at liberty to work; and I have to acknowledge
my obligations to him for his assistance in all the experiments and observations
made with it. This magnet is composed of about 450 bar magnets, each 15 in.
long, 1 in. wide, and half and inch thick, arranged in a box so as to present at
one of its extremities two external poles (Fig. 25.6). These poles projected horizontally 6 in. from the box, were each 12 in. high and 3 in. wide. They were
9 in. apart; and when a soft iron cylinder, three quarters of an inch in diameter
and 12 in. long, was put across from one to another, it required a force of nearly
100 pounds to break the contact. The pole to the left in the figure is the marked
pole[10].

---

[10] To avoid any confusion as to the poles of the magnet, I shall designate the pole pointing to the
north as the marked pole; I may occasionally speak of the north and south ends of the needle, but
do not mean thereby north and south poles. That is by many considered the true north pole of a
needle which points to the south; but in this country it is often called the south pole.

Fig. 25.6 A soft iron rod
spanning the 12 in. tall poles
of a large compound magnet.
—[*K.K.*]

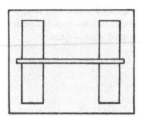

45. The indicating galvanometer, in all experiments made with this magnet, was about 8 ft from it, not directly in front of the poles, but about 16° or 17° on one side. It was found that on making or breaking the connexion of the poles by soft iron, the instrument was slightly affected; but all error of observation arising from this cause was easily and carefully avoided.

46. The electrical effects exhibited by this magnet were very striking. When a soft iron cylinder 13 in. long was put through the compound hollow helix, with its ends arranged as two general terminations (39), these connected with the galvanometer, and the iron cylinder brought in contact with the two poles of the magnet (Fig. 25.6), so powerful a rush of electricity took place that the needle whirled round many times in succession[11].

47. Notwithstanding this great power, if the contact was continued, the needle resumed its natural position, being entirely uninfluenced by the position of the helix (30). But on breaking the magnetic contact, the needle was whirled round in the opposite direction with a force equal to the former.

48. A piece of copper plate wrapped once round the iron cylinder like a socket, but with interposed paper to prevent contact, had its edges connected with the wires of the galvanometer. When the iron was brought in contact with the poles, the galvanometer was strongly affected.

49. Dismissing the helices and sockets, the galvanometer wire was passed over, and consequently only half round the iron cylinder (Fig. 25.7); but even then a strong effect upon the needle was exhibited, when the magnetic contact was made or broken.

50. As the helix with its iron cylinder was brought towards the magnetic poles, but without making contact, still powerful effects were produced. When the helix, without the iron cylinder, and consequently containing no metal but copper, was approached to, or placed between the poles (44), the needle was thrown 80°, 90°, or more, from its natural position. The inductive force was of course greater, the nearer the helix, either with or without its iron cylinder, was brought to the poles; but otherwise the same effects were produced; and the effects of approximation and removal were the reverse of each other (30).

---

[11] A soft iron bar in the form of a lifter to a horse-shoe magnet, when supplied with a coil of this kind round the middle of it, becomes, by juxta-position with a magnet, a ready source of a brief but determinate current of electricity.

Fig. 25.7 A wire attached to a
galvanometer and looped over
a soft iron rod placed between
the poles of a compound
magnet. —[*K.K.*]

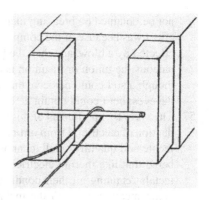

51. When a bolt of copper corresponding to the iron cylinder was introduced, no greater effect was produced by the helix than without it. But when a thick iron wire was substituted, the magneto-electric induction was rendered sensibly greater.

52. The direction of the electric current produced in all these experiments with the helix, was the same as that already described (38) as obtained with the weaker bar magnets.

53. A spiral containing 14 ft of copper wire, being connected with the galvanometer, and approximated directly towards the marked pole in the line of its axis, affected the instrument strongly; the current induced in it was in the reverse direction to the current theoretically considered by M. Ampère as existing in the magnet (38), or as the current in an electro-magnet of similar polarity. As the spiral was withdrawn, the induced current was reversed.

54. A similar spiral had the current of 80 pairs of 4-in. plates sent through it so as to form an electro-magnet, and then the other spiral connected with the galvanometer (53) approximated to it; the needle vibrated, indicating a current in the galvanometer spiral the reverse of that in the battery spiral (18. 26) On withdrawing the latter spiral, the needle passed in the opposite direction.

55. Single wires, approximated in certain directions towards the magnetic pole, had currents induced in them. On their removal, the currents were inverted. In such experiments the wires should not be removed in directions different to those in which they were approximated; for then occasionally complicated and irregular effects are produced, the causes of which will be very evident in the fourth part of this paper.

56. All attempts to obtain chemical effects by the induced current of electricity failed, though the precautions before described (22), and all others that could be thought of, were employed. Neither was any sensation on the tongue, or any convulsive effect upon the limbs of a frog, produced. Nor could charcoal or fine wire be ignited (133). But upon repeating the experiments more at leisure at the Royal Institution, with an armed loadstone belonging to Professor Daniell and capable of lifting about 30 pounds, a frog was very powerfully convulsed each time magnetic contact was made. At first the convulsions could

not be obtained on breaking magnetic contact; but conceiving the deficiency of effect was because of the comparative slowness of separation, the latter act was effected by a blow, and then the frog was convulsed strongly. The more instantaneous the union or disunion is effected, the more powerful the convulsion. I though also I could perceive the sensation upon the tongue and the flash before the eyes; but I could obtain no evidence of chemical decomposition.

57. The various experiments of this section prove, I think, most completely the production of electricity from ordinary magnetism. That its intensity should be very feeble and quantity small, cannot be considered wonderful, when it is remembered that like thermo-electricity it is evolved entirely within the substance of metals retaining all their conducting power. But an agent which is conducted along metallic wires in the manner described; which, whilst so passing possesses the peculiar magnetic actions and forces of a current of electricity; which can agitate and convulse the limbs of a frog; and which, finally, can produce a spark by its discharge through charcoal (32), can only be electricity. As all the effects can be produced by ferruginous electro-magnets (34), there is no doubt that arrangements like the magnets of Professors Moll, Henry, Ten Eyke, and others, in which as many as 2000 pounds have been lifted, may be used for these experiments; in which case not only a brighter spark may be obtained, but wires also ignited, and, as the current can pass liquids (23), chemical action be produced. These effects are still more likely to be obtained when the magneto-electric arrangements explained in the fourth section are excited by the powers of such apparatus.

58. The similarity of action, almost amounting to identity, between common magnets and either electro-magnets or volta-electric currents, is strikingly in accordance with and confirmatory of M. Ampère's theory, and furnishes powerful reasons for believing that the action is the same in both cases; but, as a distinction in language is still necessary, I propose to call the agency thus exerted by ordinary magnets, *magneto-electric* or *magnelectric* induction (26).

59. The only difference which powerfully strikes the attention as existing between volta-electric and magneto-electric induction, is the suddenness of the former and the sensible time required by the latter; but even in this early state of investigation there are circumstances which seem to indicate, that upon further inquiry this difference will, as a philosophical distinction, disappear (68).

25.3   Study Questions

QUES. 25.1. What was the inspiration for Faraday's first series of experimental researches? And what were the results?

a) Regarding terminology, in what two senses does Faraday use the term "induction"?

b) What do you think Faraday means by "Ampère's beautiful theory." And what did Ampere's experiments with electric currents suggest to Faraday?

c) What, then, did Faraday hope to accomplish? Was he successful? And what phenomenon did Faradays' work allow him to understand?

QUES. 25.2. Can electrical currents be induced by nearby electrical currents?

a) What is a galvanometer? A voltaic battery? How were these employed in Faraday's helical coil apparatus.
b) How many helices did Faraday's apparatus employ? How many independent and complete circuit loops did these form? And to what was each circuit loop attached?
c) Was Faraday's initial experiment successful? How, then, did he modify his apparatus? What, finally, was he able to observe? How did Faraday interpret his results?
d) Can a steel needle be magnetized without using a magnet? Under what conditions? Why is this significant?
e) Does it matter what type of material the helices are coiled about? For example, if the wooden cylinder is exchanged for a pasteboard cylinder which is either hollow or filled with a copper rod or an iron rod?

QUES. 25.3. Can collateral currents exert a permanent inducing power upon one another?

a) Describe Faraday's zig-zag wire apparatus. What purpose did the voltaic battery and the galvanometer serve?
b) What happened when the wires approached one another? When they receded from one another? How is this different than his previous work with coils?
c) How did Faraday measure the direction of the induced current? In particular, why do you think he placed an additional voltaic battery in the circuit containing the galvanometer?

QUES. 25.4. What is meant by the term volta-electric induction?

a) What is frictional electricity, and how does it differ from voltaic electricity? Are the effects produced by the discharge of frictional electricity the same as that produced by the flow of voltaic electricity?
b) What new law of induction does Faraday propose? Why do you think he introduces a new term to describe this law?

QUES. 25.5. How was Faraday's experiment with the iron ring similar to his previous experiments with helical coils? How were they different? Why is the use of iron significant?

QUES. 25.6. Can an ordinary magnet induce an electric current?

a) Describe the components of the apparatus shown in Fig. 25.3.
b) What happens when contact is made between the magnets and the iron rod? When contact is broken? When the polarity of the magnets is reversed?
c) What, then, is the precise relationship between the current in the helix and the orientation of the magnets?

QUES. 25.7. Can a stationary magnet induce an electric current?

a) Describe the components of the apparatus shown in Fig. 25.5. What happens when the magnet is stationary within the coil? When it is thrust in?
b) How can this effect be made more pronounced? Does it matter how deeply the magnet is thrust into the coil?

QUES. 25.8. Describe Faraday's induction experiments with Dr. Knight's powerful magnet. In particular, how did each experiment complement Faraday's previous work?

QUES. 25.9. Do volta-electric and magneto-electric currents have the same cause? Do they have the same nature or substance?

a) Apart from using a galvanometer, what alternative techniques did Faraday employ in order to infer the existence of an induced electric current? Were any of these successful?
b) Why do you suppose Faraday appeals to these demonstrations? What would it imply if these demonstrations finally proved unsuccessful?
c) What notable difference exists between volta-electric currents and magneto-electric currents? Does Faraday believe this difference is significant? Why?

25.4   Exercises

EX. 25.1 (ELECTRO-MOTIVE FORCE AND OHMS' LAW).   Chemical reactions within a voltaic battery generate an *electro-motive force* which acts on the ions in solution, separating them and causing them to accumulate on the terminals of the battery. The electro-motive force of a voltaic battery, ε, may be defined as the work, W, done in pushing a unit positive charge, q, from the negative to the positive terminal:

$$\varepsilon = W/q. \tag{25.1}$$

The charge separation caused by the electro-motive force results in an electric *tension*, or electric *potential*, between the poles, typically measured in *volts*. Notice that the electro-motive force is not technically a force since it is not measured in Newtons, but rather in volts (or joules per coulomb).

If an electrically conducting material, such as a copper wire, is placed between the terminals of the battery, then the electric potential difference induces an electric current through the connecting wire. According to *Ohm's law*,[12] the electric current,

---

[12] Georg Ohm, a German physicist, published the empirically discovered law named after him in 1827. An English translation of parts of this paper can be found in Magie, W. F. (Ed.), *A Source Book in Physics*, Harvard University Press, Cambridge, Massachusetts, 1963, pp. 456–472.

I, is proportional to this electric potential difference, V, between the ends of the wire:

$$V = IR. \tag{25.2}$$

Here R is the electrical resistance of the wire. The SI units of electrical resistance is, appropriately, the *ohm*, represented by the greek letter Ω. The electrical resistance of a wire depends on its length, L, its cross-sectional area, A, and the electrical *resistivity* of the material out of which it is made:

$$R = \rho L/A. \tag{25.3}$$

The electrical *resistivity* of a material, ρ, is the inverse of the material's electrical *conductivity*, σ; the SI unit of conductivity is the *siemen*. If the electrical resistance of a wire connecting the terminals of a battery is too small (for instance if the wire is quite thick) then the wire is said to "short" the terminals of the battery. That is, the electric potential across the battery terminals drops since the chemical reactions within the battery cannot keep up with the current drawn from the terminals through the wire.

As an exercise, look up the electrical resistivity of tungsten. Then use the above equations to calculate how much electrical current would be drawn through (a) a single 100-cm long 28-gauge tungsten wire connected between the terminals of a 9-V battery, (b) each of two identical 100-cm long 28-gauge tungsten wires connected side-by-side (in parallel) between the terminals of a 9-V battery, and (c) each of two identical 100-cm long 28-gauge tungsten wires connected end-to-end (in series) between the terminals of a 9-V battery. (d) What would happen if the 9-V battery was unable to source this level of electric current?

EX. 25.2 (MAGNETO-ELECTRIC INDUCTION LABORATORY). In this laboratory exercise, we will study electromagnetic induction using a galvanometer, a bar magnet, concentric helical coils of wire, a low-voltage power supply, and some connecting wires.[13] Connect one terminal of the galvanometer to the power supply and the other terminal, through a large resistance, to the other terminal of the power supply. The added resistance will prevent a short. If you have a low-current (*i.e.* safe) power supply, then you may use *yourself* as the resistance by grasping the ends of the connecting wires with your hands so that you become part of the circuit! Increase the voltage until you see a deflection on the galvanometer and record the relationship between the current direction and the deflection; this amounts to calibrating our galvanometer.

Now, remove the galvanometer from the power supply and connect its terminals instead to a secondary coil (the larger helical coil with the greater number of turns). In this way, you can measure any current which is induced in the secondary coil. Move a bar magnet into and out of the secondary coil and record the direction and

---

[13] This equipment is readily available from most scientific supply companies. For example, I have used Sargent Welch's primary and secondary coils (Item #CP32989-00), a galvanometer (Item #WLS30663-32) and an Elenco Regulated DC Power Supply (Item #CP32787-00).

magnitude of any induced current in the coil as a function of the position, speed and polarity of the magnet with respect to the coil. Does it matter whether you move the magnet or the coil? Draw appropriate diagrams and state, for instance, whether moving the north pole of the magnet into the coil induces a clockwise or a counterclockwise current around the helix as viewed from a clearly articulated orientation.

Next, attach a primary coil (the smaller coil) in series with the power supply. Turn on the power supply and study the behavior of a magnetic compass needle when situated near the ends of the primary coil. If the results are difficult to discern, place an iron core inside the primary coil. Does the primary coil behave like a bar magnet? Finally, move the primary coil in and out of the secondary coil, as you did previously with the bar magnet. What do you notice? From your experimental results, draw some general conclusions regarding the magneto-electric induction of current. Are your conclusions consistent with those of Faraday?

EX. 25.3 (ELECTRO-MOTIVE FORCE AND FARADAY'S LAW). Perhaps Faraday's most notable discovery is that an electro-motive force is generated around a loop of wire when the magnetic field through the loop is changed.[14] In other words, a changing magnetic field through the loop acts like a voltaic battery. This phenomenon is expressed mathematically by *Faraday's law*:

$$\varepsilon = -\frac{d\Phi}{dt} \tag{25.4}$$

The negative sign in Eq. 25.4 indicates the direction of the induced current.[15] The *magnetic flux*, Φ, is a measure of the amount of magnetic field which is passing perpendicularly through the loop. It is equal to the product of the magnetic field strength, B, and the cross-sectional area of the loop, A, which the magnetic field penetrates:[16]

$$\Phi = B \cdot A. \tag{25.5}$$

---

[14] We were introduced (anachronistically) to the concept of the magnetic field in Ex. 7.4; we will have much more to say about this in Chap. 28. Suffice it to say for the ti being that a magnetic field is a medium for communicating forces between magnetic bodies.

[15] See the discussion of Lenz's law in Ex. 25.4.

[16] In Eq. 25.5, we have tacitly assumed that the magnetic field strength is the same throughout the entire area of the loop, A. In many cases, however, the magnetic field is non-uniform. Then we must sum the contributions to the magnetic flux at the location of each of N tiny area elements, ΔA_i, which comprise the total area of the loop:

$$\Phi = \sum_{i=1}^{N} B_i \cdot \Delta A_i.$$

In each ΔA_i, the magnetic field strength, B_i, is approximately uniform. And in the limit that $N \to \infty$ and $\Delta A_i \to 0$, the above sum may be written as a surface integral over infinitesimal area segments; see Appendix A.

Now suppose that a copper wire is bent into a circular loop. A bar magnet on a rotating shaft flips about in front of the copper loop such that the magnetic flux through the loop varies according to

$$\Phi(t) = B_o \cos(2\pi t/T) \cdot \pi R^2.$$

Here, B_o denotes the magnetic field strength; suppose that its value is 1 mT. The factor πR^2 represents the area of the loop; suppose the loop radius is one centimeter. The cosine term indicates that the magnetic flux through the copper loop is periodically varying as the magnet flips about its axis once every T seconds; suppose that the period is 1 ms.

a) First, make a graph of the magnetic flux through the loop as a function of time. Be sure to label the axes and include some numbers to provide a sense of scale.
b) Derive an expression, in terms of the variables provided, for the electro-motive force acting on the loop. Overlay your function $\varepsilon(t)$ on your previous graph. What do you notice about the relative phases of $\Phi(t)$ and $\varepsilon(t)$?
c) What is the maximum value of the electro-motive force? What is the minimum value? Is it ever zero? If so, when?
d) If the most powerful magnet you had on hand was a 500 mT magnet, what variables could you modify (and by how much) for your spinning-magnet and wire-loop apparatus to act as a 60 Hz, 170-V (peak-to-peak) generator? (HINT: Could you use more loops of wire?)

Ex. 25.4 (LENZ'S LAW). According to Faraday's law, when the magnetic flux through a closed loop changes, an electro-motive force is generated which tends to create an electric current in the loop. But in which way does the current flow? A simple way of remembering the direction is *Lenz's law*:

> *The induced current in the loop is always in a direction so as to create its own magnetic flux through the loop which opposes the change in flux which induced the current.*

Recall the configuration of the magnetic field around a wire.[17] Then use Lenz's law to answer the following questions.

a) Consider Fig. 25.3 in the text. Suppose that the top bar magnet has its north pole on the left and the bottom bar magnet has its north pole on the right. In which direction will a current be induced when (i) contact is initially made between the bar magnets and the iron rod, (ii) a few moments after the initial contact, and (iii) when contact is broken.
b) Consider Fig. 25.5 in the text. Suppose that the cylindrical magnet has its north pole on the right. In which direction will a current be induced (i) before the magnet is inserted into the tube? (ii) as the magnet is slowly thrust into the tube? (iii) when the motion of the magnet is stopped inside the tube? (iv) when the magnet is drawn back out of the tube?

---

[17] The magnetic field around a long straight wire obeys the right hand rule; see Ex. 7.4 in the present volume.

25.5 Vocabulary

1. Induction
2. Electro-magnet
3. Concise
4. Helix
5. Interpose
6. Calico
7. Galvanometer
8. Ascertain
9. Periphrasis
10. Ferruginous
11. Terrestrial

Chapter 26
Arago's Mysterious Wheel

*I hoped to make the experiment of M. Arago a new source of
electricity; and did not despair, by reference to terrestrial
magneto-electric induction, of being able to construct a new
electrical machine.*

—Michael Faraday

26.1 Introduction

After a number of failed attempts, Faraday was able to generate electric currents
from ordinary magnetism. His experimental procedure was explained at the outset of
his first series of *Experimental Researches in Electricity*. In particular, Faraday discovered that an electrical current could be momentarily induced in a coil of copper
wire (i) when an electrical current in a second, nearby coil, was changed abruptly,
or (ii) when an iron rod, around which the coil had been wound, was abruptly magnetized or demagnetized. Faraday refers to these phenomena as *volta-electric* and
magneto-electric induction, respectively, and he carefully articulated the direction
of the induced electric currents (see Fig. 25.4).[1]

In the reading selection that follows, which is a continuation of his First Series
of his *Experimental Researches in Electricity*, Faraday begins to develop a general
law which governs *magneto-electric induction*.[2] He begins by describing an experimental puzzle: Arago had recently demonstrated that an ordinary magnetic compass
needle, when suspended by a thin string above a spinning copper disc, experiences a
torque which causes it to twist in the same direction as the rotating disc. Why might
this be? After all, copper is not a magnetic material like iron, so it should exert
no force whatsoever on the suspended needle. Faraday perceives that this strange
effect might be due to electrical currents induced within the spinning copper disc.

---

[1] In Exs. 25.3 and 25.4 of the previous chapter, the phenomenon of magneto-electric induction was
expressed in terms of (what are today known as) *Faraday's Law* (Eq. 25.4) and *Lenz's law*.

[2] For the sake of brevity, I have omitted §3 of Faraday's text. In this omitted section, entitled
New Electrical State or Condition of Matter, Faraday speculates about the nature of the so-called
electro-tonic state which arises within a material so as to drive the induced electrical currents.

© Springer International Publishing Switzerland 2016
K. Kuehn, *A Student's Guide Through the Great Physics Texts,*
Undergraduate Lecture Notes in Physics, DOI 10.1007/978-3-319-21816-8_26

Fig. 26.1 In Arago's apparatus, a suspended compass needle rotates in the direction of a copper disk spinning beneath it; a glass plate separates them to prevent air currents. This image is from p. 391 of Larden, W., *Electricity for public schools and colleges*, London: Longman's, 1891

So he constructs another apparatus with which he might explore whether electrical currents are in fact generated in a copper disk which is spinning near a powerful magnet (Fig. 26.1).

26.2   Reading: Faraday, *Experimental Researches in Electricity*

Faraday, M., *Experimental Researches in Electricity*, vol. 1, Taylor and Francis, London, 1839. First Series, beginning at §4.

26.2.1   Explication of Arago's Magnetic Phenomena

81. If a plate of copper be revolved close to a magnetic needle, or magnet, suspended in such a way that the latter may rotate in a plane parallel to that of the former, the magnet tends to follow the motion of the plate; or if the magnet be revolved, the plate tends to follow the motion of the plate; and the effect is so powerful that magnets or plates of many pounds weight may be thus carried round. If the magnet and plate be at rest relative to each other, not the slightest effect, attractive or repulsive, or of any kind, can be observed between them (62.). This is the phenomenon discovered by M. Arago; and he states that the effect takes place not only with all metals, but with solids, liquids, and even gases, *i.e.* with all substances (130.).

82. Mr. Babbage and Sir John Herschel, on conjointly repeating the experiments in this country[3], could obtain the effects only with excellent conductors of electricity. They refer the effect to magnetism induced in the plate by the magnet; the pole of the latter causing an opposite pole in the nearest part of the plate, and round this a more diffuse polarity of its own kind (120.). The essential circumstance in producing the rotation of the suspended magnet is, that the substance revolving below it shall acquire and lose its magnetism in a finite time, and not instantly (124.). This theory refers the effect to an attractive force, and is not agreed to by the discoverer, M. Arago, nor by M. Ampère, who quote against it the absence of all attraction when the magnet and metal are at rest (62. 126.), although the induced magnetism should still remain; and who, from experiments made with a long dipping needle, conceive the action to be always repulsive (125.).

83. Upon obtaining electricity from magnets by the means already described (36. 46.), I hoped to make the experiment of M. Arago a new source of electricity; and did not despair, by reference to terrestrial magneto-electric induction, of being able to construct a new electrical machine. Thus stimulated, numerous experiments were made with the magnet of the Royal Society at Mr. Christie's house, in all of which I had the advantage of his assistance. As many of these were in the course of the investigation superseded by more perfect arrangements, I shall consider myself at liberty to rearranged them in a manner calculated to convey most readily what appears to me to be a correct view of the nature of the phenomenon.

84. The magnet has been already described (44.). To concentrate the poles, and bring them nearer to each other, two iron or steel bars, each about 6 or 7 in. long, 1 in. wide, and half an inch thick, were put across poles as in Fig. 26.2, and being supported by twine from slipping, could be placed as near to or as far from each other as was required. Occasionally two bars of soft iron were employed, so bent that when applied, one to each pole, the two smaller resulting poles were vertically over each other, either being the uppermost at pleasure.

85. A disc of copper, 12 in. in diameter, and about one fifth of an inch in thickness, fixed upon a brass axis, was mounted in frames so as to be revolved either vertically or horizontally, its edge being at the same time introduced more or less between the magnetic poles (Fig. 26.2). The edge of the plate was well amalgamated for the purpose of obtaining a good but movable contact; a part round the axis was also prepared in a similar manner.

86. Conductors or collectors of copper and lead were constructed so as to come in contact with the edge of the copper disc (85.), or with other forms of plates hereafter to be described (101.). These conductors were about 4 in. long, one third of an inch wide, and one fifth of an inch thick; one end of each was slightly grooved, to allow of more exact adaptation to the somewhat convex

---

[3] Philosophical Transactions, 1825. p. 467.

Fig. 26.2 A copper disk free
to rotate around a brass axle
inserted between the poles of
a compound magnet. —[*K.K.*]

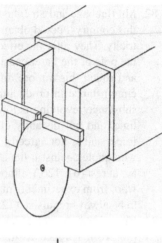

Fig. 26.3 A galvanometer.
—[*K.K.*]

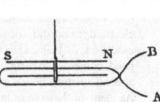

edge of the plates, and then amalgamated. Copper wires, one sixteenth of an
inch in thickness, attached, in the ordinary manner, but convolutions to the
other ends of these conductors, passed away to the galvanometer.

87. The galvanometer was roughly made, yet sufficiently delicate in its indications. The wire was of copper covered with silk, and made 16 or 18
convolutions. Two sewing-needles were magnetized and passed through a
stem of dried grass parallel to each other, but in opposite directions, and about
half an inch apart; this system was suspended by a fibre of un-spun silk, so
that the lower needle should be between the convolutions of the multiplier,
and the upper above them. The latter was by much the most powerful magnet,
and gave terrestrial direction to the whole; Fig. 26.3 represents the direction
of the wire and of the needles when the instrument was placed in the magnetic
meridian; the ends of the wires are marked *A* and *B* for convenient reference
hereafter. The letters *S* and *N* designate the south and north ends of the needle when affected merely by terrestrial magnetism; the end *N* is therefore the
marked pole (44.). The whole instrument was protected by a glass jar, and
stood, as to position and distance relative to the large magnet, under the same
circumstances as before (45.).

88. All these arrangements being made, the copper disk was adjusted as in
Fig. 26.2, the small magnetic poles being about half an inch apart, and the
edge of the plate inserted about half their width between them. One of the galvanometer wires was passed twice or thrice loosely round the brass axis of the
plate, and the other attached to a conductor (86.), which itself was retained

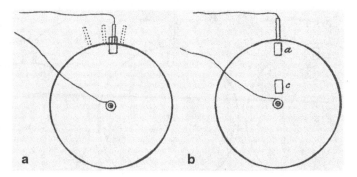

Fig. 26.4 (a) Electrical contact was made from the galvanometer to both the axle and to the edge of the rotating copper disk. (b) The copper disk could be raised or lowered so that the poles of the magnet could straddle the disk at positions *a* or *c*. —[*K.K.*]

by the hand in contact with the amalgamated edge of the disc at the part immediately between the magnetic poles. Under these circumstances all was quiescent, and the galvanometer exhibited no effect. But the instant the plate moved, the galvanometer was influenced, and by revolving the plate quickly the needle could be deflected 90° or more.

89. It was difficult under the circumstances to make the contact between the conductor and the edge of the revolving disc uniformly good and extensive; it was also difficult in the first experiments to obtain a regular velocity of rotation: both these causes tended to retain the needle in a continual state of vibration; but no difficulty existed in ascertaining to which side it was deflected, or generally, about what line it vibrated. Afterwards, when the experiments were made more carefully, a permanent deflection of the needle of nearly 45° could be sustained.

90. Here therefore was demonstrated the production of a permanent current of electricity by ordinary magnets (57.).

91. When the motion of the disc was reversed, every other circumstance remaining the same, the galvanometer needle was deflected with equal power as before; but the deflection was on the opposite side, and the current of electricity evolved, therefore, the reverse of the former.

92. When the conductor was placed on the edge of the disc a little to the right or left, as in the dotted positions Fig. 26.4a, the current of electricity was still evolved, and in the same direction as the first (88. 91.). This occurred to a considerable distance, *i.e.* 50 or 60° on each side of the place of the magnetic poles. The current gathered by the conductor and conveyed to the galvanometer was of the same kind on both sides of the place of greatest intensity, but gradually diminished in force from that place. It appeared to be equally powerful at equal distances from the place of the magnetic poles, not being affected in that respect by the direction of the rotation. When the rotation of the disc was reversed, the direction of the current of electricity was reversed also; but the other circumstances were not affected.

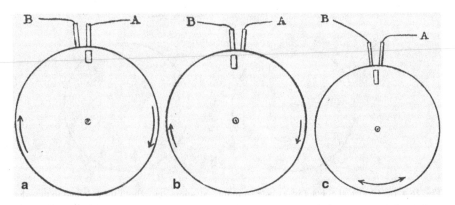

Fig. 26.5 Both wires from the galvanometer could also be put into electrical contact with the perimeter of the spinning disk, with the poles of the magnet nearer the right contact (**a**), the left contact (**b**), or midway between the two contacts (**c**). —[*K.K.*]

93. On raising the plate, so that the magnetic poles were entirely hidden from each other by its intervention (*a*. Fig. 26.4b), the same effects were produced in the same order, and with equal intensity as before. On raising it still higher, so as to bring the place of the poles to *c*, still the effects were produced and apparently with as much power as at first.

94. When the conductor was held against the edge as if fixed to it, and with it moved between the poles, even though but for a few degrees, the galvanometer needle moved and indicated a current of electricity, the same as that which would have been produced if the wheel had revolved in the same direction, the conductor remaining stationary.

95. When the galvanometer connexion with the axis was broken, and its wires made fast to two conductors, both applied to the edge of the copper disc, then currents of electricity were produced, presenting more complicated appearances, but in perfect harmony with the above results. Thus, if applied as in Fig. 26.5a, a current of electricity through the galvanometer was produced; but if their place was a little shifted, as in Fig. 26.5b, a current in the contrary direction resulted; the fact being, that in the first instance that galvanometer indicated the difference between a strong current through *A* and a weak one through *B*, and in the second, of a weak current through *A* and a strong one through *B* (92.), and therefore produced opposite deflections.

96. So also when the two conductors were equidistant from the magnetic poles, as in Fig. 26.5c, no current at the galvanometer was perceived, whichever way the disc was rotated, beyond what was momentarily produced by irregularity of contact; because equal currents in the same direction tended to pass into both. But when the two conductors were connected with one wire, and the axis with the other wire (Fig. 26.6), then the galvanometer showed a current according with the direction of rotation (91.); both conductors now acting consentaneously, and as a single conductor did before (88.).

Fig. 26.6 One contact might even be attached to the axle while the other is split into two, which are equidistant from the poles of the magnet. —[*K.K.*]

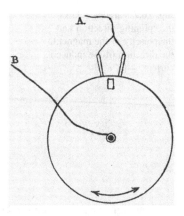

97. All of these effects could be obtained when only one of the poles of the magnet was brought near to the plate; they were of the same kind as to direction &c., but by no means so powerful.

98. All care was taken to render these results independent of the earth's magnetism, or of the mutual magnetism of the magnet and the galvanometer needles. The contacts were made in the magnetic equator of the plate, and at other parts; the plate was placed horizontally, and the poles vertically; and other precautions were taken. But the absence of any interference of the kind referred to, was readily shown by the want of all effect when the disc was removed from the poles, or the poles from the disc; every other circumstance remaining the same.

99. The relation of the current of electricity produced, to the magnetic pole, to the direction of rotation of the plate, &c. &c., may be expressed by saying, that when the unmarked pole (44. 84.) is beneath the edge of the plate, and the latter revolves horizontally, screw-fashion, the electricity which can be collected at the edge of the plate nearest to the pole is positive. As the pole of the earth may mentally be considered the unmarked pole, this relation of the rotation, the pole and the electricity evolved, is not difficult to remember. Or if, in Fig. 26.7, the circle represent the copper disc revolving in the direction of the arrows, and *a* the outline of the unmarked pole placed beneath the plate, then the electricity collected at *b* and the neighbouring parts is positive, whilst that collected at the centre *c* and other parts is negative (88.). The currents in the plate are therefore from the centre by the magnetic poles towards the circumference.

100. If the marked pole be placed above, all other things remaining the same, the electricity at *b*, Fig. 26.7, is still positive. If the marked pole be placed below, or the unmarked pole above, the electricity is reversed. If the direction of revolution in any case is reversed, the electricity is also reversed.

101. It is now evident that the rotating plate is merely another form of the simpler experiment of passing a piece of metal between the magnetic poles in a rectilinear direction, and that in such cases currents of electricity are produced

Fig. 26.7 Faraday relates
the spinning direction and
the polarity of the magnet to
the direction of the induced
current. —[*K.K.*]

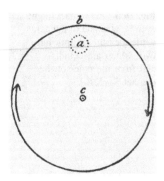

at right angles to the direction of the motion, and crossing it at the place of
the magnetic pole or poles. This was sufficiently shown by the following simple experiment: A piece of copper plate one-fifth of an inch thick, one inch
and a half wide, and 12 in. long, being amalgamated at the edges, was placed
between the magnetic poles, whilst the two conductors from the galvanometer were held in contact with its edges; it was then drawn through between the
poles of the conductors in the direction of the arrow, Fig. 26.8; immediately the
galvanometer needle was deflected, its north or marked end passed eastward,
indicating that the wire *A* received negative and the wire *B* positive electricity;
and as the marked pole was above, the result is in perfect accordance with the
effect obtained by the rotary plate (99.).

102. On reversing the motion of the plate, the needle at the galvanometer was
deflected in the opposite direction, showing an opposite current.

103. To render evident the character of the electrical current existing in various
parts of the moving copper plate, differing in their relation to the inducing
poles, one collector (86.) only was applied at the part to be examined near to
the pole, the other being connected with the end of the plate as the most neutral
place: the results are given at Fig. 26.9a–d, the marked pole being above the
plate. In Fig. 26.9a, *B* received positive electricity; but the plate moving in the
same direction, it received on the opposite side, Fig. 26.9b, negative electricity;
reversing the motion of the latter, as in Fig. 26.9d, *B* received positive electricity; or reversing the motion of the first arrangement, that of Fig. 26.9a–c,
B received negative electricity.

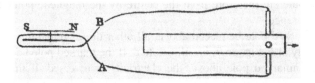

Fig. 26.8 An electrical current can be detected when a copper plate, placed beneath the marked
pole of a stationary magnet, is slid between two electrical contacts whose other ends are attached
to a galvanometer.—[*K.K.*]

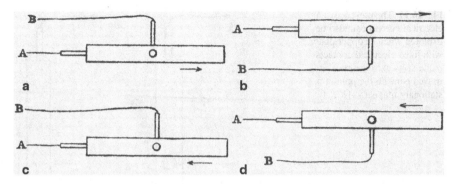

Fig. 26.9 The electrical contacts can be made at various locations so as to measure the direction of the current in the moving copper plate.—[*K.K.*]

Fig. 26.10 The magnetic pole need not be placed directly above the moving copper plate.—[*K.K.*]

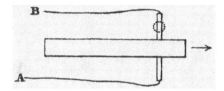

104. When the plates were previously removed sideways from between the magnets, as in Fig. 26.10, so as to be quite out of the polar axis, still the same effects were produced, though not so strongly.

105. When the magnetic poles were in contact, and the copper plate was drawn between the conductors near to the place, there was but very little effect produced. When the poles were opened by the width of a card, the effect was somewhat more, but still very small.

106. When an amalgamated copper wire, one eighth of an inch thick, was drawn through between the conductors and poles (101.), it produced a very considerable effect, though not so much as the plates.

107. If the conductors were held permanently against any particular parts of the copper plates, and carried between the magnetic poles with them, effects the same as those described were produced, in accordance with the results obtained with the revolving disk (94.).

108. On the conductors being held against the ends of the plates, and the plates then passed between the magnetic poles, in a direction transverse to their length, the same effects were produced (Fig. 26.11a). The parts of the plates towards the end may be considered either as mere conductors, or as portions of metal in which the electrical current is excited, according to their distance and the strength of the magnet; but the results were in perfect harmony with those before obtained. The effect was as strong as when the conductors were held against the sides of the plate (101.).

109. When the mere wire from the galvanometer, connected so as to form a complete circuit, was passed through the poles, the galvanometer was affected; and upon passing it to and fro, so as to make the alternate impulses pronounced

Fig. 26.11 The presence of an
electrical current can also be
explored when a copper plate,
with fixed electrical contacts
made at its extreme ends, is
moved beneath the pole of a
stationary magnet.—[*K.K.*]

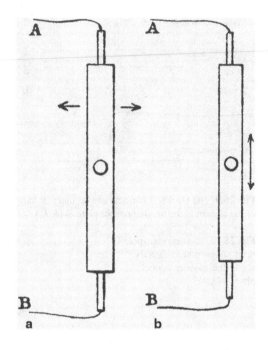

correspond with the vibrations of the needle, they could be increased 20 or 30° on each side of the magnetic meridian.

110. Upon connecting the ends of a plate of metal with the galvanometer wires, and then carrying it between the poles from end to end (as in Fig. 26.11b), in either direction, no effect whatever was produced upon the galvanometer. But the moment the motion became transverse, the needle was deflected.

111. These effects were also obtained from electro-magnetic poles, resulting from the use of copper helices or spirals, either alone or with iron cores (34. 54.). The directions of the motions were precisely the same; but the action was much greater when the iron cores were used than without.

112. When a flat spiral was passed through edgeways between the poles, a curious action at the galvanometer resulted; the needle first went strongly one way, but then suddenly stopped, as if it struck against some solid obstacle, and immediately returned. If the spiral were passed through from above downwards, or from below upwards, still the motion of the needle was in the same direction, then suddenly stopped, and then was reversed. But on truing the spiral half-way round, *i.e.* edge for edge, then the directions of the motions were reversed, but still were suddenly interrupted and inverted as before. This double action depends upon the halves of the spiral (divided by a line passing through its centre perpendicular to the direction of its motion) acting in opposite directions; and the reason why the needle went to the same side, whether the spiral passed by the poles in the one or the other direction, depended upon the circumstance, that upon changing the motion, the direction of the wires in the approaching half of the spiral was changed also. The effects, curious as they appear when witnessed, are immediately referable to the action of single wires (40. 109.).

113. Although the experiments with the revolving plate, wires, and plates of metal, were first successfully made with the large magnet belonging to the Royal Society, yet they were all ultimately repeated with a couple of bar magnets 2 ft long, 1 in. and a half wide, and half an inch thick; and, by rendering the galvanometer (87.) a little more delicate, with the most striking results. Ferro-electric-magnets, as those of Moll, Henry, &c. (57.), are very powerful. It is very essential, when making experiments on different substances, that thermo-electric effects (produced by contact of the fingers, &c.) be avoided, or at least appreciated and accounted for; they are easily distinguished by their permanency, and their independence of the magnets.

26.3   Study Questions

QUES. 26.1. How can Arago's magnetic phenomenon be used to generate a sustained electric current?

a) What is Arago's magnetic phenomenon? How did Babbage and Herschel explain it? And what is the weakness of this explanation?
b) Describe Faraday's apparatus. In particular, how did Faraday make electrical contact with the spinning copper disk? And how did he determine whether a current was induced in the disk?
c) Under what conditions did Faraday observe an induced current in the disk? Was the current transitory or sustained? And how does he summarize its behavior?
d) How does Faraday finally explain the behavior of Arago's wheel? To what simpler system does he compare the spinning wheel? Is this a good comparison?

26.4   Vocabulary

1. Induction
2. Electro-magnet
3. Galvanometer
4. Supersede
5. Amalgamated
6. Convex
7. Convolution
8. Terrestrial
9. Quiescent
10. Ascertain
11. Consentaneously
12. Inference
13. Magnetic meridian
14. Thermo-electric

Chapter 27
Faraday's Law

> *The law which governs the evolution of electricity by magneto-electric induction, is very simple, although rather difficult to express.*
>
> —Michael Faraday

27.1 Introduction

In his *Explication of Arago's Magnetic Phenomena* (§4 in the First Series of *Experimental Researches in Electricity*), Faraday described how he was able to generate a sustained electric current using only a bar magnet and a spinning copper disk. This he did by first touching the ends of a conducting wire to the disk, one to its axis and one to its periphery. When the pole of a magnet was aimed at a point on the disk between the two wire contacts, an electrical current flowed thought the wire when the disk was spun. This curious observation had profound consequences; it led directly to the development of the practical electric generator. For example, a constant electric current can be generated by a magnet placed before the face of a conducting disk being spun by a water- or wind-mill. But why does this work? What is the physical law governing this phenomenon? To answer this questions, Faraday considers a simpler system: a single conducting wire passing before the face of a bar magnet.

27.2 Reading: Faraday, *Experimental Researches in Electricity*

Faraday, M., *Experimental Researches in Electricity*, vol. 1, Taylor and Francis, London, 1839. First Series, §4 continued.

114. The relation which holds between the magnetic pole, the moving wire or metal, and the direction of the current evolved, *i.e.* the law which governs the evolution of electricity by magneto-electric induction, is very simple, although rather difficult to express. If in Fig. 27.1 *PN* represents a horizontal wire passing by a marked magnetic pole, so that the direction of its motion shall

© Springer International Publishing Switzerland 2016
K. Kuehn, *A Student's Guide Through the Great Physics Texts*,
Undergraduate Lecture Notes in Physics, DOI 10.1007/978-3-319-21816-8_27

Fig. 27.1 Faraday's explanation of the sense of the
electric current generated in
a wire passing before the
marked (*north*) face of a bar
magnet.—[*K.K.*]

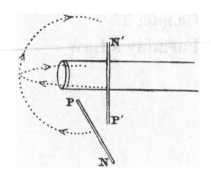

coincide with the curved line proceeding from below upwards; or if its motion parallel to itself be in a line tangential to the curved line, but in the general direction of the arrows; or if it pass the pole in other directions, but so as to cut the magnetic curves[1] in the same general direction, or on the same side as they would be cut by the wire if moving along the dotted curved line;—then the current of electricity in the wire is from P to N. If it be carried in the reverse directions, the electric current will be from N to P. Or if the wire be in the vertical position, figured P' N', and it will be carried in similar directions, coinciding with the dotted horizontal curve so far, as to cut the magnetic curves on the same side with it, the current will be from P' to N'. If the wire be considered a tangent to the curved surface of the cylindrical magnet, and it be carried round that surface into any other position, or if the magnet itself be revolved on its axis, so as to bring any part opposite to the tangential wire,— still, if afterwards the wire be moved in the directions indicated, the current of electricity will be from P to N; or if it be moved in the opposite direction, from N to P; so that as regards the motions of the wire past the pole, they may be reduced to two, directly opposite to each other, one of which produces a current from P to N, and the other from N to P.

115. The same holds true of the unmarked pole of the magnet, except that if it be substituted from the one in the figure, then, as the wires are moved in the direction of the arrows, the current of electricity would be from N to P, and as they move in the reverse direction, from P to N.

116. Hence the current of electricity which is excited in metal when moving in the neighbourhood of a magnet, depends for its direction altogether upon the relation of the metal to the resultant of magnetic action, or to the magnetic curves, and may be expressed in a popular way thus; Let AB (Fig. 27.2) represent a cylinder magnet, A being the marked pole, and B the unmarked pole; let PN be a silver knife-blade resting across the magnet with its edge upward, and with its marked or notched side towards the pole A; then in whatever direction

---

[1] By magnetic curves, I mean the lines of magnetic forces, however modified by the juxtaposition of poles, which would be depicted by iron filings; or those to which a very small magnetic needle would form a tangent.

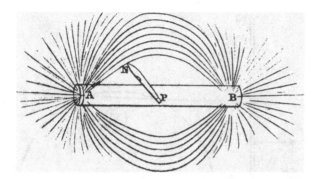

Fig. 27.2 An electric current arises in a knife blade cutting through the magnetic field lines surrounding a bar magnet.—[*K.K.*]

or position this knife be moved edge foremost, either about the marked or the unmarked pole, the current of electricity produced will be from *P* to *N* provided the intersected curves proceeding from *A* abut upon the notched surface of the knife, and those from *B* upon the un-notched side. Or if the knife be moved with its back foremost, the current will be from *N* to *P* in every possible position and direction, provided the intersected curves abut on the same surfaces as before. A little model is easily constructed, by using a cylinder of wood for a magnet, a flat piece for the blade, and a piece of thread connecting one end of the cylinder with the other, and passing through a hole in the blade, for the magnetic curves: this readily gives the result of any possible direction.

117. When the wire under induction is passing by an electro-magnetic pole, as for instance one end of a copper helix traversed by the electric current (34), the direction of the current in the approaching wire is the same with that of the current in the parts or sides of the spirals nearest to it, and in the receding wire the reverse of that in the parts nearest to it.

118. All these results show that the power of inducing electric currents is circumferentially excited by a magnetic resultant or axis of power, just as circumferential magnetism is dependent upon and is exhibited by an electric current.

119. The experiments described combine to prove that when a piece of metal (and the same may be true of all conducting matter) is passed either before a single pole, or between the opposite poles of a magnet, or near electro-magnetic poles, whether ferruginous or not, electrical currents are produced across the metal transverse to the direction of motion; and which therefore, in Arago's experiments, will approximate towards the direction of radii. If a single wire be moved like the spoke of a wheel near a magnetic pole, a current of electricity is determined through it from one end towards the other. If a wheel be imagined, constructed of a great number of these radii, and this revolved near a pole, in the manner of the copper disc (85), each radius will have a current produced in it as it passes by the pole. If the radii be supposed to be in contact

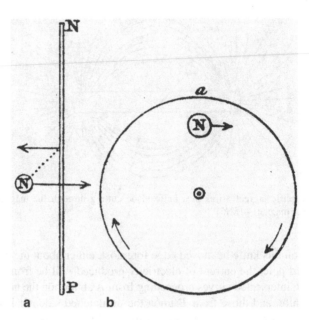

Fig. 27.3 The generation of electric current in a moving conducting wire **a** or a spinning copper disk **b** near the marked (*north*) pole of a magnet.—[*K.K.*]

laterally, a copper disk results, in which the directions of the currents will be generally the same, being modified only by the coaction which can take place between the particles, now that they are in metallic contact.

120. Now that the existence of these currents is known, Arago's phenomena may be accounted for without considering them as due to the formation in the copper of a pole of the opposite kind to that approximated, surrounded by a diffuse polarity of the same kind (82); neither is it essential that the plate should acquire and lose its state in a finite time; nor on the other hand does it seem necessary that any repulsive force should be admitted as the cause of the rotation (82).

121. The effect is precisely of the same kind as the electro-magnetic rotations which I had the good fortune to discover some years ago.[2] According to the experiments then made, which have since been abundantly confirmed, if a wire (PN, Fig. 27.3a) be connected with the positive and negative ends of a voltaic battery so that the positive electricity shall pass from P to N, and a marked magnetic pole N be placed near the wire between it and the spectator, the pole will move in a direction tangential to the wire, *i.e.* towards the right, and the wire will move tangentially towards the left, according to the directions of the arrows. This is exactly what takes place in the rotation of a plate beneath

---

[2] Quarterly Journal of Science, vol. xii. pp. 74, 186, 283, 416.

a magnetic pole; for let N (Fig. 27.3b) be a marked pole above the circular plate, the latter being rotated in the direction of the arrow: immediately currents of positive electricity set from the central parts in the general direction of the radii by the pole to the parts of the circumference a on the other side of that pole (99. 119), and are therefore exactly in the same relation to it as the current in the wire (PN, Fig. 27.3a) and therefore the pole in the same manner moves to the right hand.

122. If the rotation of the disc be reversed, the electric currents are reversed (91), and the pole therefore moves to the left hand. If the contrary pole be employed, the effects are the same, *i.e.* in the same direction, because currents of electricity the reverse of those described, are produced, and by reversing both poles and currents, the visible effects remain unchanged. In whatever position the magnetic axis be placed, provided the same pole be applied to the same side of the plate, the electric current produced is in the same direction, in consistency with the law already stated (114, &c); and thus every circumstance regarding the direction of the motion may be explained.

123. These currents are discharged or return in the parts of the plate on each side of and more distant from the place of the pole, where, of course, the magnetic induction is weaker: and when the collecters are applied, and a current of electricity is carried away to the galvanometer, the deflection there is merely a repetition, by the same current or part of it, of the effect of rotation in the magnet over the plate itself.

124. It is under the point of view just put forth that I have ventured to say it is not necessary that the plate should acquire and lose its state in a finite time (120); for if it were possible for the current to be fully developed in the instant *before* it arrived at its state of nearest approximation to the vertical pole of the magnet, instead of opposite to or a little beyond it, still the relative motion of the pole and plate would be the same, the resulting force being tangential instead of direct.

125. But it is possible (though not necessary for the rotation) that time may be required for the development of the maximum current in the plate, in which case the resultant of all the forces would be in advance of the magnet when the plate is rotated, or in the rear of the magnet when the latter is rotated, and many of the effects with pure electro-magnetic poles tend to prove this is the case. Then, the tangential force may be resolved into two others, one parallel to the plane of rotation, and the other perpendicular to it; the former would be the force excited in making the plate revolve with the magnet, or the magnet with the plate; the latter would be a repulsive force, and is probably that, the effects of which M. Arago has also discovered (82).

126. The extraordinary circumstances accompanying this action, which has seemed so inexplicable, namely, the cessation of all phenomena when the magnet and metal are brought to rest, now receives a full explanation (82); for then the electrical currents which cause the motion, cease altogether.

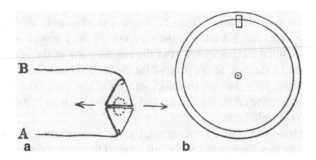

Fig. 27.4 The induction of current is hindered when a film of paper interrupts an otherwise connected copper plate **a** or disk **b** moving past the pole of a magnet.—[*K.K.*]

127. All the effects of solution of metallic continuity, and the consequent diminution of power described by Messrs. Babbage and Herschel,[3] now receive their natural explanation, as well also as the resumption of power when the cuts were filled but by metallic substances, which, though conductors of electricity, where themselves very deficient in the power of influencing magnets. And new modes of cutting the plate may be devised, which shall almost entirely destroy its power. Thus, if a copper plate (91) be cut through at about a fifth or sixth of its diameter from the edge, so as to separate a ring from it, and this ring be again fastened on, but with a thickness of paper intervening (Fig. 27.4a), and if Arago's experiment be made with this compound plate so adjusted that the section shall continually traverse opposite the pole, it is evident that the magnetic currents will be greatly interfered with, and the plate probably lose much of its effect.[4] An elementary result of this kind was obtained by using two piece of thick copper, shaped as in Fig. 27.4a. When the two neighbouring edges were amalgamated and put together, and the arrangement passed between the poles of the magnet, in a direction parallel to these edges, a current was urged through the wires attached to the outer angles, and the galvanometer became strongly affected; but when a single film of paper was interposed, and the experiment repeated, no sensible effect could be produced.

128. A section of this kind could not interfere much with the induction of magnetism, supposed to be of the nature ordinarily received by iron.

129. The effect of rotation or deflection of the needle, which M. Arago obtained by ordinary magnets, M. Ampère succeeded in procuring by electro-magnets. This is perfectly in harmony with the results relative to volta-electric and magneto-electric induction described in this paper. And by using flat spirals of copper wire, through which electric currents were sent, in place of ordinary

---

[3] *Philosophical Transactions*, 1825, p. 481.

[4] This experiment has actually been made by Mr. Christie, with the results here described, and is recorded in the *Philosophical Transactions* for 1827, p. 82.

magnetic poles (111), sometimes applying a single one to one side of the rotating plate, and sometimes two to opposite sides, I obtained the induced currents
of electricity from the plate itself, and could lead them away to, and ascertain
their existence by, the galvanometer.

130. The cause which has now been assigned for the rotation in Arago's experiment,
namely, the production of electrical currents, seems abundantly sufficient in all
cases where the metal, or perhaps even other conductors, are concerned; but
with regard to such bodies as glass, resins and above all, gases, it seems impossible that currents of electricity capable of producing these effects should be
generated in them. Yet Arago found that the effects in question were produced
by these and all bodies tried (81). Messrs. Babbage and Herschel, it is true,
did not observe them with any substance not metallic, except carbon, in a
highly conducting state (82). Mr. Harris has ascertained their occurrence with
wood, marble, freestone and annealed glass, but obtained no effect with sulphuric acid and saturated solution of sulphate of iron, although these are better
conductors of electricity than the former substances.

131. Future investigations will no doubt explain these difficulties, and decide the
point whether the retarding or dragging action spoken of is always simultaneous with electric currents.[5] The existence of the action in metals, only whilst
the currents exist, *i.e.*, whilst motion is given (82, 88), and the explication
of the repulsive action observed by M. Arago (82, 125), are the strong reasons for referring it to this cause; but it may be combined with others which
occasionally act alone.

132. Copper, iron, tin, zinc, lead, mercury, and all the metals tried, produced electrical currents when passed between the magnetic poles: the mercury was put
into a glass tube for the purpose. The dense carbon deposited in coal gas
retorts, also produced the current, but ordinary charcoal did not. Neither could
I obtain any sensible effects with brine, sulphuric acid, saline solutions, &c.,
whether rotated in basins, on inclosed in tubes and passed between the poles.

133. I have never been able to produce any sensation upon the tongue by the wires
connected with the conductors applied to the edges of the revolving plate (88)
or slips of metal (101). Nor have I been able to heat a fine platina wire, or
produce a spark, or convulse the limbs of a frog. I have failed also to produce
any chemical effects by electricity thus evolved (22, 56).

134. As the electric current in the revolving copper plate occupies but a small
space, proceeding by the poles and being discharged right and left at very
small distances comparatively; and as it exists in a thick mass of metal possessing almost the highest conducting power of any, and consequently offering
extraordinary facility for its production and discharge; and as, notwithstanding

---

[5] Experiments which I have since made convince me that this particular action is always due to the
electrical currents formed; and they supply a test by which it may be distinguished from the action
of ordinary magnetism, or any other cause, including those which are mechanical or irregular,
producing similar effects.

this, considerable currents may be drawn off which can pass through narrow wires, 40, 50, or even 100 ft long; it is evident that the current existing in the plate itself must be a very powerful one, when the rotation is rapid and the magnet strong. This is also abundantly proved by the obedience and readiness with which a magnet 10 or 12 pounds in weight follows the motion of the plate and will strongly twist up the cord by which it is suspended.

135. Two rough trials were made with the intention of constructing magneto-electric machines. In one, a ring 1 in. and a half broad and 12 in. external diameter, cut from a thick copper plate, was mounted so as to revolve between the poles of the magnet and represent a plate similar to those formerly used (101), but of interminable length; the inner and outer edges were amalgamated, and the conductors applied one to each edge, at the place of the magnetic poles. The current of electricity evolved did not appear by the galvanometer to be stronger, if so strong, as that from the circular plate (88).

136. In the other, small thick disks of copper or other metal, half an inch in diameter, were revolved rapidly near to the poles, but with the axis of rotation out of the polar axis; the electricity evolved was collected by conductors applied as before to the edges (86). Currents were procured, but of strength much inferior to that produced by the circular plate.

137. The latter experiment is analogous to those made by Mr. Barlow with a rotating iron shell, subject to the influence of the earth.[6] The effects then obtained have been referenced by Messrs. Babbage and Herschel to the same cause as that considered as influential in Arago's experiment;[7] but it would be interesting to know how far the electric current which might be produced in the experiment would account for the deflexion of the needle. The mere inversion of a copper wire six or seven times near the poles of the magnet, and isochronously with the vibrations of the galvanometer needle connected with it, was sufficient to make the needle vibrate through an arc of 60° or 70°. The rotation of a copper shell would perhaps decide the point, and might even throw light upon the more permanent, though somewhat analogous effects obtained by Mr. Christie.

138. The remark which has already been made respecting iron (66), and the independence of the ordinary magnetical phenomena of that substance, and the phenomena now described of magneto-electric induction in that and other metals, was fully confirmed by many results of the kind detailed in this section. When an iron plate similar to the copper one formerly described (101) was passed between the magnetic poles, it gave a current of electricity like the copper plate, but decidedly of less power; and in the experiments upon the induction of electric currents (9), no difference in the kind of action between iron and other metals could be perceived. The power therefore of an iron plate to drag a magnet after it, or to intercept magnetic action, should be carefully distinguished from the similar power of such metals as silver, copper,

---

[6] *Philosophical Transactions*, 1825, p. 317.

[7] *Ibid.* 1825, p. 485.

&c. &c. in as much as in the iron by far the greater part of the effect is due to what may be called ordinary magnetic action. There can be no doubt that the cause assigned by Messrs. Babbage and Herschel in explication of Arago's phenomena is true when iron is the metal used.

139. The very feeble powers which were found by those philosophers to belong to bismuth and antimony, when moving, of affecting the suspended magnet, and which has been confirmed by Mr. Harris, seem at first disproportionate to their conducting powers; whether it be so or not must be decided by future experiment (73).[8] These metals are highly crystalline, and probably conduct electricity with different degrees of facility in different directions, and it is not unlikely that where a mass is made up of a number of crystals heterogeneously associated, an effect approaching to that of actual division may occur (127); or the currents of electricity may become more suddenly deflected at the confines of smaller crystalline arrangements, and so be more readily and completely discharged within the mass.

Royal Institution, November 1831.

Note.—In consequence of the long period which has intervened between the reading and printing of the foregoing paper, accounts of the experiments have been dispersed, and through a letter of my own to M. Hachette, have reached France and Italy. That letter was translated (with some errors), and read to the Academy of Sciences at Paris, 26th December, 1831. A copy of it in *Le Temps* of the 28th December quickly reached Signor Nobili, who, with Signor Antinori, immediately experimented upon the subject, and obtained many of the results mentioned in my letter; others they could not obtain or understand, because of the brevity of my account. These results by Signori Nobili and Antinori have been embodied in a paper dated 31st January 1832, and printed and published in the number of the *Antologia* dated November 1831, (according at least to the copy of the paper kindly sent me by Signor Nobili). It is evident the work could not have been then printed; and though Signor Nobili, in his paper, has inserted my letter as the text of his experiments, yet the circumstance of back date has caused many here, who have heard of Nobili's experiments by report only, to imagine his results were anterior to, instead of being dependent upon, mine.

I may be allowed under these circumstances to remark, that I experimented on this subject several years ago, and have published results. (See Quarterly Journal of Science for July 1825. p. 338.) The following is also an excerpt from my notebook, dated November 28, 1825: "Experiments on inductions by connecting wire of

---

[8] I have since been able to explain these differences, and prove, with several metals, that the effect is in the order of the conducting power; for I have been able to obtain, by magneto-electric induction, currents of electricity which are proportionate in strength to the conducting power of the bodies experimented with (211).

voltaic-battery:—a battery of four troughs, ten pairs of plates, each arranged side by side—the poles connected by a wire about 4 ft long, parallel to which was another similar wire separated from it only by two thicknesses of paper, the ends of the latter were attached to a galvanometer:—exhibited no action, &.c &c. &c.—Could not in any way render induction evident from the connecting wire." The cause of failure at that time is now evident (79).

M.F., April 1832.

27.3   Study Questions

QUES. 27.1. What is the law which governs magneto-electric induction in a wire when moving in the neighborhood of a magnet?

a) Which direction is the electric current in each of the moving wires of Fig. 27.1? What happens if the motion is reversed? If the pole of the magnet is reversed? How does Faraday express this law "in a popular way"?
b) What happens when a loop of wire approaches a magnetic pole, whether of a bar magnet or an energized helical electro-magnet?
c) If a spoked wheel is spun before the marked pole of a magnet, in which direction will the current be induced in the spoke(s) as they pass before the pole? How is this related to Faraday's spinning copper disk experiment?

QUES. 27.2. What effect, if any, does a current-carrying wire have on a nearby magnet? And what effect, in turn, does the magnet have upon a nearby current-carrying wire?

QUES. 27.3. Do insulating materials experience magneto-electric induction when moving in front of the pole of a magnet? Out of what material, then, did Faraday build his first magneto-electric machine?

27.4   Exercises

EX. 27.1 (THE LORENTZ FORCE LAW AND FARADAY'S LAW). Faraday found that when a conducting wire is passed near the pole of a magnet, an electric current is generated in the wire. This is depicted schematically in Fig. 27.1. In the present exercise, we will explore the connection between this observation and the Lorentz force law. Recall that, according to the Lorentz force law, a magnetic field exerts a force on a nearby electric current.[9] More generally, a magnetic field exerts a force

---

[9] We came across the Lorentz force law as a means of understanding Ampére's observation that current carrying wires repel one another; see Chap. 8, and especially Ex. 8.2, in the present volume.

on any electric charge in motion. Since an electric current consists of individual charged particles in motion, the Lorentz force may be expressed in terms of the speed of the particle, v, and the strength of the magnetic field, B:

$$F = qvB \qquad (27.1)$$

Here, q is the charge of the moving particle. The direction of the Lorentz force acting on the moving particle is perpendicular to the plane formed by the magnetic field and the velocity. It is given by another *right hand rule*.[10] If the index finger of the right hand is extended in the direction of the velocity of the particle, and the remaining fingers of the right hand are bent so as to point in the direction of the magnetic field, then the thumb of this hand, when extended, will point in the direction of the Lorentz force acting on a positively charged particle. If the charged particle is negative, then the direction of the Lorentz force is reversed.[11]

Consider, now, a cylindrical copper rod which is oriented along the x-axis of a certain cartesian coordinate system. Suppose the rod is moving with a uniform velocity, v, along the y-axis of this coordinate system through a uniform magnetic field, B, which is directed upward, along the z-axis.

a) From a microscopic perspective, the rod itself consists of an equal number of positive and negative charges, all moving with the rod along the y-axis through the magnetic field. Are the Lorentz forces acting on the positive and the negative charges within the rod in the same direction? At which end will the positive charges accumulate? What about the negative charges?

b) Suppose, for the moment, that the positive charges within the rod are free to move in response to the Lorentz force. Write down a mathematical expression for the work, W, done by the Lorentz force in moving a positive charge, q, along the entire length of the rod, L. (HINT: Recall Eq. 10.1.)

c) Since the Lorentz force does work in separating the positive and negative charges in the moving rod, the moving rod acts like a voltaic battery whose electro-motive force is given by Eq. 25.1. From this perspective, write down a mathematical expression for the potential difference, ΔV, which develops between the ends of the rod as a result of its motion through the magnetic field. (ANSWER: $\Delta V = vBL$)

---

[10] The present right hand rule must not be confused with the other right-hand rule, articulated in Chap. 7, which was used to determine the direction of the magnetic field surrounding a current-carrying wire.

[11] The direction and magnitude of the Lorentz force can be summarized compactly using the cross product notation of vector algebra:

$$\vec{F} = q\vec{v} \times \vec{B}. \qquad (27.2)$$

Here, q is a scalar quantity representing the charge, and $\vec{v}$ and $\vec{B}$ are vector quantities representing the velocity and magnetic field. The cross-product is discussed in more detail in Appendix A.

Fig. 27.5 A copper rod
sliding across parallel rails
connected by a tungsten bar
in the presence of a magnetic
field

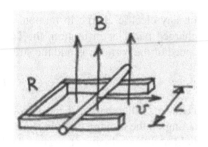

d) Suppose, now, that as the rod moves with velocity v, it slides along a set of parallel copper rails which are separated by a distance L and which are connected
by a tungsten bar whose resistance is R (see Fig. 27.5). Based on your previous considerations, what is the direction and magnitude of the electrical current
driven by the electro-motive force through the circuit formed by the copper rod,
the rails and the tungsten bar?

e) The induced current in the loop passes through the tungsten bar, which then heats
up. The rate of heat generation when a current, I, passes through a resistance, R,
is given by the *Joule heating formula*

$$P = I^2 R. \qquad (27.3)$$

Here, P is the rate at which electrical energy is transformed into heat; the SI units
of power is Joules per second (or Watts).[12] What is the rate of heat generation in
the tungsten wire? (ANSWER: $P = (vBL)^2/R$)

f) We have seen that an electric current is induced in the circuit due to the Lorentz
force law. But the induced electric current itself also feels a Lorentz force. In
which direction does this Lorentz force act on the induced current in the moving
copper rod? Does it act to speed up, or slow down, the rod? What force, then,
must be applied to the rod in order to keep it moving at a constant speed? And
what would happen if this force were not continually applied to the rod?

g) What is the rate of work done by the force which is required to keep the rod
moving at a constant speed? Is this equal to the rate of energy dissipation in the
tungsten wire? Is this surprising? Why or why not?

h) So far, we have approached this problem from the perspective of the Lorentz
force law. Let us instead look at it from the perspective of Faraday's law
(Eq. 25.4). As the copper rod slides along the rails, the magnetic flux through
the loop increases. Using the speed of the copper rod, what is the rate of change
of magnetic flux through the loop? And how does this allow us to calculate the
electro-motive force generated in the loop?

---

[12] See Helmholtz's discussion of the connection between electrical energy and heat in Chap. 11 of
the present volume.

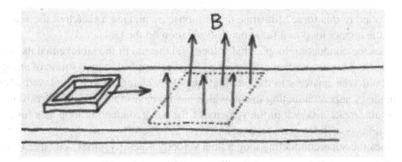

Fig. 27.6 A square loop sliding across a frictionless surface toward a region having a uniform magnetic field

i) Is the electro-motive force calculated using Faraday's law the same as the electro-motive force calculated using the Lorentz force law? Which method of calculation is better?

EX. 27.2 (ST. LOUIS MOTOR LABORATORY). Attach a St. Louis Motor to a DC power supply.[13] Increase the voltage and observe the motor spin. Using what you have learned thus far, try to explain how the motor works. Draw diagrams and be as precise as possible in your explanation. (a) In which direction are magnetic field(s) of the permanent magnets directed? (b) In which way do the electric currents in the coils flow? (c) In which way do the force(s) act on the coils at each instant? (d) Why does the shaft keep spinning?

EX. 27.3 (EDDY CURRENTS, MAGNETIC DAMPING, AND THE CONSERVATION OF ENERGY). When a conducting material, such as a copper plate or loop, moves through a non-uniform magnetic field, electrical currents are generated in the material. As we have seen in Ex. 27.1, the creation of these so-called *eddy currents* can be understood from the perspective of either the Lorentz force law (Eq. 27.1) or Faraday's law (Eq. 25.1). In this exercise we will explore the nature of eddy currents in both conducting and superconducting materials. Suppose a square loop of copper wire slides across the surface of a frictionless table. As it slides, it passes over a rectangular region of the table in which a uniform magnetic field is directed upward, as depicted in Fig. 27.6.

a) Describe the electric current generated in the loop (i) before it reaches the magnetic field, (ii) as it is entering the magnetic field, (iii) while it is entirely within the region of the magnetic field, (iv) as it is exiting the magnetic field, and (v) after it is entirely out of the magnetic field.
b) Recall that any induced current in the loop itself feels a Lorentz force. What is the net force acting on the loop in each of these regions? How does the loop

[13] The St. Louis Motor (Item #CP79945-00) and Elenco Regulated DC Power Supply (Item #CP32787-00) are available from Sargent Welch, Buffalo Grove, Illinois.

respond to this force? Illustrate this response by making a sketch of the velocity of the copper loop as a function of its location on the table.

c) Does the conducting loop's final velocity (at the end of the table) equal its initial velocity? Is your answer consistent with the principle of conservation of energy?

d) Would your answers to the previous questions change if you employed a loop made of superconducting metal, which has no measurable electrical resistance? Again, make a sketch of the velocity of the superconducting loop as a function of time as it slides across the table.

e) Does the superconducting loop's final velocity equal its initial velocity? Is your answer consistent with the principle of conservation of energy?

Ex. 27.4 (FALLING MAGNET PUZZLE). Drop a steel (or other non-magnetic metal) object through a long hollow aluminum or copper tube. Next, drop a similarly sized magnetic object down the same tube. Describe your observations, and clearly explain why this happens. Is the energy of this system conserved as the magnet falls through the tube?

Ex. 27.5 (ARAGO'S MAGNETIC PHENOMENON). Recall the inspiration for Faraday's experiments. Arago noticed that a magnetic compass needle suspended above a spinning copper disk begins to spin in the same direction as the disk. Using what you have learned about the magneto-electric induction of currents (and the forces which they experience) try to explain Arago's magnetic phenomenon.

27.5 Vocabulary

1. Magneto-electric induction
2. Circumferential
3. Lateral
4. Cessation
5. Amalgamate
6. Anneal
7. Explication
8. Brine
9. Procure
10. Brevity
11. Render

Chapter 28
The Magnetic Field

I am now about to leave the strict line of reasoning for a time,
and enter upon a few speculations respecting the physical
character of the lines of force, and the manner in which they
may be supposed to be continued through space.

—Michael Faraday

28.1 Introduction

Until the time of Faraday, the action of magnets and electric charges were described
almost exclusively in terms of their measured forces upon other nearby magnets
and electric charges. Indeed, this was the disposition of many of the founders of
the sciences of electricity and magnetism—men such as Coulomb, Biot, Poisson
and Ampère.[1] For example, Ampère describes his strictly empirical approach to
formulating a law of magnetic forces in the following way.[2]

> I have relied solely on experimentation to establish the laws of the phenomena and from
> them I have derived the formula which alone can represent the forces which are produced;
> I have not investigated the possible cause of these forces, convinced that all research of this
> nature must proceed from pure experimental knowledge of the laws and from the value,
> determined solely by deduction from these laws, of the individual forces in the direction
> which is, of necessity, that of a straight line drawn through the material points between
> which the forces act.

---

[1] See, for example, the section entitled "Biot and the Principles of Laplacian Physics," starting
on p. 118 of Hofmann, J. R., *André-Marie Ampère: Enlightenment and Electrodynamics*, Cambridge Science Biographies, Cambridge University Press, 1996. Also see James Clerk Maxwell's
discussion of *action-at-a-distance* and *mediated action* in Chap. 30 of the present volume.

[2] These quotes are from Ampère's "On the mathematical theory of electrodynamic phenomena,
experimentally deduced." See pp. 155–200 of Tricker, R. A. R., *Early Electrodynamics*, Pergamon,
New York, 1965.

© Springer International Publishing Switzerland 2016
K. Kuehn, *A Student's Guide Through the Great Physics Texts,*
Undergraduate Lecture Notes in Physics, DOI 10.1007/978-3-319-21816-8_28

Ampère then proceeds to contrast his own approach, which rejected all hypothetical causes, with that of Oersted, who imagined an *electric conflict* swirling in circles around a current-carrying wire.[3]

> It does not appear that this approach, the only one which can lead to results which are free of all hypothesis, is preferred by physicists in the rest of Europe like it is by Frenchmen; the famous scientist who first saw the poles of a magnet transported by the action of a conductor in directions perpendicular to those of the wire, concluded that electrical matter revolved about it and pushed the poles along with it.

Like many of the French scientists, Michael Faraday was himself a very meticulous and well-respected experimentalist. But like Oersted, he made a decided break with the French empirical tradition by explaining electric and magnetic actions in terms of invisible—but nonetheless real—electric and magnetic fields. In his 1852 article entitled "On the Physical Character of the Lines of Magnetic Force," published in *The London, Edinburgh and Dublin Philosophical Magazine and Journal of Science*, Faraday attempts to justify and defend his bold speculations. As you explore the text of this article, you might consider the following questions. Do you think that magnetic fields *actually exist*? If so, in what sense? Do they exist in the same sense that your hand exists? That light and gravity exist? That justice exists? That God exists? Or is the magnetic field perhaps just a useful (albeit disposable) conceptual hypotheses which we have inherited from the creative mind of Faraday?

28.2   Reading: Faraday, *On the Physical Character of the Lines of Magnetic Force*

Faraday, M., *Experimental Researches in Electricity*, vol. 3, Taylor and Francis, London, 1855. Reprinted from the *Philosophical Magazine* for June 1852.

NOTE.—The following paper contains so much of a speculative and hypothetical nature, that I have thought it more fitted for the pages of the Philosophical Magazine than those of the Philosophical Transactions. Still it is so connected with, and dependent upon former researches, that I have continued the system and series of paragraph numbers from them to it. I beg, therefore, to inform the reader, that those in the body of the text refer chiefly to papers already published, or ordered for publication, in the Philosophical Transactions; and that they are not quite essential to him in the reading of the present paper, unless he is led to a serious consideration of its contents. The paper, as is evident, follows Series XXVIII. and XXIX., now printing in the Philosophical Transactions, and depends much for its experimental support upon the more strict results and conclusions contained in them.

---

[3] See Oersted's *Experiments on the Effect of a Current of Electricity on the Magnetic Needle* in Chap. 7 of the present volume.

3243. I have recently been engaged in describing and defining the lines of magnetic
force (3070.), *i.e.* those lines which are indicated in a general manner by
the disposition of iron filings or small magnetic needles, around or between
magnets; and I have shown, I hope satisfactorily, how these lines may be
taken as exact representants of the magnetic power, both as to disposition
and amount; also how they may be recognized by a moving wire in a manner altogether different in principle from the indications given by a magnetic
needle, and in numerous cases with great and peculiar advantages. The definition then given had no reference to the physical nature of the force at the
place of action, and will apply with equal accuracy whatever that may be;
and this being very thoroughly understood, I am now about to leave the strict
line of reasoning for a time, and enter upon a few speculations respecting
the physical character of the lines of force, and the manner in which they
may be supposed to be continued through space. We are obliged to enter into
such speculations with regard to numerous natural powers, and, indeed, that
of gravity is the only instance where they are apparently shut out.

3244. It is not to be supposed for a moment that speculations of this kind are useless, or necessarily hurtful, in natural philosophy. They should ever be held as
doubtful, and liable to error and to change; but they are wonderful aids in the
hands of the experimentalist and mathematician. For not only are they useful in rendering the vague idea more clear for the time, giving it something
like a definite shape, that it may be submitted to experiment and calculation; but they lead on, by deduction and correction, to the discovery of new
phænomena, and so cause an increase and advance of real physical truth,
which, unlike the hypothesis that lead to it, becomes fundamental knowledge not subject to change. Who is not aware of the remarkable progress in
the development of the nature of light and radiation in modern times, and the
extent to which that progress has been aided by the hypotheses both of emission and undulation? Such considerations form my excuse for entering now
and then upon speculations; but though I value them highly when cautiously
advanced, I consider it as an essential character of a sound mind to hold them
in doubt; scarcely giving them the character of opinions, but esteeming them
merely as probabilities and possibilities, and making a very broad distinction
between them and the facts and laws of nature.

3245. In the numerous cases of force acting at a distance, the philosopher has
gradually learned that it is by no means sufficient to rest satisfied with the
mere fact, and has therefore directed his attention to the manner in which
the force is transmitted across the intervening space; and even when he can
learn nothing sure of the manner, he is still able to make clear distinction in
different cases, by what may be called the affections of the lines of power;
and thus, by these and other means, to make distinctions in the nature of the
lines of force of different kinds of power as compared with each other, and
therefore between the powers to which they belong. In the action of gravity for instance, the line of force is a straight line as far as we can test it
by the resultant phænomena. It cannot be deflected, or even affected, in its

course. Neither is the action in one line at all influenced, either in direction or amount, by a like action in another line; *i.e.* one particle gravitating toward another particle has exactly the same amount of force in the same direction, whether it gravitates to the one alone or towards myriads of other like particles, exerting in the latter case upon each one of them a force equal to that which it can exert upon the single one when alone: the results of course can combine, but the direction and amount of force between any two given particles remain unchanged. So gravity presents us with the simplest case of attraction; and appearing to have no relation to any physical process by which the power of the particles is carried on between them, seems to be a pure case of attraction or action at a distance, and offers therefore the simplest type of the cases which may be like it in that respect. My object is to consider how far magnetism is such an action at a distance; or how far it may partake of the nature of other powers, the lines of which depend, for the communication of force, upon intermediate physical agencies (3075.).

3246. There is one question in relation to gravity, which, if we could ascertain or touch it, would greatly enlighten us. It is, whether gravitation requires *time*. If it did, it would show undeniably that a physical agency existed in the course of the line of force. It seems equally impossible to prove or disprove this point; since there is no capability of suspending, changing, or annihilating the power (gravity), or annihilating the matter in which the power resides.

3247. When we turn to radiation phænomena, then we obtain the highest proof, that though nothing ponderable passes, yet the lines of force have a physical existence independent, in a manner, of the body radiating, or of the body receiving the rays. They may be turned aside in their course, and they deviate from a straight into a bent or a curved line. They may be affected in their nature so as to be turned on their axis, or else to have different properties impressed on different sides. Their sum of power is limited; so that if the force, as it issues from its source, is directed on to or determined upon a given set of particles, or into another direction, without being proportionately removed from the first. The lines have no dependence upon a second or reacting body, as in gravitation; and they require time for their propagation. In all these things they are in marked contrast with the lines of gravitating force.

3248. When we turn to the electric force, we are presented with a very remarkable general condition intermediate between the conditions of the two former cases. The power (and its lines) here requires the *presence* of two or more acting particles or masses, as in the case of gravity; and cannot exist with one only, as in the case of light. But though two particles are requisite, they must be in an *antithetical* condition in respect of each other, and not, as in the case of gravity, alike in relation to the force. The power is now dual; there it was simple. Requiring two or more particles like gravity, it is unlike gravity in that the power is limited. One electro-particle cannot affect a second, third and fourth, as much as it does the first; to act upon the latter its power must be proportionately removed from the former, and this limitation appears to

exist as a necessity in the dual character of the force; for the two states, or
places, or directions of force must be equal to each other.

3249. With the electric force we have both the static and dynamic state. I use these
words merely as names, without pretending to have a clear notion of the
physical condition which they seem meaningly to imply. Whether they are
two fluids or one, or any fluid of electricity, or such a thing as may be rightly
called a current, I do not know; still there are well-established electric conditions and effects which the words *static, dynamic,* and *current* are generally
employed to express; and with this reservation they express them as well as
any other. The lines of force of the *static* condition of electricity are present in
all cases of induction. They terminate at the surfaces of the conductors under
induction, or at the particles of non-conductors, which, being electrified, are
in that condition. They are subject to inflection in their course (1215. 1230.),
and may be compressed or rarefied by bodies of different inductive capacities (1252. 1277.); but they are in those cases affected by the intervening
matter; and it is not certain how the line of electric force would exist in relation to a perfect vacuum, *i.e.* whether it would be a straight line, as that of
gravity is assumed to be, or curved in such a manner as to show something
like physical existence separate from the mere distant actions of the surfaces
or particles bounding or terminating the induction. No condition of *quality*
or *polarity* has as yet been discovered in the line of static electric force; nor
has any relation of *time* been established in respect of it.

3250. The lines of force of dynamic electricity are either limited in their extent,
as in the lowering by discharge, or otherwise of the inductive condition of
static electricity; or endless and continuous, as closed curves in the case of
a voltaic circuit. Being definite in their amount for a given source, they can
still be expanded, contracted, and deflected almost to any extent, according
to the nature and size of the media through which they pass, and to which
they have a direct relation. It is probable that matter is always essentially
present; but the hypothetical æther may perhaps be admitted here as well as
elsewhere. No condition of quality or polarity has as yet been recognised in
them. In respect of *time*, it has been found, in the case of a Leyden discharge,
that time is necessary even with the best conductors; indeed there is reason
to think it is as necessary there as in the cases dependent on bad conducting
media, as, for instance, in the lightening flash.

3251. Three great distinctions at least may be taken among these cases of the exertion of force at a distance; that of gravitation, where propagation of the force
by physical lines through the intermediate space is not supposed to exist;
that of radiation, where the propagation does exist, and where the propagating line or ray, once produced, has existence independent either of its
source, or termination; and that of electricity, where the propagating process
has intermediate existence, like a ray, but at the same time depends upon
both extremities of the line of force, or upon conditions (as in the connected
voltaic pile) equivalent to such extremities. Magnetic action at a distance
has to be compared with these. It may be unlike any of them; for who shall

say we are aware of all the physical methods or forms under which force is communicated? It has been assumed, however, by some, to be a pure case of force at a distance, and so like that of gravity; whilst others have considered it as better represented by the idea of streams of power. The question at present appears to be, whether the lines of magnetic force have or have not a physical existence; if they have, whether such physical existence has a static or dynamic form (3075. 3156. 3172. 3173.).

3252. The lines of magnetic force have not as yet been affected in their *qualities, i.e.* nothing analogous to the polarization of a ray of light or heat has been impressed on them. A relation between them and the rays of light when polarized has been discovered (2146.);[4] but it is not of such a nature as to give proof as yet, either that the lines of magnetic force have a separate existence, or that they have not; though I think the facts are in favour of the former supposition.[5] The investigation is an open one, and very important.

3253. No relation of *time* to the lines of magnetic force has as yet been discovered. That iron requires *time* for its magnetization is well known. Plücker says the same is the case for bismuth, but I have not been able to obtain the effect showing this result. If that were the case, then mere space with its æther ought to have a similar relation, for it comes between bismuth and iron (2787.); and such a result would go far to show that the lines of magnetic force have a separate physical existence. At present, such results as we have cannot be accepted as in any degree proving the point of *time*; though if that point were proved they would most probably come under it. It may be as well to state, that in the case also of the moving wire or conductor (125. 3076.), time is required.[6] There seems no hope of touching the investigation by any method like those we are able to apply to a ray of light, or to the current of the Leyden discharge; but the mere statement of the problem may help towards its solution.

3254. If an action in *curved* lines or directions could be proved to exist in the case of the lines of magnetic force, it would also prove their physical existence external to the magnet on which they might depend; just as the same proof applies in the case of static electric induction.[7] But the simple disposition of the lines, as they are shown by iron particles, cannot as yet be brought in proof of such a curvature, because they may be dependent upon the presence of these particles and their mutual action on each other and the magnets; and it is possible that attractions and repulsions in right lines might produce the same arrangement. The results therefore obtained by the moving wire

---

[4] *Philosophical Transactions*, 1846, p. 1.

[5] In this paragraph, Faraday is refers to his recent discovery that the plane of polarization of light is rotated when propagating through a magnetic field; see Tyndall's explanation of the *Faraday rotation* of light in Chap. 24 of the present volume.—[*K.K.*]

[6] Experimental Researches, 8vo edition, vol. ii. pp. 191, 195.

[7] Philosophical Transactions, 1838, p. 16.

(3076. 3176.),[8] are more likely to supply data fitted to elucidate this point, when they are extended, and the true magnetic relation of the moving wire to the space which it occupies is fully ascertained.

3255. The *amount* of the lines of magnetic force, or the force which they represent, is clearly limited, and therefore quite unlike the force of gravity in that respect (3245); and this is true, even though the force of a magnet in free space must be conceived of as extending to incalculable distances. This limitation in amount of force appears to be intimately dependent upon the dual nature of the power, and is accompanied by a displacement or removability of it from one object to another, utterly unlike anything which occurs in gravitation. The lines of force abutting on one end or pole of a magnet may be changed in their direction almost at pleasure (3238.), though the original seats of their further parts may otherwise remain the same. For, by bringing fresh terminals of power into presence, a new disposition of the force upon them may be occasioned; but though these may be made, either in part or entirely, to receive the external power, and thus alter its direction, no change in the amount of the force is thus produced. And this is the case in strict experiments, whether the new bodies introduced are soft iron or magnets (3218. 3223.).[9] In this respect, therefore, the lines of magnetic force and of electric force agree. Results of this kind are well shown in some recent experiments on the effect of iron, when passing by a copper wire in the magnetic field of a horseshoe magnet (2129. 3130.), and also by the action of iron and magnets on each other (3218. 3223.).

3256. It is evident, I think, that the experimental data are as yet insufficient for a full comparison of the various lines of power. They do not enable us to conclude, with much assurance, whether the magnetic lines of force are analogous to those of gravitation, or direct actions at a distance; or whether, having a physical existence, they are more like in their nature to those of electric induction or the electric current. I incline at present to the latter view more than to the former, and will proceed to the further and future elucidation of the subject.

3257. I think I have understood that the mathematical expression of the laws of magnetic action at a distance is the same as that of the laws of static electric actions; and it has been assumed at times that the supposition of north and south magnetisms, spread over the poles or respective ends of a magnet, would account for all its external actions on other magnets or bodies. In either the static or dynamic view, or in any other view like them, the exertion of the magnetic forces outwards, at the poles or ends of the magnet, must be an essential condition. Then, with a given bar-magnet, can these forces exist without a mutual relation of the two, or else a relation to the contrary magnetic forces of equal amount originating in other sources? I do not believe they can; because, as I have shown in recent researches, the sum

---

[8] Ibid. 1852.

[9] *Philosophical Transactions*, 1852.

of the lines of force is equal for any section across them taken anywhere externally between the poles (3109.). Besides that, there are many other experimental facts which show the relation and connexion of the forces at one pole to those at the other;[10] and there is also the analogy with static electrical induction, where the one electricity cannot exist without relation to, equality with, and dependence on the other. Every dual power appears subject to this law as a law of necessity. If the opposite magnetic forces could be independent of each other, then it is evident that *a charge* with one magnetism only is possible; but such a possibility is negatived by every known experiment and fact.

3258. But supposing this necessary relation, which constitutes polarity, to exist, then how is it sustained or permitted in the case of an independent bar-magnet situated in free space? It appears to me, that the outer forces at the poles can only have relation to each other by *curved* lines of force through the surrounding space; and I cannot conceive curved lines of force without the conditions of a physical existence in that intermediate space. If they exist, it is not by a succession of particles, as in the case of static electric induction (1215. 1231.), but by the condition of space free from such material particles. A magnet placed in the middle of the best vacuum we can produce, and whether that vacuum be formed in a space previously occupied by paramagnetic or diamagnetic bodies, acts as well upon a needle as if it were surrounded by air, water or glass; and therefore these lines exist in such a vacuum as well as where there is matter.

3259. It may perhaps be said that there is no proof of any outer lines of force, in the case of a magnet, except when the objects employed experimentally to show these lines, as a magnetic needle, soft iron, a moving wire, or a crystal of bismuth, are present; that these bodies in fact, cause and develope the lines; just as in the case of gravity no idea of a line of gravitating force, in respect of a particle of matter by itself, can be formed: the idea exists only when a second particle is concerned. We are dealing, however, with a dual power; and we know that we cannot call into action, by magnetic induction upon soft iron or by electric currents, or otherwise, one magnetism without the other. Supposing, therefore, a bar of soft iron, or another bar-magnet, when brought end on and near to the first magnet, did by that approach develope the external force, the power which then only would become external should produce a corresponding external force of the contrary kind at the opposite extremity, or should not. If the first case occurs, it should be accompanied by the development of lines of force equivalent to it *within* the magnet. But I think we know, now, that in a very hard and perfect magnet there is no change of this kind (3223.). The outer and the inner lines of force remain the

---

[10] The manner in which a large powerful magnet deranges, overpowers, and even inverts the magnetism of a similar magnet, when it is brought near it in different directions without touching it, presents a number of such cases.

same in amount, whether the secondary magnet or the soft iron is present or away. It is the *disposition* only of the outer lines that is changed; their sum, and therefore the existence, remains the same. If the second case occurs, then the magnet, if broken in half under induction, should present in its fragments case of absolute magnetic charge, or charge with one magnetism only (3257. 3261.).

3260. Or if it be imagined for a moment, that the two polarities of the bar-magnet are in relation to each other, but that whilst there is no external object to be acted upon they are related to each other through the magnet itself (an idea very difficult to conceive after the experimental demonstration of the course of the lines as closed curves (3117. 3230.)), still it would follow, that upon the forces being determined externally, a change in the sum of force both within and without the magnet should be caused. We can now, however, take cognizance of both these partitions of force: and it appears that, with a good magnet, whether alone or under the influence of soft iron or other magnets of fourfold strength, the sum of forces without (3223.), and therefore also within (3117. 3121.) the magnet remains the same.

3261. If the northness and southness be considered so far independent of each other as to be compared to two fluids diffused over the two ends of the magnet (like the two electricities over a polarised conductor), then breaking the magnet in half ought to leave the two parts, one absolutely or differentially north in character and the other south. Each should not be both north and south in equality of proportion, considering only the external force. But this never happens. If it be said that the new fracture renders manifest, externally, two new poles, opposite in kind but equal in force (which is the fact), because of the necessity of the case, then the same necessity exists also for the dependence and relation of the original poles of the original magnet, no matter what or where the first source of the power may be. But in that case the *curved lines* of force between the poles of the original magnet follow as a consequence; and the curvature of these lines appears to me to indicate their physical existence.

3262. If the magnetic poles in a bar-magnet be supposed to exert some kind of power internally, backward, as if they were centres of force, both within and without the magnet, by which they are able, upon the breaking of the magnet, to develope the contrary poles and their force then that power cannot be the identical portion which is at the same time exerted externally; and if not the same, then when the magnet is broken, the two halves ought to have a degree of north or south charge. They ought not to be determinate magnets having equipotential poles. But they are so; and we may *break a hard magnet in half* whilst opposed to another powerful magnet which ought most to disturb the forces, and yet the broken halves are perfect magnets, equivalent in their polarities, just as if, when they were made by breaking, the dominant magnet was away. The power at the old poles is neither increased nor diminished, but remains in amount and in polar direction unchanged.

3263. Falling back, therefore, upon the case of a hard, well-made and well-charged straight bar-magnet, subject only to its own powers, it appears to me that we must either deny the joint external relation of the poles, and consider them as having no mutual tendency towards or action upon each other, or else admit that there is such an action exerted in or transmitted on through *curved* lines. To deny such an action, would be to set up a distinction between the action of the north end of a bar upon its south end, and its action upon the south end of the other magnets, which, in the face of all the old experiments, and the new ones made with the moving wire (3076.), it appears to me impossible to admit. To acknowledge the action in curved lines, seems to me to imply at once that the lines have a physical existence. It may be a vibration of the hypothetical æther, or a state or tension of that æther equivalent to either a dynamic or a static condition; or it may be some *other state*, which though difficult to conceive, may be equally distinct from the supposed non-existence of the line of gravitating force, and the independent and separate existence of the line of radiant force (3251.). Still the existence of the state does not appear to me to be mere assumption or hypothesis, but to follow in some degree as a consequence of the known condition of the force concerned, and the facts dependent on it.

3264. I have not referred in the foregoing considerations to the view I have recently supported by experimental evidence, that the lines of force, considered simply as representants of the magnetic power (3117.), are closed curves, passing in one part of their course through the magnet, and in the other part through the space around it. These lines are identical in their nature, qualities and amount, both within the magnet and without. If to these lines as formerly defined (3071.), we add the idea of physical existence, and then reconsider such of the cases which have just been mentioned as come under the new idea, it will be seen at once that the probability of curved external lines of force, and therefore of the physical existence of the lines, is as great, and even far greater, than before. For now no back action in the magnet could be supposed; and the external relation and dependence of the polarities (3257. 3263.) would, if it were possible, be even more necessary than before. Such a view would tend to give, but not necessarily, a dynamic form to the idea of magnetic force; and its close relation to dynamic electricity is well known (3265.). This I will proceed to examine; but before doing so, will again look for a moment at static electric induction, as a case of the dual powers in mutual dependence by curved lines of force, but with these lines terminated, and not existing as closed circuits. An electric conductor polarized by induction, or an insulated, unconnected, rectilineal, voltaic battery presents such a case, and resembles a magnet in the disposition of the external lines of force. But the sustaining action (as regards the induction) being dependent upon the necessary relation of the opposite dual conditions of the force, is external to the conductor, or the battery; and in such a case, if the conductor or battery be separated in the middle, no charge appears there, nor any origin of new lines of inductive force. This is, no doubt, a consequence

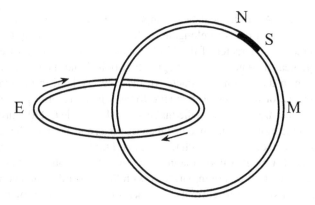

Fig. 28.1 The relative orientation of a ring of electric current and the lines of magnetic force it produces.—[*K.K.*]

of the fact, that the lines of static inductive force are not continued internally; and, at the same time, a cause why the two divided portions remain in opposite states or absolutely charged. In the magnet such a division *does* develope new external lines of force; which being equal in amount to those dependent on the original poles, shows that the lines of force are continuous through the body of the magnet, and with that continuity gives the necessary reason why no absolute charge of northness or southness is found in the two halves.

3265. The well-known relation of the electric and magnetic forces may be thus stated. Let two rings, in planes at right angles to each other, represent them, as in Fig. 28.1. If a current of electricity be sent round the ring E in the direction marked, then lines of magnetic force will be produced, correspondent to the polarity indicated by a supposed magnetic needle placed at NS, or in any part of the ring M to which such a needle may be supposed to be shifted. As these rings represent the lines of electro-dynamic force and of magnetic force respectively, they will serve for a standard of comparison. I have elsewhere called the electric current, or the line of electro-dynamic force, "an axis of power having contrary forces exactly equal in amount in contrary directions" (517.). The line of magnetic force may be described in *precisely the same terms*; and these two axes of power, considered as right lines, are perpendicular to each other; with this additional condition, which determines their mutual direction, that they are separated by a right line perpendicular to both. The meaning of the words above, when applied to the electric current, is precise, and does not imply that the forces are contrary *because* they are in reverse directions, but are *contrary in nature*; the turning one round, end for end, would not at all make it resemble the other; a consideration which may have influence with those who admit electric fluids, and endeavour to decide whether there are one or two electricities.

3266. When these two axes of power are compared, they have some remarkable correspondences, especially in relation to their position at right angles to each other. As a physical fact, Ampère[11] and Davy[12] have shown, that an electric current tends to elongate itself; and, so far, that may be considered as marking a character of the *electric* axis of power. When a free magnetic needle near the end of a bar-magnet first points and then tends to approach it, I see in the action a character of the contrary kind in the *magnetic* axis of power; for the lines of magnetic force, which, according to my previous researches, are common to the magnet and the needle (3230.), are shortened, first by the motion of the needle when it points, and again by the action which causes the needle to approach the magnet. I think I may say, that all the other actions of a magnet upon magnets, or soft iron, or other paramagnetic and diamagnetic bodies, are in harmony with the same effect and conclusions.

3267. Again:—like electric currents, or lines of force, or axes of power, when placed side by side, attract each other. This is well known and well seen, when wires carrying such currents are placed parallel to each other. But like magnetic axes of power or lines of force repel each other: the parallel case to that of the electric currents is given, by placing two magnetic needles side by side with like poles in the same direction; and by the use of iron filings, numerous pictorial representations (3234.) of the same general result may be obtained.

3268. Now these effects are not merely *contrasts* continued through two or more different relations, but they are contrasts which *coincide* when the position of the two axes of power at right angles to each other are considered (1659. 3265.). The tendency to *elongate* in the electric current, and the tendency to *lateral* separation of the magnetic lines of force which surround that current, are both tendencies in the same direction, though they seem like contrasts, when the two axes are considered out of their relation of mutual position; and this, with other considerations to be immediately referred to, probably points to the intimate physical relation, and it may be, to the oneness of condition of that which is apparently two powers or forms of power, electric and magnetic. In that case many other relations, of which the following are some forms, will merge in the same result. Thus, unlike magnetic lines, when end on, repel each other, as when similar poles are face to face; and unlike electric currents, if placed in the same relation stop each other; or if raised in intensity, when thus made static, repel each other. Like electric currents or lines of force, when end on to each other, coalesce; like magnetic lines of force similarly placed do so too (3266. 3295.). Like electric currents, end to end, do not add their sums; but whilst there is no change in quantity, the intensity is increased. Like magnetic lines of force, similarly placed do not increase each other, for the power then also remains the same (3218.): perhaps some effect correspondent to the gain of intensity in the former case may be produced, but there is none as yet distinctly recognised. Like electric

---

[11] *Ann. de Chim.* 1822, vol. xxi. p. 47.

[12] *Phil. Trans.* 1823, p. 153.

currents, side by side, add the quantities together; a case supplied either by uniting several batteries by their like ends, or comparing a large plate battery with a small one. Like magnetic lines of force do the same (3232.).

3269. The mutual relation of the magnetic lines of force and the electric axis of power has been known ever since the time of Oersted and Ampère. This, with such considerations as I have endeavoured to advance, enables us to form a guess or judgement, with a certain degree of probability, respecting the nature of the lines of magnetic force. I incline to the opinion that they have a physical existence correspondent to that of their analogue, the electric lines; and having that notion, am further carried on to consider whether they have a probable dynamic condition, analogous to that of the electric axis to which they are so closely, and perhaps, inevitably related, in which case the idea of magnetic currents would arise; or whether they consist in a state of tension (of the æther?) round the electric axis, and may therefore be considered as static in their nature. Again and again the idea of an electro-tonic state (60. 1114. 1661. 1729. 1733.) has been forced on my mind; such a state would coincide and become identified with that which would then constitute the physical lines of magnetic force. Another consideration tends in the same direction. I formerly remarked that the magnetic equivalent to *static* electricity was not known; for if the undeveloped state of electric force correspond to the like undeveloped condition of the magnetic force, and if the electric current or axis of electric power correspond to the lines of magnetic force or axis of magnetic power, then there is no known magnetic condition which corresponds to the static state of the electric power (1734.). Now assuming that the physical lines of magnetic force are currents, it is very unlikely that such a link should be naturally absent; more unlikely, I think, than that the magnetic condition should depend upon a state of tension; the more especially as under the latter supposition, the lines of magnetic power would have a physical existence as positively as in the former case, and the curved condition of the lines, which seems to me such a necessary admission, according to the natural facts, would become a possibility.

3270. The considerations which arise during the contemplation of the phænomena and laws that are made manifest in the mutual action of magnets, currents of electricity, and *moving conductors* (3084. &c.), are, I think, altogether in favour of the physical existence of the lines of magnetic force. When only a single magnet is employed in such cases, and the use of iron or paramagnetic bodies is dismissed, then there is no effect of attraction or repulsion or any ordinary magnetic result produced. The phænomena may all very fairly be looked upon as purely electrical, for they are such in character; and if they coincide with magnetic actions (which is no doubt the case), it is probably because the two actions are one. But being considered as electrical actions, they convey a different idea of the condition of the field where they occur, to that involved in the thought of magnetic action at a distance. When a copper wire is placed in the neighbourhood of a bar-magnet, it does not, as far as we are aware (by the evidence of a magnetic needle or other means), disturb in the least degree the disposition of the magnetic forces, either in itself or in surrounding space. When it is moved across the lines of force, a current of electricity is developed in it, or tends to be developed; and I have every reason to believe, that if we could employ a perfect conductor, and obtain a perfect result, it would be the full equivalent to the force, electric or magnetic, which is exerted in the place occupied by the conductor. But, as I have elsewhere observed (3172.), this current, having its full and equivalent relation to the magnetic force, can hardly be conceived of as having its entire foundation in the mere fact of motion. The motion of an external body, otherwise physically indifferent, and having no relation to the magnet, could not beget a physical relation such as that which the moving wire presents. There must, I think, be a previous state, a state of tension or a static state, as regards the wire, which, when motion is superadded, produces a dynamic state or current of electricity. This state is sufficient to constitute and give a physical existence to the lines of magnetic force, and permit the occurrence of curvature or its equivalent external relation of poles, and also the various other conditions, which I conceive are incompatible with mere action at a distance, and which yet do exist amongst magnetic phænomena.

3271. All the phænomena of the moving wire seem to be to show the physical existence of an atmosphere of power about a magnet, which, as the power is antithetical, and marked in its direction by the lines of magnetic force, may be considered as disposed in sphondyloids, determined by the lines or rather shells of force.[13] As the wire intersects the lines within a given sphondyloid external to the magnet, a current of electricity is generated, and that current is definite and the same for any or every intersection of the given sphondyloid. At the same time, whether the wire be quiescent or in motion, it does not cause derangement, or expansion, or contraction of the lines of force; the state of the power in the neighbouring or other parts of the sphondyloid remaining sensibly the same (3176.).

---

[13] The lines of magnetic force have been already defined (3071.). They have also been traced, as I think, and shown to be closed curves passing in one part of the course through the magnet to which they belong, and in the other part, through space (3117.). If, in the case of a straight bar-magnet, any one of these lines, E, be considered as revolving round the axis of the magnet, it will describe a surface; and as the line itself is a closed curve, the surface will form a tube round the axis and inclose a solid form. Another line of force, F, will produce a similar result. The sphondyloid body may be either that contained by the surface of revolution E, or that contained between the two surfaces of E and F, and which, for the sake of brevity, I have (by the advice of a kind friend) called simply *sphondyloid*. The parts of the solid described, which are within and without the magnet, are in power equivalent to each other. When it is needful to speak of them separately, they are easily distinguished as the inner and outer sphondyloids; the surface of the magnet being then part of the bounding surface.

3272. The old experiment of a wire when carrying an electric current[14] moving
round a magnetic pole, or of a current being produced in the same wire when
it is carried per force round the same pole (114.), shows the electrical dependence of the magnet and the wire, both when the current is employed from
the first, and when it is generated by the motion. It coincides in principle
with the results already quoted, and it includes, experimentally, all currents
of electricity, whatever the medium in which they occur, even up to that due
to the discharge of the Leyden jar or that between the electrodes of the voltaic
battery. I think it also indicates the state of magnetic or electric tension in the
surrounding space, not only when that space is occupied by metal or a wire,
but also by air and other bodies; for whatever be the state in one case, it is
probably general and therefore common to all (3173.).

28.3   Study Questions

QUES. 28.1. Why does Faraday engage in speculation regarding the physical existence of magnetic lines of force? What, according to Faraday, are the limits and
dangers of such speculation?

QUES. 28.2. How does Faraday distinguish the nature of gravity from that of light?
In particular, which of these (i) acts in a straight line,(ii) acts instantaneously,
(iii) may be polarized, (iv) can be separated from its source, (v) depends upon a
second, reacting, body for its existence? What do these distinctions imply about the
physical existence of gravity, and of light?

QUES. 28.3. Do electric lines of force physically exist?

a) In what sense are electric fields like, or unlike, gravity and light?
b) What is the nature of electric lines of force in the case of static electricity?
   Under what conditions does static electricity arise? Does each line of force have
   a beginning and an end, or are they closed loops?
c) How are the lines of electric force affected by the presence, or absence, of nearby
   matter? Can they be separated from their source? Can they be polarized like
   light? Do they act instantaneously?
d) What is the nature of the electric lines of force associated with dynamic
   electricity?
e) What, then, does Faraday conclude about the physical existence of electric lines
   of force? Are his conclusions justified?

QUES. 28.4. Is the physical existence of magnetic lines of force a "mere assumption
or hypothesis"?

a) Can lines of magnetic force be polarized like light? Do they act instantaneously?

---

[14] *Experimental Researches*, 8vo edition, vol. ii. p. 127.

b) Why is Faraday skeptical that the pattern of iron filings around a bar magnet provides ample proof for the existence of magnetic lines of force?

c) For a given bar magnet, are the number of lines of magnetic force limited, or unlimited? What governs the number of lines issuing from a particular magnet?

d) Do lines of magnetic force terminate? If so, where? If not, what is the relationship between the lines of force and magnetic poles?

e) Is a "charge" with only one magnetism, then, possible? What does this imply about the curvature of magnetic lines of force—and about their physical existence?

f) Do magnets require a surrounding medium in order to act on nearby magnets? Does an ambient medium affect a magnet's action in any way?

g) How does Faraday answer the objection that, just as in the case of gravity (which Faraday argues does *not* exist physically) a test object (*e.g.* a magnet) is required to show lines of magnetic force?

h) Are the number, and quality, of the lines of magnetic force equal inside and outside a magnet?

i) Do the lines of electric force outside a voltaic cell terminate at the poles of the cell, or do they continue through the cell? Do the lines of magnetic force outside a horseshoe magnet terminate at the poles of the magnet, or do they continue through the magnet?

j) What happens if a voltaic cell is divided in half? If a magnet is divided in half? What does this imply about the difference between the concept of positive and negative electricity, on the one hand, and the concept of "north-ness" and "south-ness", on the other hand?

k) What does Faraday mean when he says that the axes of electric and magnetic power are perpendicular to one another? And if a ring of electric force is typically accompanied by an electric current, is a ring of magnetic force similarly accompanied by motion? Or is it static in nature? Is there, then, a magnetic analogue to static electricity?

QUES. 28.5. Is it true that the relative motion of a copper wire and a magnet cannot, by itself, explain the generation of electric current?

a) Does the mere presence of a copper wire affect any lines of magnetic force which are in its vicinity?

b) Does a magnet induce an electric current in a stationary wire? In a moving wire? Does the moving wire itself derange the lines of magnetic force near a magnet?

c) What do you think: does the generation of a current by a moving wire imply the physical existence of an "atmosphere of power" about a nearby magnet? In what shape is Faraday's proposed "atmosphere of power" about a straight bar magnet?

28.4   Exercises

EX. 28.1 (SPECULATION ESSAY).   Before speculating about the physical character of
the lines of force, Faraday states that we "are obliged to enter into such speculations
with regard to numerous natural powers". Would you agree? And regarding such
speculation, would you agree that there is "a very broad distinction between them
and the facts and laws of nature"?

EX. 28.2 (ELECTRIC FIELD MAPPING LABORATORY).   In this laboratory exercise, we
will explore the configuration of the electric field around various charged bodies.
You will be using an apparatus called an Overbeck Field Mapping apparatus.[15]
Essentially, it consists of a base plate of insulating material to which can be attached
various field plates with differently shaped pairs of conductors. These various field
plates are attached to the underside of the base plate. The pairs of conductors on a
field plate are attached, *via* conducting bars, to the opposite terminals of a power
supply so as to charge them oppositely. A special probe is then used to measure
the voltage at various locations between the pair of conductors on a particular field
plate. By using the probe to measure the voltage in the region between the conductors, one can map the electric field in the region between them on the field plate.
The probe itself consists of a wand having two terminals. One terminal is attached
to a point of known electric potential (voltage); the other terminal is touched to a
point in the region of the field plate between the conductors. A galvanometer is then
used to measure the current between the terminals as follows: (i) if the point on
the plate is at a higher potential than the point of known voltage, then an electrical
current will flow from the plate to the point of known voltage, (ii) if the point on
the plate is at a lower potential than the point of known voltage, then an electrical
current will flow from the point of known voltage to the plate, and (iii) if the point
on the plate is at the same potential as the point of known voltage, then no electrical
current will flow between the plate to the point of known voltage. So by moving
the wand around on the field plate, one can map out several equipotential lines in
the regions separating the conductors. These can be sketched on a sheet of paper
fixed to the top of the base plate. The electric field lines can then be constructed by
drawing lines perpendicular to the equipotential lines. You should do this for each
of the field plates, and comment on the relationship between the configuration of the
charges and the shape of the electric field lines.

---

[15] CENCO Overbeck Electric Field Mapping Apparatus (Model No. CP79587-00), Sargent Welch,
Buffalo Grove, IL.

28.5   Vocabulary

1. Deduction
2. Radiation
3. Myriad
4. Antithetical
5. Dynamic
6. Polarity
7. Bismuth
8. Elucidate
9. Analogous
10. Paramagnetic
11. Diamagnetic
12. Manifest
13. Insulate
14. Conducting
15. Quiescent
16. Ponderable

Chapter 29
Paramagnetism and Diamagnetism

> *The lines from all the sources tend to coalesce, to pass through*
> *the best conductors, and to contract in length.*
>
> —Michael Faraday

29.1 Introduction

The nature of magnetism is a topic to which Faraday devotes considerable attention
in his experimental researches. For example, in his 20th and 21st Series, Faraday had
discovered that while some substances (such as iron, nickel and cobalt) are strongly
attracted to the poles of a magnet, other substances (such as Bismuth, Phosphorus, and, to a lesser extent, even water and gold) are repelled.[1] Faraday referred
to the former substances as *paramagnetic* and the latter as *diamagnetic*, so as to
indicate their respective orientations when suspended between the poles of a powerful electromagnet—either parallel to or across the axis of the magnet. Such careful
experimental work eventually led Faraday to the concept of what is now known as
the magnetic field—an "atmosphere of power" which surrounds a magnet and which
acts on other nearby magnets. In his article entitled "On the Physical Character of the
Lines of Magnetic Force," Faraday argues that magnetic fields are not mere mathematical constructions. Rather, like beams of light, they have a real physical existence. The curvature of these lines of force, in Faraday's estimation, is evidence of
their physical existence. After all, things that do not physically exist cannot be bent.

As we now continue to study Faraday's article in the present chapter, we will
consider the configuration and intensity of the magnetic field lines in the vicinity of
magnets of various shapes and sizes. For example, which is more powerful: a tall,
thin bar magnet or a short, fat one? More generally, is it the volume or the surface of
a magnet that is the seat of its power? Perhaps most interesting is Faraday's analysis
of how the type of substance placed in the vicinity of a permanent magnet affects the
configuration of the lines of magnetic force. Such considerations inspire Faraday to
develop a general rule governing the attraction and repulsion of magnets.

---

[1] See sections 2243–2453 in Faraday, M., *Experimental Researches in Electricity*, vol. 3, Taylor
and Francis, London, 1855.

© Springer International Publishing Switzerland 2016
K. Kuehn, *A Student's Guide Through the Great Physics Texts*,
Undergraduate Lecture Notes in Physics, DOI 10.1007/978-3-319-21816-8_29

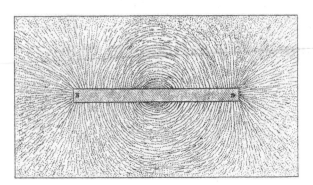

Fig. 29.1 The lines of force surrounding an ordinary bar magnet.—[*K.K.*]

29.2   Reading: Faraday, *On the Physical Character of the Lines of Magnetic Force*

Faraday, M., *Experimental Researches in Electricity*, vol. 3, Taylor and Francis, London, 1855. Reprinted from the *Philosophical Magazine* for June 1852.

3273. I will now venture for a time to assume the physical existence of the external lines of magnetic force, for the purpose of considering how the idea will accord with the general phænomena of magnetism. The magnet is evidently the sustaining power, and in respect of its internal condition or that of its particles, there is no idea put forth to represent it which at all approaches in probability and beauty to that of Ampère (1659.). Its analogy with the helix is wonderful; nevertheless there is, as yet, a striking experimental distinction between them; for whereas a magnet can never raise up a piece of soft iron to a state more than equal to its own, as measured by the moving wire (3219.), a helix carrying a current can develope in an iron core magnetic lines of force, of a 100 or more times as much power as that possessed by itself, when measured by the same means. In every point of view, therefore, the magnet deserves the utmost exertions of the philosopher for the development of its nature, both as a magnet and also as a source of electricity, that we may become acquainted with the great law under which the apparent anomaly may disappear, and by which all these various phænomena presented to us shall become *one*.

3274. The physical lines of force, in passing out of the magnet into space, present a great variety of conditions as to form (3238.). At times their refraction is very sudden, leaving the magnet at right, or obtuse, or acute angles, as in the case of a hard well-charged bar-magnet, Fig. 29.1; in other cases the change of form of the line in passing from the magnet into space is more gradual, as in the circular plate or globe-magnet, Figs. 29.2, 29.3a, and b.

Fig. 29.2 The lines of force surrounding a spherical magnet.—[*K.K.*]

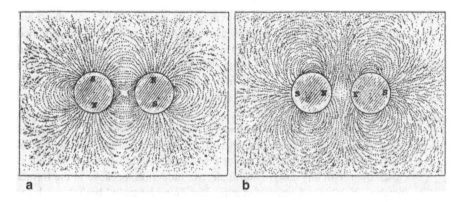

Fig. 29.3 The lines of force surrounding two nearby spherical magnets which are (**a**) anti-parallel and (**b**) whose like poles face each other.—[*K.K.*]

Here the form of the magnet as the source of the lines has much to do with the result; but I think the condition and relation of the surrounding medium has an essential and evident influence, in a manner I will endeavour to point out presently. Again, this refraction of the lines is affected by the relative difference of the nature of the magnet and the medium or space around it; as the difference is greater, and therefore the transition is more sudden, so the line of force is more instantaneously bent. In the case of the earth, both the nature of its substance and also its form, tend to make the refractions of the line of force at its surface very gradual; and accordingly the line of dip does not sensibly vary under ordinary circumstances, at the same place, whether it be observed upon the surface or above or below it.

3275. Though the physical lines of force of a magnet may, and must be considered as extending to infinite distance around it as long as the magnet is absolutely alone (3110.), yet they may be condensed and compressed into a very small local space, by the influence of other systems of magnetic power. This is indicated by Fig. 29.4. I have no doubt, after the experimental results given

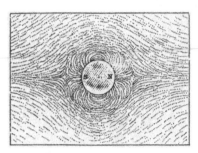

Fig. 29.4 The lines of force surrounding a small spherical magnet which is itself in the vicinity of a larger magnet.—[*K.K.*]

in Series XXVIII, respecting definite magnetic action (3109.), that the sphondyloid representing the total power, which in the experiment that supplied the figure had a sectional area of not 2 in.^{2} in surface, would have equal power upon the moving wire, with that infinite sphondyloid which would exist if the small magnet were in free space.

3276. The magnet, with its surrounding sphondyloid of power, may be considered as analogous in its condition to a voltaic battery immersed in water or any other electrolyte; or to a gymnotus (1773. 1784.) or torpedo, at the moment when these creatures, at their own will, fill the surrounding fluid with lines of electric force. I think the analogy with the voltaic battery so placed, is closer than with any case of *static* electric induction, because in the former instance the physical lines of electric force may be traced both through the battery and its surrounding medium, for they form continuous curves like those I have imagined within and without the magnet. The direction of these lines of electric force may be traced, experimentally, many ways. A magnetic needle freely suspended in the fluid will show them in and near to the battery, by standing at right angles to the course of the lines. Two wires from a galvanometer will show them; for if the line joining the two ends in the fluid be at right angles to the lines of electric force (or the currents), there will be no action at the galvanometer; but if oblique or parallel to these lines, there will be deflection. A plate, or wire, or ball of metal in the fluid will show the direction, provided any electrolytic action can go on against it, as when a little acetate of lead is present in the medium, for then the electrolysis will be a maximum in the direction of the current or line of force, and nothing at all in the direction at right angles to it. The same ball will disturb and inflect the lines of electric force in the surrounding fluid, just as I have considered the case to be with paramagnetic bodies amongst magnetic lines (2806. 2821. 2874.). No one I think will doubt that as long as the battery is in the fluid, and has its extremities in communication by the fluid, lines of electric force having a physical existence occur in every part of it, and the fluid surrounding it.

3277. I conceive that when a magnet is in free space, there is such a medium (magnetically speaking) around it. That a vacuum has its own magnetic relations of attraction and repulsion is manifest from former experimental results (2787.); and these place the vacuum in relation to material bodies, not at either extremity of the list, but in the *midst* of them, as, for instance, between gold and platina (2399.), having other bodies on either side of it. What that surrounding magnetic medium, deprived of all material substance, may be, I cannot tell, perhaps the æther. I incline to consider this outer medium as *essential* to the magnet; that it is that which relates the external polarities to each other by curved lines of power; and that these must be so related as a matter of necessity. Just as in the case of the battery above, there is no line of force, either in or out of the battery, if this relation be cut off by removing or intercepting the conducting medium;—or in that of static electric induction, which is impossible until this related state be allowed (1169.);[2]—so I conceive, that without this external mutually related condition of the poles, or a related condition of them to other poles sustained and rendered possible in like manner, a magnet could not exist; an absolute northness or southness, being as impossible as an absolute or an unrelated state of positive or negative electricity (1178.).

3278. In this view of a magnet, the medium or space around it is as essential as the magnet itself, being a part of the true and complete magnetic system. There are numerous experimental results which show us that the relation of the lines to the surrounding space can be varied by occupying it with different substances; just as the relation of a ray of light to the space through which it passes can be varied by the presence of different bodies made to occupy that space, or as the lines of electric force are affected by the media through which either induction or conduction takes place. This variation in regard to the magnetic power may be considered as depending upon the mutual relation of the two external polarities, or to carry onwards the physical line of force; and I have on a former occasion in some degree considered it and its consequences, using the phrase *magnetic conduction* to represent the physical effect (2797.) produced by the presence either of paramagnetic or diamagnetic bodies.

3279. When, for instance, a piece of cold iron (3129.) or nickel (3240.) is introduced into the magnetic field, previously occupied by air or being even mere space, there is a concentration of lines of force on to it, and more power is transmitted through the space thus occupied than if the paramagnetic body were not there. The lines of force therefore converge on to or diverge from it, giving what I have called conduction polarity (2818.); and this is the whole effect produced as regards the amount of the power; for not the slightest addition to, or diminution of, that external to the magnet is made (3218. 3223.). A new disposition of the force arises; for some passes now where it did not

---

[2] Philosophical Magazine, March 1843; or Experimental Researches, 8vo, vol. ii, p. 279.

pass before, being removed from places where it was previously transmitted. Supposing that the magnet was inclosed in a surrounding solid mass of iron, then the effect of its superior conducting power would be to cause a great contraction inwards of the sphere of external action, and of the various sphondyloids, which we may suppose to be identified in different parts of it. A magnetic needle, if it could be introduced into the iron medium, would indicate extreme diminution, if not apparent annihilation, of the external power of this magnet; but the moving wire would show that it was there present to its full extent (3152. 3162.) in a very concentrated condition, just as it shows it in the very body of a magnet (3116.); and the power within the magnet, it being a hard and perfect one, would remain the same.

3280. The reason why a magnetic needle would fail as a correct indicator of the amount of power present in a given space is, that when perfect, it, because of the necessary condition of hardness, cannot carry on through its mass more lines of force than it can excite (3223.). But because of the coalescence of like lines of force end on (3226.), such a needle, when surrounded by a bad magnetic conductor, determines on to itself many of the lines which would otherwise pass elsewhere, has a high magnetic polarity, and is affected in proportion; every experiment, as far as I can perceive, tending to show that the attractions and repulsions are merely consequences of the tendency which the lines of physical magnetic force have to shorten themselves (3266.). So when the magnetic needle is surrounded by a medium gradually increasing in conducting power, it seems to show less and less force in its neighbourhood, though in reality the force is increasing there more and more. We can easily conceive a very hard and feebly charged magnetic needle surrounded by a medium, as soft iron, better than itself in conducting power, *i.e.* carrying on by conduction more lines of force than the needle could determine or carry on by its state of charge (3298.). In that case I conceive it would, if free to move, point feebly in the iron, because of the coalescence of the lines of force, but would be repelled bodily from the chief magnet, in analogy with the action on a diamagnetic body. As I have before stated, the principle of the moving wire can be applied successfully in those cases where that of the magnetic needle fails (3255.).

3281. If other paramagnetic bodies than iron be considered in their relation to the surrounding space, then their effects may be assumed as proportionate to the conducting power. If the surrounding medium were hard steel, the contraction of the sphondyloid of power would be much less than with iron; and the effects, in respect of the magnetic needle, would occur in a limited degree. If a solution of protosulphate of iron were used, the effect would occur in a very much less degree. If a solution were prepared and adjusted so as to have no paramagnetic or diamagnetic relation (2422.), it would be the same to the lines of force as free space. If a diamagnetic body were employed, as water, glass, bismuth or phosphorus, the extent of action of the sphondyloids would expand (3279.); and a magnetic needle would appear to increase in intensity of action, though placed in a region having a smaller amount of magnetic force passing across it than before (3155.). Whether in any of these cases, even in that of iron, the body acting as a conductor has a state induced upon its particles for the time like that of a magnet in the corresponding state, is a question which I put upon a former occasion (2833.); but I leave its full investigation and decision for a future time.

3282. The circumstances dependent upon the shape and size of magnets appear to accord singularly well with the view I am putting forth of the action of the surrounding medium. If there be a function in that medium equivalent to conduction, involving differences in conduction in different cases, that of necessity implies also a reaction or resistance. The differences could not exist without. The analogous case is presented to us in every part by the electric force. When, therefore, a magnet, in place of being a bar, is made into a horseshoe form, we see at once that the lines of force and the sphondyloids are greatly distorted or removed from their former regularity; that a line of maximum force from pole to pole grows up as the horseshoe form is more completely given; that the power gathers in, or accumulates about this line, just because the badly conducting medium, *i.e.* the space or air between the poles, is shortened. A bent voltaic battery in its surrounding medium (3276.), or a gymnotus curved at the moment of its peculiar action (1785.), present exactly the like results.

3283. The manner in which the keeper or sub-magnet, when in place, reduces the power of the magnet in the space or air around is evident. It is the substitution of an excellent conductor for a poor one; far more of the power of the magnet is transmitted through it than through the same space before, and less, there-fore, in other places. If a horseshoe magnet be charged to saturation with its keeper on, and its power be then ascertained, removing the keeper will cause the power to fall. This will be (according to the hypothesis) because the iron keeper could, by its conduction, sustain higher external conditions of the magnetic force, and therefore the *magnet* could take up and sustain a higher level of charge. The case passes into that of a steel ring magnet, which being magnetized, shows no external signs of power, because the lines of force of one part are continued on by every other part of the ring; and yet when broken exhibit strong polarity and external action, because then the lines, which, being determined at a given point, were before carried on through the continuous magnet, have now to be carried on and continued through the surrounding space.

3284. These results, again, pass into the fact, easily verified partially, that if soft iron surround a magnet, being in contact with its poles, that magnet may receive a much higher charge than it can take, being surrounded with a lower paramagnetic substance, as air; also another fact, that when masses of soft iron are at the ends of a magnet, the latter can receive and keep a higher charge than without them; for these masses carry on the physical lines of force, and deliver them to a body of surrounding space; which is either widened, and therefore increased in the direction across the lines of force, or shortened in that direction parallel to them, or both; and both are circumstances which facilitate the conduction from pole to pole, and the relation of the external lines of force *within* the magnet. In the same way the armature of a natural loadstone is useful. All these effects and expedients accord with the view, that the space or medium external to the magnet is as important to its existence as the body of the magnet itself.

3285. Magnets, whether large or small, may be supersaturated, and then they fall in power when left to themselves; quickly at first if strongly supersaturated, and more slowly afterwards. This, upon the hypothesis, would be accounted for by considering the surrounding medium as unable, by its feeble magneto-conducting power, to sustain the higher state of charge. If the conducting power were increased sufficiently, then the magnet would not be supersaturated, and its power would not fall. Thus, if a magnet were surrounded by iron, it might easily be made to assume and retain a state of charge, which, if the iron were suddenly replaced by air, would instantly fall. Indeed, magnets can only be supersaturated by placing them for the time under the dominion of other sources of magnetic power, or of other more favourable surrounding media than that in which they manifest themselves as supersaturated.

3286. The well-known result, that small bar-magnets are far stronger in proportion to their size than larger similar magnets, harmonizes and *sustains* that view of the action of the external medium which has now been given. A sewing-needle can be magnetized far more strongly than a bar 12 in. long and an inch in diameter; and the reason under the view taken is, that the excited system in the magnet (correspondent to the voltaic battery in the analogy quoted (3276.)) is better sustained by the necessary conjoint action of the surrounding medium in the case of the small magnet. For as the imperfect magneto-conducting power of that medium (or the consequent state of tension into which it is thrown) acts back upon the magnet, (3282.), so the smaller the sum of exciting force in the centre of the magnetic sphondyloids, the better able will the surrounding medium be to do its part in sustaining the resultant of force. It is very manifest, that if the 12 in. bar be conceived of as subdivided into sewing-needles, and these be separated from each other, the whole amount of exciting force acts upon, and is carried onwards in closed magnetic curves, by a very much larger amount of external surrounding medium than when they are all accumulated in the single bar.

3287. The results which have been observed in the relation of *length* and *thickness* of a bar-magnet, harmonize with the view of the office of the external medium now urged. If we take a small, well-proportioned, saturated magnet, as a sewing-needle; alone, it has, as just stated, such relation to the surrounding space as to have its high condition sustained; if we place a second like magnetic needle by the side of the first, the surrounding space of the two is scarcely enlarged, it is not at all improved in conducting character, and yet it has to sustain double the internal exciting magnetic force exerted when there was one needle only (3232.); this must react back upon the magnets, and cause a reduction of their power. The addition of a third needle repeats the effect; and if we conceive that successive needles are added until the bundle is an inch thick, we have a result which will illustrate the effect of a thickness too large, and disproportionate to the length.

3288. On the other hand, if we assume two such needles similarly placed in a right line at a distance from each other, each has its surrounding system of curves occupying a certain amount of space; if brought together, by unlike poles, they form a magnet of double the length; the external lines of force coalesce (3226.), those at the faces of contact nearly disappear; those which proceed from the extreme poles coalesce externally, and form one large outer system of force, the lines of which have a greater length than the corresponding lines of either of the two original needles. Still, by the supposition that the magnets are perfectly hard and invariable, the exciting force within remains, or tends to remain the same (3227.) in quantity, there is nothing to increase it. The increase in length, therefore, of the external circuit, which acts as a resisting medium upon the internal action, will tend to diminish the force of the whole system. Such would be the case if a voltaic battery surrounded by distilled water, as the analogous illustration (3276.), could be elongated in the water, and so its poles be removed further apart; and though in the case of magnets previously charged, some effect equivalent to intensity of excitement may be produced by conjoining several together on end, yet the diminished sustentation of power externally appears to follow as a consequence of the increased distance of the extreme poles, or external, mutually dependent parts. Static electric induction also supplies a correspondent and illustrative case.

3289. The usual case in which the influence of length and thickness becomes evident, is not, however, always or often that of the juxtaposition of magnets already as highly charged as they can be, but rather that of a bar about to be charged. If two bars, alike in steel, hardness, &c., one an inch long and the tenth of an inch in diameter, and the other of the same length by five-tenths of an inch in diameter, be magnetized to supersaturation, the latter, though it contains 25 times the steel of the former, will not retain 25 times the power, for the reasons already given (3287.); the surrounding medium not being able to sustain external lines of force to that amount. But if a third bar, 2 in. long and also five-tenths in diameter, be magnetized at the same time, it can receive much more power than the second one. A natural reason for this presents itself by hypothesis; for the limitation of power in the two cases is not in the magnets themselves, but in the external medium. The shorter magnet has contact and connexion with that medium by a certain amount of surface; and just what power the medium outside that surface can support, the magnet will retain. Make the magnet as long again, and there is far more contact and relation with the surrounding medium than before; and therefore the power which the magnet can retain is greater. If there were such limited points of resulting action in the magnet as is often understood by the word *poles*, then such a result could hardly be the case, on my view of the physical actions. But such poles do not exist. Every part of the surface of the magnet, so to say, is pouring forth externally lines of magnetic force, as may be seen in Figs. 29.1, 29.2, 29.3a, and b (3274.). The larger the magnet, to a certain extent, and the larger the amount of external conducting medium in contact with it, the more freely is this transmission made. If the second magnet, being an inch long, be conceived to be charged to its full amount, and then, whilst in free space, could have half an inch of iron added to its length at each end, we see and know that many of the lines of force originally issuing from that part of its surface still left in contact with the air at the equatorial part, would now move internally towards the ends, and issue at a part of the soft iron surface; indicating the manner in which the tension would be relieved by this better conducting medium at the ends, and by increased surface of contact with the surrounding bad conductor of air or space. The thick, short magnet could evidently excite and carry on physical lines of magnetic power far more numerous than those which the space about it can receive and convey from pole to pole; and the increase in the length of the magnet may go on advantageousely, until the increasing sum of power, sustainable by the increasing medium in the circuit, is equal to that which the magnet can sustain or transmit internally; for all the lines of power, wherever they issue from the magnet, have to pass through its equator; and in this way the equator or thickness of the magnet becomes related to its length. So the advantageous increase in length of the bar is limited by the increasing resistance within, and especially at the equator of the bar; and the increase in breadth, by the increasing resistance (for increasing powers) of the external surrounding medium (3287.).

3290. It is very interesting to observe the results obtained when an attempt is made to magnetize, regularly, a thin steel wire, about 15 or 20 in. in length, and 0.05 of an inch in diameter. It can hardly be effected by bars; and when the wire is afterwards examined by filings (3234.), it is found to have irregular and consecutive poles, which vary as the magnetization is repeated with the same wire, as if they broke out suddenly by a rupture of something like unstable equilibrium; the effects apparently being chiefly referable to the cause now assigned. Again, when a magnet is made out of a thin, hard, steel plate, whose length is 10 or 12 times its width, it is well known how the lines of force issue from it in greatest abundance at the extreme angles, and then at the edges; and how a spot on the face gives exit to a much smaller number of lines than a like spot on the edge, at the same distance from the magnetic equator. Iron filings show such results readily, and so also do the vibrations of a magnetic needle, and likewise the revolutions of a wire ring (3212.). Now this state of the plate-magnet is precisely that which would be expected from the hypotheses of the necessary and dependent state of the magnet on the medium surrounding it.

3291. The mutual dependence of a magnet and the external medium, assumed in the view now put forth, bears upon, and may probably explain, numerous observations of the apparently superficial character of the magnetism of iron and magnets in different cases. If a hard steel bar be magnetized by touch of other magnets, both the vicinity of the superficial parts of the bar to the exciting magnet in the first instance, and afterwards to the surrounding sustaining medium, will tend to cause the magnetism to be superficial in the bar. If a small magnet or a horseshoe bar be surrounded by a thick shell of iron as its external medium, the inner surface of iron, or that nearest to the magnet, with its neighboring parts, will convey on more power than the parts further away. If a thick iron core be placed in a helix carrying a feeble or moderate electric current, it is the part of the core nearest to the helix which becomes most highly charged. Probably many other like results may appear, or be hereafter devised, and may greatly help to assist the discussion of the question of physical lines of force now under consideration.

3292. When, in place of considering the medium external to a magnet as homogeneous or equal in magnetic power, we make it variable in different parts, then the effects in it appear to me still to be in perfect accordance with the notion of physical lines of magnetic force, which, being present externally, are definite in direction and amount. The series of substances at our command which affect the surrounding space in this respect do not present a great choice of successive steps; but having iron, nickel and cobalt, very high as paramagnetic bodies, we then possess hard steel, as very far beneath them; next, perhaps, oxides of iron, and so on by solutions of the magnetic metals to oxygen, water, glass, bismuth and phosphorus, in the diamagnetic direction. Taking the magnetic force of the earth as supplying the source of power, and placing a globe of iron or nickel in the air, we see by the pointing of a small magnetic needle (or in another case, by the use of iron filings (3240.)), the deflected course of the lines of force as they enter into and pass out of the sphere, consequent upon the conducting power of the paramagnetic body. These have been described in their forms in another place (3238.). If we take a large bar-magnet, and place a piece of soft iron, about half the width of the magnet, and three or four times as long as it is wide, end on to, and about its own width from one pole, and covering that with paper, then observe the forms of the lines of force by iron filings; it will be seen how beautifully those issuing from the magnet converge, by fine inflections, on to the iron, entering by a comparatively small surface, and how they pass out in far more diffuse streams by a much larger surface at the further part of the bar, Fig. 29.5a. If we take several pieces of iron, cubes for instance, then the lines of force which are altogether outside of them, may be seen undergoing successive undulations in contrary directions, Fig. 29.5b. Yet in all these cases of the globe, bar and cubes, I, at least, am satisfied that a section across the same lines of force in any part of their course, however or whichever way deflected, would yield the same amount of effect (3109. 3218.); at the same time this effect of deflection is not only consistent with, but absolutely suggests the idea of a physical line of force.

3293. Then the manner in which the power disappears in such cases to an ordinary magnetic needle is perfectly consistent. A little needle held by the side of the soft bar described above (3292.), indicates much less magnetic power than

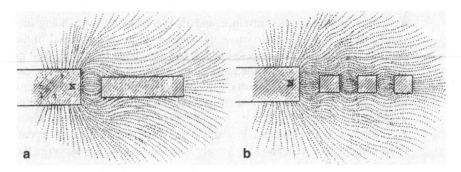

Fig. 29.5 The lines of force in the vicinity of a rod of iron (**a**) or three cubes of iron (**b**) placed near the pole of a bar magnet.—[*K.K.*]

if the iron were away. If held in a hole made in the iron, it is almost indifferent to the magnet; yet what power remains shows that the lines through the air in the hole are in the same general direction as those through the neighbouring iron. These effects are perfectly well known, no doubt; and my object is only to show that they are consistent with, and support the idea of external media having magnetic conducting power. But these apparent destructions of power, and even far more anomalous cases (2868. 3155.), are fully accounted for by the hypothesis; and the force absolutely unaffected in amount is found, experimentally, by the moving wire. I have had occasion before to refer to the modification of the magnetic force (in relation to the magnetic needle), where, its absolute quantity being the same, it passes across better or worse conductors, and I have temporarily used the words *quantity* and *intensity* (2866. 2868. 2870.). I would, however, rather not attempt to limit or define these or such like terms now, however much they may be wanted, but wait until what is at present little more than suggestion, may have been canvassed, and if true in itself, may have received assurance from the opinions or testimony of others.

3294. The association of magnet with magnet, and all the effects then produced (3218.) are in harmony, as far as I can perceive, with the idea of a physical line of magnetic force. If the magnets are all free to move, they set to each other, and then tend to approach; the great result being, that the lines from all the sources tend to coalesce, to pass through the best conductors, and to contract in length. When there are several magnets in presence and in restrained conditions, the lines of force, which they present by filings, are most varied and beautiful (3238.); but all are easily read and understood by the principles I have set forth. As the power is definite in amount, its removability from place to place, according to the changing disposition of the magnets, or the introduction of better or worse conductors into the surrounding media, becomes a perfectly simple result.

3295. As magnets may be looked upon as the habitations of bundles of lines of force, they probably show us the tendencies of the physical lines of force where they occur in the space around; just as electric currents, when conducted by solid wires, or when passing, as the Leyden or the voltaic spark, through air or a vacuum, are alike in their essential relations. In that case, the repulsion of magnets when placed side by side, indicates the lateral tendency of separation of lines of magnetic force (3267.). The effect, however, must be considered in relation to the simultaneous gathering up of the terrestrial lines of force in the surrounding space upon each magnet, and also the tendency of each magnet to secure its own independent external medium. The effect coincides with, and passes into that of the lateral repulsion of balls of iron in a previously equal magnetic field (2814.); which again, by a consideration of the action in two directions, *i.e.* parallel to and across the magnetic axis, links the phænomena of separation with those of attraction.

3296. When speaking of magnets, in illustration of the question under consideration, I mean magnets perfect in their kind, *i.e.* such as are very hard and hold their charge, so that there shall be neither internal reaction of discharge or development (3224.), nor any external change, except what may depend upon such absolute and permanent loss of exciting power as is consequent upon an over-ruling change of the external relations. Heterogeneous magnets which might allow of irregular variations of power, are out of present consideration.

3297. With regard to the great point under consideration, it is simply, whether the lines of magnetic force have a *physical existence* or not? Such a point may be investigated, perhaps even satisfactorily, without our being able to go into the further questions of how they account for magnetic attraction and repulsion, or even by what condition of space, æther or matter, these lines consist. If the extremities of a straight bar-magnet, or if the polarities of a circular plate of steel (3274.), are in magnetic relation to each other externally (3257.), then I think the existence of *curved* lines of magnetic force must be conceded (3258. 3263.);[3] and if that be granted, then I think that the physical nature of the lines must be granted also. If the external relation of the poles or polarity is denied, then, as it appears to me, the internal relation must be denied also; and with it a vast number of old and new facts (3070. &c.) will be left without either theory, hypothesis, or even a vague supposition to explain them.

3298. Perhaps both magnetic attraction and repulsion, in all forms and cases, resolve themselves into the differential action (2757.) of the magnets and substances which occupy space, and modify its magnetic power. A magnet first originates lines of magnetic force; and then, if present with another magnet, offers in one position a very free conduction of the new lines, like a paramagnetic body; or if restrained in the contrary position, resists their passage, and resembles a highly diamagnetic substance. So, then, a source of magnetic lines being present, and also magnets or other bodies affecting and varying the conducting power of space, those bodies which can convey

---

[3] See for a case of curved lines the inclosed and compressed system of forces belonging to the central circular magnet, Fig. 29.4 (3275.).

onwards the most force, may tend, by differential actions, with the others present, to take up the position in which they can do so the most freely, whether it is by pointing or by approximation; the best conductor passing to the place of strongest action (2757.), whilst the worst retreats from it, and so the effects both of attraction and repulsion be produced. The tendency of the lines of magnetic force to shorten (3266. 3280.) would be consistent with such a notion. The result would occur whether the physical lines of force were supposed to consist in a dynamic or static state (3269.).

3299. Having applied the term *line of magnetic force* to an abstract idea, which I believe represents accurately the nature, condition, direction, and comparative amount of the magnetic forces, without reference to any physical condition of the force, I have now applied the term *physical line of force* to include the further idea of their physical nature. The first set of lines I *affirm* upon the evidence of strict experiment (3071. &c.). The second set of lines I advocate, chiefly with a view of stating the question of their existence; and though I should not have raised the argument unless I had thought it both important and likely to be answered ultimately in the affirmative, I still hold the opinion with some hesitation, with as much, indeed, as accompanies any conclusion I endeavour to draw respecting points in the very depths of science, as for instance, regarding one, two or no electric fluids; or the real nature of a ray of light, or the nature of attraction, even that of gravity itself, or the general nature of matter.

Royal Institution, March 6, 1852.

29.3 Study Questions

QUES. 29.1. What are the shapes and properties of the lines of magnetic force?

a) Can one permanent magnet be used to create another permanent magnet with more power than itself? Can a helical coil of current-carrying wire be used to create a magnet with more power than itself?
b) What determines the shape of the lines of magnetic force which proceed from the surface of a magnet? Does the shape of the magnet matter? The properties of the surrounding medium?
c) Do the lines extend to infinite distances from a magnet?

QUES. 29.2. How can one observe lines of electric force?

a) Why does an electric current accompany a line of electric force? What two examples of such currents does Faraday provide?
b) How can one observe lines of electric force using a compass? Using a galvanometer?
c) Does the presence of an obstruction, such as a plate or a ball of metal, affect the shape of the lines of electric force in the fluid surrounding it?

QUES. 29.3. Does empty space possess magnetic properties?

a) To what other material(s) does Faraday compare empty space? In what sense is it in the *midst* of these other materials?
b) Does Faraday know what the magnetic medium filling space *is*? Is this medium *essential* to the power of the magnet itself? How is the medium between the poles of a magnet like the conducting medium between the poles of a voltaic cell?
c) Is Faraday correct in saying that without this medium, the magnet could not exist?

QUES. 29.4. Does the substance in the space around a magnet have an effect upon the shape or the number of lines of magnetic force in this space?

a) Does a paramagnetic substance, such as iron or nickel, alter the shape of the lines of magnetic force? If so, how? Does it change the total number of lines of magnetic force?
b) What would happen if a magnet were completely encased in a solid mass of iron? What would happen to the concentration, and to the number, of the lines of magnetic force? Would it change the magnet's effect upon a nearby compass needle?
c) What is the natural tendency of lines of magnetic force? How, then, does Faraday re-interpret the phenomenon of magnetic attractions and repulsions?
d) What would happen to the concentration of the lines of magnetic force if a magnet were surrounded by a diamagnetic substance, such as water, glass, bismuth or phosphorous? Would this change the number of lines of magnetic force going through the magnet?

QUES. 29.5. How is magnetic saturation related to the size of the magnet and the medium in which it is situated?

a) What does it mean for a magnet to be saturated? To be supersaturated?
b) Whose magnetization is stronger *in proportion to its size*: a small or a large magnet?
c) If two magnetic needles which are saturated are brought side by side, will each retain its former state of saturation? If not, will their individual magnetisms be increased or decreased? What if they are brought end to end?
d) Which is more important in determining the maximum strength of a magnet, its volume or its surface area? What does this have to do with the ability of the surrounding medium to sustain lines of magnetic force?

QUES. 29.6. What is the overall tendency of lines of magnetic force? What principle governs the arrangement of a system which includes at least one magnet and a number of other substances? How will such a system tend to order itself? And how can this account for the preferred location and orientation of paramagnetic and diamagnetic substances?

29.4 Exercises

Ex. 29.1 (MAGNETIC FIELD SKETCHING). Make a careful sketch of the magnetic field lines in the vicinity of each of the following systems. Be sure to indicate the direction of the magnetic field lines using arrows, and remember that every magnetic field line you draw must form a closed loop! (a) a long, slender bar magnet, (b) a horseshoe magnet, (c) two cubical bar magnets whose north poles are facing one another and which are separated by a distance equal to one side length, (d) a "Helmholtz coil" arrangement of current-carrying loops, (e) a magnetic quadrupole arrangement, (f) a "Tokamak" arrangement, in which a helical coil of current-carrying wire is bent around into the shape of a torus, (g) a tiny sphere of iron suspended one diameter above the north face of a short, broad bar magnet, (h) a tiny sphere of bismuth suspended one diameter above the north face of a short, broad bar magnet, and (i) a tiny sphere of superconducting niobium suspended one diameter above the north face of a short, broad bar magnet.

29.5 Vocabulary

1. Venture
2. Anomaly
3. Obtuse
4. Acute
5. Sphondyloid
6. Gymnotus
7. Electrolytic
8. Paramagnetic
9. Diamagnetic
10. Supersaturated
11. Conjoint
12. Manifest
13. Sustentation
14. Juxtaposition
15. Advantageous
16. Coalesce
17. Superficial

Chapter 30
Action-at-a-Distance

> To a person ignorant of the properties of air, the transmission of
> force by means of that invisible medium would appear as
> unaccountable.
>
> —James Clerk Maxwell

30.1 Introduction

James Clerk Maxwell (1831–1879) was born in Edinburgh, Scotland.[1] As an adolescent, Maxwell's father enrolled him at the Edinburgh Academy, where he excelled in drawing, geometry, and English verse. Recognizing his son's aptitudes, his father began to bring young Maxwell to meetings of the Edinburgh Society of Arts and the Royal Society, and by the age of 15 he had contributed a mathematical paper to the *Proceedings of the Edinburgh Royal Society*. Maxwell went on to attended the University of Edinburgh between 1847 and 1850, where he carried out original laboratory experiments on polarization and chemistry, attended philosophy lectures by William Hamilton, perused high-level scientific works such as Young's *Lectures*, Fourier's *Theory of Heat* and Newton's *Optics*, and published two mathematical papers "On the theory of rolling curves" and "On the Equilibrium of Elastic Solids." He entered Peterhouse College in Cambridge, where he spent one term, before migrating to Trinity College. He graduated in 1854 and attained a Fellowship and Lectureship at Trinity until 1856. He then served as Professor of Natural Philosophy at Marischal College in Aberdeen for 3 years before moving to a position as Professor of Physics and Astronomy at Kings College in London until 1865. During his time in London, Maxwell was occupied with formulating his electromagnetic theory of light. In 1871, Maxwell was elected an Honorary Fellow of Trinity and became Professor of Experimental Physics at Cambridge. In Cambridge, he oversaw the erection and furnishing of the Cavendish Laboratory.

---

[1] Much of the bibliographical information in this introduction was gleaned from Maxwells' obituary published in Tait, P. G., James Clerk Maxwell, *Proceedings of the Royal Society of Edinburgh*, *10*, 331–339, 1878–1880 and also from Campbell, L., and W. Garnett, *The Life of James Clerk Maxwell*, Macmillan and Co., London, 1882.

© Springer International Publishing Switzerland 2016
K. Kuehn, *A Student's Guide Through the Great Physics Texts*,
Undergraduate Lecture Notes in Physics, DOI 10.1007/978-3-319-21816-8_30

Fig. 30.1 An etching of
James Clerk Maxwell by
George J. Stodart based on
the photograph by Fergus of
Greenack

Maxwell is best known for his work on the kinetic theory of gasses and his electromagnetic theory of light. Regarding the former topic, Maxwell argued that the
macroscopic properties of a gas, like pressure and temperature, may be understood
in terms of the velocity distribution of the particles comprising the gas.[2] Regarding
the latter topic, Maxwell drew together disparate lines of research—from electricity,
magnetism and the wave theory of light—to argue that "light is an electromagnetic disturbance in the form of waves propagated through the electromagnetic field
according to electromagnetic laws." Maxwell's electromagnetic theory of light, presented to the Royal Society in 1864 as "A Dynamical Theory of the Electromagnetic
Field," was inspired by Faraday's conception of electric and magnetic forces as
manifestations of strains arising within a subtle medium pervading space. Under-
lying Maxwell's theory is the idea that electrically charged bodies do not act upon
one another immediately, that is, without any intervening medium. Rather, forces
between bodies are mediated by the luminferous æther. The conflict between natural philosophers who advocated "action-at-a-distance" and those who advocated
"mediated action" is the topic of the reading selection included in the present chapter. It was first published in the Proceedings of the Royal Institution of Great Britain
in 1873. Whose arguments do you find most convincing? (Fig. 30.1).

30.2   Reading: Maxwell, *On Action at a Distance*

Maxwell, J. C., On action at a distance, *Proceedings of the Royal Institution of Great
Britain*, *VII*, 48–49, 1873–1875.

I have no new discovery to bring before you this evening. I must ask you to go over
very old ground, and to turn your attention to a question which has been raised again
and again ever since men began to think.

---

[2] The kinetic theory of gasses will be discussed in more detail in volume IV.

The question is that of the transmission of force. We see that two bodies at a distance from each other exert a mutual influence on each other's motion. Does this mutual action depend on the existence of some third thing, some medium of communication, occupying the space between the bodies, or do the bodies act on each other immediately, without the intervention of anything else?

The mode in which Faraday was accustomed to look at phenomena of this kind differs from that adopted by many other inquirers, and my special aim will be to enable you to place yourselves at Faraday's point of view, and to point out the scientific value of that conception of *lines of force* which, in his hands, became the key to the science of electricity.

When we observe one body acting on another at a distance, before we assume that this action is direct and immediate, we generally inquire whether there is any connection between the two bodies; and if we find strings, or rods, or mechanism of any kind, capable of accounting for the observed action between the bodies, we prefer to explain the action by means of these intermediate connections, rather than to admit the notion of direct action at a distance.

Thus, when we ring a bell by means of a wire, the successive parts of the wire are first tightened and then moved, till at last the bell is rung at a distance by a process in which all the intermediate particles of the wire have taken part one after the other. We may ring a bell at a distance in other ways, as by forcing air into a long tube, at the other end of which is a cylinder with a piston which is made to fly out and strike the bell. We may also use a wire; but instead of pulling it, we may connect it at one end with a voltaic battery, and at the other with an electro-magnet, and thus ring the bell by electricity.

Here are three different ways of ringing a bell. They all agree, however, in the circumstance that between the ringer and the bell there is an unbroken line of communication, and that at every point of this line some physical process goes on by which the action is transmitted from one end to the other. The process of transmission is not instantaneous, but gradual; so that there is an interval of time after the impulse has been given to one extremity of the line of communication, during which the impulse is on its way, but has not reached the other end.

It is clear, therefore, that in many cases the action between bodies at a distance may be accounted for by a series of actions between each successive pair of a series of bodies which occupy the intermediate space; and it is asked, by the advocates of mediate action, whether, in those cases in which we cannot perceive the intermediate agency, it is not more philosophical to admit the existence of a medium which we cannot at present perceive, than to assert that a body can act at a place where it is not.

To a person ignorant of the properties of air, the transmission of force by means of that invisible medium would appear as unaccountable as any other example of action at a distance, and yet in this case we can explain the whole process, and determine the rate at which the action is passed on from one portion to another of the medium.

Why then should we not admit that the familiar mode of communicating motion by pushing and pulling with our hands is the type and exemplification of all action between bodies, even in cases in which we can observe nothing between the bodies which appears to take part in the action?

Here for instance is a kind of attraction with which Professor Guthrie has made us familiar. A disk is set in vibration, and is then brought near a light suspended body, which immediately begins to move towards the disk, as if drawn towards it by an invisible cord. What is this cord? Sir W. Thomson has pointed out that in a moving fluid the pressure is least where the velocity is greatest. The velocity of the vibratory motion of the air is greatest nearest the disk. Hence the pressure of the air on the suspended body is less on the side nearest the disk than on the opposite side, the body yields to the greater pressure, and moves toward the disk.

The disk, therefore, does not act where it is not. It sets the air next it in motion by pushing it, this motion is communicated to more and more distant portions of the air in turn, and thus the pressures on opposite sides of the suspended body are rendered unequal, and it moves towards the disk in consequence of the excess of pressure. The force is therefore a force of the old school—a case of *vis a tergo*—a shove from behind.

The advocates of the doctrine of action at a distance, however, have not been put to silence by such arguments. What right, say they, have we to assert that a body cannot act where it is not? Do we not see an instance of action at a distance in the case of a magnet, which acts on another magnet not only at a distance, but with the most complete indifference to the nature of the matter which occupies the intervening space? If the action depends on something occupying the space between the two magnets, it cannot surely be a matter of indifference whether this space is filled with air or not, or whether wood, glass, or copper, be placed between the magnets.

Besides this, Newton's law of gravitation, which every astronomical observation tends only to establish more firmly, asserts not only that the heavenly bodies act on one another across immense intervals of space, but that two portions of matter, the one buried a 1000 miles deep in the interior of the earth, and the other a 100,000 miles deep in the body of the sun, act on one another with precisely the same force as if the strata beneath which each is buried had been non-existent. If any medium takes part in transmitting this action, it must surely make some difference whether the space between the bodies contains nothing but this medium, or whether it is occupied by the dense matter of the earth or of the sun.

But the advocates of direct action at a distance are not content with instances of this kind, in which the phenomena, even at first sight, appear to favour their doctrine. They push their operations into the enemy's camp, and maintain that even when the action is apparently the pressure of contiguous portions of matter, the contiguity is only apparent—that a space *always* intervenes between the bodies which act on each other. They assert, in short, that so far from action at a distance being impossible, it is the only kind of action which ever occurs, and that the favourite old *vis a tergo* of the schools has no existence in nature, and exists only in the imagination of schoolmen.

The best way to prove that when one body pushes another it does not touch it, is to measure the distance between them. Here are two glass lenses, one of which is pressed against the other by means of a weight. By means of the electric light we may obtain on the screen an image of the place where the one lens presses against the other. A series of coloured rings is formed on the screen. These rings were first observed and first explained by Newton. The particular colour of any ring depends on the distance between the surfaces of the pieces of glass. Newton formed a table of the colours corresponding to different distances, so that by comparing the colour of any ring with Newton's table, we may ascertain the distance between the surfaces at that ring. The colours are arranged in rings because the surfaces are spherical, and therefore the interval between the surfaces depends on the distance from the line joining the centres of the spheres. The central spot of the rings indicates the place where the lenses are nearest together, and each successive ring corresponds to an increase of about the 4000th part of a millimètre in the distance of the surfaces.

The lenses are now pressed together with a force equal to the weight of an ounce; but there is still a measurable interval between them, even at the place where they are nearest together. They are not in optical contact. To prove this, I apply a greater weight. A new colour appears at the central spot, and the diameters of all the rings increase. This shews that the surfaces are now nearer than at first, but they are not yet in optical contact, for if they were, the central spot would be black. I therefore increase the weights, so as to press the lenses into optical contact.

But what we call optical contact is not real contact. Optical contact indicates only that the distance between the surfaces is much less than a wavelength of light. To shew that the surfaces are not in real contact, I remove the weights. The rings contract, and several of them vanish at the centre. Now it is possible to bring two pieces of glass so close together, that they will not tend to separate at all, but adhere together so firmly, that when torn asunder the glass will break, not at the surface of contact, but at some other place. The glasses must then be many degrees nearer than when in mere optical contact.

Thus we have shewn that bodies begin to press against each other whilst still at a measurable distance, and that even when pressed together with great force they are not in absolute contact, but may be brought nearer still, and that by many degrees.

Why, then, say the advocates of direct action, should we continue to maintain the doctrine, founded only on the rough experience of a pre-scientific age, that matter cannot act where it is not, instead of admitting that all the facts from which our ancestors concluded that contact is essential to action were in reality cases of action at a distance, the distance being too small to be measured by their imperfect means of observation?

If we are ever to discover the laws of nature, we must do so by obtaining the most accurate acquaintance with the facts of nature, and not by dressing up in philo-sophical language the loose opinions of men who had no knowledge of the facts which throw the most light on these laws. And as for those who introduce ætherial, or other media, to account for these actions, without any direct evidence of the exis-tence of such media, or any clear understanding of how the media do their work, and who fill all space three or four times over with æthers of different sorts, why the less these men talk about their philosophical scruples about admitting action at a distance the better.

If the progress of science were regulated by Newton's first law of motion, it would be easy to cultivate opinions in advance of the age. We should only have to compare the science of to-day with that of 50 years ago; and by producing, in the geometrical sense, the line of progress, we should obtain the science of 50 years hence.

The progress of science in Newton's time consisted in getting rid of the celestial machinery with which generations of astronomers had encumbered the heavens, and thus "sweeping cobwebs off the sky."

Though the planets had already got rid of their crystal spheres, they were still swimming in the vortices of Descartes. Magnets were surrounded by effluvia, and electrified bodies by atmospheres, the properties of which resembled in no respect those of ordinary effluvia and atmospheres.

When Newton demonstrated that the force which acts on each of the heavenly bodies depends on its relative position with respect to the other bodies, the new theory met with violent opposition from the advanced philosophers of the day, who described the doctrine of gravitation as a return to the exploded method of explaining everything by occult causes, attractive virtues, and the like.

Newton himself, with that wise moderation which is characteristic of all his speculations, answered that he made no pretence of explaining the mechanism by which the heavenly bodies act on each other. To determine the mode in which their mutual action depends on their relative position was a great step in science, and this step Newton asserted that he had made. To explain the process by which this action is effected was a quite distinct step, and this step Newton, in his *Principia*, does not attempt to make.

But so far was Newton from asserting that bodies really do act on one another at a distance, independently of anything between them, that in a letter to Bentley, which has been quoted by Faraday in this place, he says:—

> It is inconceivable that inanimate brute matter should, without the mediation of something else, which is not material, operate upon and affect other matter without mutual contact, as it must do if gravitation, in the sense of Epicurus, be essential and inherent in it... That gravity should be innate, inherent, and essential to matter, so that one body can act upon another at a distance, through a vacuum, without the mediation of anything else, by and through which their action and force may be conveyed from one to another, is to me so great an absurdity, that I believe no man who has in philosophical matters a competent faculty of thinking can ever fall into it.

Accordingly, we find in his *Optical Queries*, and in his letters to Boyle, that Newton had very early made the attempt to account for gravitation by the means of the pressure of a medium, and that the reason he did not publish these investigations "proceeded from hence only, that he found he was not able, from experiment and observation, to give a satisfactory account of this medium, and the manner of its operation in producing the chief phenomena of nature.[3]"

The doctrine of direct action at a distance cannot claim for its author the discoverer of universal gravitation. It was first asserted by Roger Cotes, in his preface to the *Principia*, which he edited during Newton's life. According to Cotes, it is by experience that we learn that all bodies gravitate. We do not learn in any other way that

---

[3] Maclaurin's *Account of Newton's Discoveries*.

they are extended, movable, or solid. Gravitation, therefore, has as much right to be considered an essential property of matter as extension, mobility, or impenetrability.

And when the Newtonian philosophy gained ground in Europe, it was the opinion of Cotes rather than that of Newton that became most prevalent, till at last Boscovich propounded his theory, that matter is a congeries of mathematical points, each endowed with the power of attracting or repelling the others according to fixed laws. In his world, matter is unextended, and contact is impossible. He did not forget, however, to endow his mathematical points with inertia. In this some of the modern representatives of his school have thought that he "had not quite got so far as the strict modern view of 'matter' as being but an expression for modes or manifestations of 'force'.[4]"

But if we leave out of account for the present the development of the ideas of science, and confine our attention to the extension of its boundaries, we shall see that it was most essential that Newton's method should be extended to every branch of science to which it was applicable—that we should investigate the forces with which bodies act on each other in the first place, before attempting to explain *how* that force is transmitted. No men could be better fitted to apply themselves exclusively to the first part of the problem, than those who considered the second part quite unnecessary.

Accordingly Cavendish, Coulomb, Poisson, the founders of the exact sciences of electricity and magnetism, paid no regard to those old notions of "magnetic effluvia" and "electric atmospheres," which had been put forth in the previous century, but turned their undivided attention to the determination of the law of force, according to which electrified and magnetized bodies attract or repel each other. In this way the true laws of these actions were discovered, and this was done by men who never doubted that the action took place at a distance, without the intervention of any medium, and who would have regarded the discovery of such a medium as complicating rather than as explaining the undoubted phenomena of attraction.

We have now arrived at the great discovery of Oersted of the connection between electricity and magnetism. Oersted found that an electric current acts on a magnetic pole, but that it neither attracts it nor repels it, but causes it to move round the current. He expressed this by saying that "the electric conflict acts in a revolving manner."

The most obvious deduction from this new fact was that the action of the current on the magnet is not a push-and-pull force, but a rotatory force, and accordingly many minds were set a-speculating on vortices and streams of æther whirling round the current.

But Ampère, by a combination of mathematical skill with experimental ingenuity, first proved that two electric currents act on one another, and then analysed this action into the resultant of a system of push-and-pull forces between the elementary parts of these currents.

---

[4] Review of Mrs Somerville, *Saturday Review*, Feb. 13, 1869.

The formula of Ampère, however, is of extreme complexity, as compared with Newton's law of gravitation, and many attempts have been made to resolve it into something of greater apparent simplicity.

I have no wish to lead you into a discussion of any of these attempts to improve a mathematical formula. Let us turn to the independent method of investigation employed by Faraday in those researches in electricity and magnetism which have made this Institution one of the most venerable shrines of science.

No man ever more conscientiously and systematically laboured to improve all his powers of mind than did Faraday from the very beginning of his scientific career. But whereas the general course of scientific method then consisted in the application of the ideas of mathematics and astronomy to each new investigation in turn, Faraday seems to have had no opportunity of acquiring a technical knowledge of mathematics, and his knowledge of astronomy was mainly derived from books.

Hence, though he had a profound respect for the great discovery of Newton, he regarded the attraction of gravitation as a sort of sacred mystery, which, as he was not an astronomer, he had no right to gainsay or to doubt, his duty being to believe it in the exact form in which it was delivered to him. Such a dead faith was not likely to lead him to explain new phenomena by means of direct attractions.

Besides this, the treatises of Poisson and Ampère are of so technical a form, that to derive any assistance from them the student must have been thoroughly trained in mathematics, and it is very doubtful if such a training can be begun with advantage in mature years.

Thus Faraday, with his penetrating intellect, his devotion to science, and his opportunities for experiments, was debarred from following the course of thought which had led to the achievements of the French philosophers, and was obliged to explain the phenomena to himself by means of a symbolism which he could understand, instead of adopting what had hitherto been the only tongue of the learned.

This new symbolism consisted of those lines of force extending themselves in every direction from electrified and magnetic bodies, which Faraday in his mind's eye saw as distinctly as the solid bodies from which they emanated.

The idea of lines of force and their exhibition by means of iron filings was nothing new. They had been observed repeatedly, and investigated mathematically as an interesting curiosity of science. But let us hear Faraday himself, as he introduces to his reader the method which in his hands became so powerful[5].

> It would be a voluntary and unnecessary abandonment of most valuable aid if an experimentalist, who chooses to consider magnetic power as represented by lines of magnetic force, were to deny himself the use of iron filings. By their employment he may make many conditions of the power, even in complicated cases, visible to the eye at once, may trace the varying direction of the lines of force and determine the relative polarity, may observe in which direction the power is increasing or diminishing, and in complex systems may determine the neutral points, or places where there is neither polarity nor power, even when they occur in the midst of powerful magnets. By their use probable results may be seen at once, and many a valuable suggestion gained for future leading experiments.

---

[5] *Exp. Res.* 3284.

30.2.1   Experiment on Lines of Force

In this experiment each filing becomes a little magnet. The poles of opposite names belonging to different fillings attract each other and stick together, and more filings attach themselves to the exposed poles, that is, to the ends of the row of filings. In this way the filings, instead of forming a confused system of dots over the paper, draw together, filing to filing, till long fibres of filings are formed, which indicate by their direction the lines of force in every part of the field.

The mathematicians saw in this experiment nothing but a method of exhibiting at one view the direction in different places of the resultant of two forces, one directed to each pole of the magnet; a somewhat complicated result of the simple law of force.

But Faraday, by a series of steps as remarkable for their geometrical definiteness as for their speculative ingenuity, imparted to his conception of these lines of force a clearness and precision far in advance of that with which the mathematicians could then invest in their own formulæ.

In the first place, Faraday's lines of force are not to be considered merely as individuals, but as forming a system, drawn in space in a definite manner so that the number of lines which pass through an area, say of 1 in.^{2}, indicates the intensity of the force acting through the area. Thus the lines of force become definite in number. The strength of a magnetic pole is measured by the number of lines which proceed from it; the electro-tonic state of a circuit is measured by the number of lines which pass through it.

In the second place, each individual line has a continuous existence in space and time. When a piece of steel becomes a magnet, or when an electric current begins to flow, the lines of force do not start into existence each in its own place, but as the strength increases new lines are developed within the magnet or current, and gradually grow outwards, so that the whole system expands from within, like Newton's rings in our former experiment. Thus every line of force preserves its identity during the whole course of its existence, though its shape and size may be altered to any extent.

I have no time to describe the methods by which every question relating to the forces acting on magnets or on currents, or to the induction of currents in conducting circuits, may be solved by the consideration of Faraday's lines of force. In this place they can never be forgotten. By means of this new symbolism, Faraday defined with mathematical precision the whole theory of electro-magnetism, in language free from mathematical technicalities, and applicable to the most complicated as well as the simplest cases. But Faraday did not stop here. He went on from the conception of geometrical lines of force to that of physical lines of force. He observed that the motion which the magnetic or electric force tends to produce is invariably such as to shorten the lines of force and to allow them to spread out laterally from each other. He thus perceived in the medium a state of stress, consisting of a tension, like that of a rope, in the direction of the lines of force, combined with a pressure in all directions a right angles to them.

This is quite a new conception of action at a distance, reducing it to a phenomenon of the same kind as that action at a distance which is exerted by means of the tension of ropes and the pressure of rods. When the muscles of our bodies are excited by that stimulus which we are able in some unknown way to apply to them, the fibres tend to shorten themselves and at the same time to expand laterally. A state of stress is produced in the muscle, and the limb moves. This explanation of muscular action is by no means complete. It gives no account of the cause of the excitement of the state of stress, nor does it even investigate those forces of cohesion which enable the muscles to support this stress. Nevertheless, the simple fact, that it substitutes as a kind of action which extends continuously along a material substance for one of which we know only a cause and an effect at a distance from each other, induces us to accept it as a real addition to our knowledge of animal mechanisms.

For similar reasons we may regard Faraday's conception of a state of stress in the electro-magnetic field as a method of explaining action at a distance by means of the continuous transmission of force, even though we do not know how the state of stress is produced.

But one of Faraday's most pregnant discoveries, that of the magnetic rotation of polarised light, enables us to proceed a step farther. The phenomenon, when analysed into its simplest elements, may be described thus:—Of two circularly polarised rays of light, precisely similar in configuration, but rotating in opposite directions, that ray is propagated with the greater velocity which rotates in the same direction as the electricity of the magnetizing current.

It follows from this, as Sir. W. Thomson has shewn by strict dynamical reasoning, that the medium when under the action of magnetic force must be in a state of rotation—that is to say, that small portions of the medium, which we may call molecular vortices, are rotating, each on its own axis, the direction of this axis being that of the magnetic force.

Here, then, we have an explanation of the tendency of the lines of magnetic force to spread out laterally and to shorten themselves. It arises from the centrifugal force of the molecular vortices.

The mode in which electromotive force acts in starting and stopping the vortices is more abstruse, though it is of course consistent with dynamical principles.

We have thus found that there are several different kinds of work to be done by the electromagnetic medium if it exists. We have also seen that magnetism has an intimate relation to light, and we know that there is a theory of light which supposes it to consist of the vibrations of a medium. How is this luminiferous medium related to our electro-magnetic medium?

It fortunately happens that electro-magnetic measurements have been made from which we can calculate by dynamical principles the velocity of propagation of small magnetic disturbances in the supposed electro-magnetic medium.

This velocity is very great, from 288 to 314 millions of metres per second, according to different experiments. Now the velocity of light, according to Foucault's experiments, is 298 millions of metres per second. In fact, the different determinations of either velocity differ from each other more than the estimated velocity of light does from the estimated velocity of the propagation of small electro-magnetic disturbance. But if the luminiferous and the electro-magnetic media occupy the same place, and transmit disturbances with the same velocity, what reason have we to distinguish the one from the other? By considering them as the same, we avoid at least the reproach of filling space twice over with different kinds of æther.

Besides this, the only kind of electro-magnetic disturbances which can be propagated through a non-conducting medium is a disturbance transverse to the direction of propagation, agreeing in this respect with what we know of that disturbance we call light. Hence, for all we know, light also may be an electro-magnetic disturbance in a non-conducting medium. If we admit this, the electro-magnetic theory of light will agree in every respect with the undulatory theory, and the work of Thomas Young and Fresnel will be established on a firmer basis than ever, when joined with that of Cavendish and Coulomb by the key-stone of the combined sciences of light and electricity—Faraday's great discovery of the electro-magnetic rotation of light.

The vast interplanetary and interstellar regions will no longer be regarded as waste places in the universe, which the Creator has not seen fit to fill with the symbols of the manifold order of His kingdom. We shall find them to be already full of this wonderful medium; so full, that no human power can remove it from the smallest portion of space, or produce the slightest flaw in its infinite continuity. It extends unbroken from star to star; and when a molecule of hydrogen vibrates in the dog-star, the medium receives the impulses of these vibrations; and after carrying them in its immense bosom for 3 years, delivers them in due course, regular order, and full tale into the spectroscope of Mr. Huggins, at Tulse Hill.

But the medium has other functions and operations besides bearing light from man to man, and from world to world, and giving evidence of the absolute unity of the metric system of the universe. Its minute parts may have rotatory as well as vibratory motion and the axes of rotation form those lines of magnetic force which extend in unbroken continuity into regions which no eye has seen, and which, by their action on our magnets, are telling us in language not yet interpreted, which is going on in the hidden underworld from minute to minute and from century to century.

And these lines must not be regarded as mere mathematical abstractions. They are the directions in which the medium is exerting a tension like that of a rope, or rather, like that of our own muscles. The tension of the medium in the direction of the earth's magnetic force is in this country one grain weight on 8 ft^2. In some of Dr. Joule's experiments, the medium has exerted a tension of 200 lbs. weight per square inch.

But the medium, in virtue of the very same elasticity by which it is able to transmit the undulations of light, is also able to act as a spring. When properly wound up, it exerts a tension, different from the magnetic tension, by which it draws oppositely electrified bodies together, produces effects through the length of telegraph wires, and when of sufficient intensity, leads to the rupture and explosion called lightning.

These are some of the already discovered properties of that which has often been called a vacuum, or nothing at all. Then enable us to resolve several kinds of action at a distance into actions between contiguous parts of a continuous substance. Whether this resolution is of the nature of explication or complication, I must leave to the metaphysicians.

30.3   Study Questions

QUES. 30.1. What is the strongest argument proposed by the advocates of mediated action?

a) What is meant by the term *mediated action*?
b) What examples are provided by Maxwell of mediated action?
c) Which famous scientists were proponents of this view?

QUES. 30.2. What is the strongest argument proposed by advocates of action-at-a-distance?

a) What is meant by the term *action-at-a-distance*?
b) What examples are provided by Maxwell of action-at-a-distance?
c) Which famous scientists were proponents of this view?

QUES. 30.3. Evaluate Faraday's concept of lines of force (*i.e.* electric and magnetic fields).

a) What was his motivation in developing the concept of lines of force?
b) What are the benefits (if any) of adopting Faraday's conception of lines of force?
c) To what does Maxwell liken the tension exerted by Faraday's lines of force?

QUES. 30.4. What does Maxwell suggest regarding the nature of light?

a) What did Faraday find when passing light through a magnetic field? What does this suggest?
b) What is significant about Maxwell's theoretical calculation of the speed of electromagnetic disturbances?
c) What are the benefits of identifying light with electromagnetic disturbances? Has Maxwell offered a convincing proof?

30.4   Exercises

EX. 30.1 (FORCES AND FIELDS).   According to Coulomb's law, when a tiny charge q_b is placed in the vicinity of another charge, q_a, charge q_b feels a force equal to

$$F_{ab} = k \frac{q_a q_b}{r^2}.$$

Here, r is the separation between the charges and k is the Coulomb constant. If one takes seriously the concept of the electric field, then one may claim that charge q_b

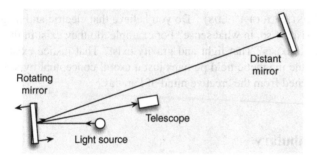

Fig. 30.2 A simplified schematic diagram depicting Foucault's speed of light apparatus

is sitting in an electric field, E_a, which is generated by charge q_a. In this view, the electric field strength, E_a, produced by charge q_a may be defined in terms of the force, F_{ab}, which q_a exerts on q_b when in its vicinity:

$$E_a \equiv F_{ab}/q_b. \tag{30.1}$$

As an exercise, what is the strength of the electric field at a distance of 1 μm from a single proton? And what are the (SI) units of this electric field? Does the electric field produced by the proton depend on the magnitude of any other charge(s) placed in its vicinity? Finally, does the electric field strength depend on the medium surrounding the proton? For example, by how much is the electric field strength changed if it is surrounded by water instead of vacuum? (HINT: the relative permittivity of water at room temperature is about 80.)

EX. 30.2 (FOUCAULT AND THE SPEED OF LIGHT). In order to measure the speed of light, one must precisely measure the time that it takes light to travel a known distance.[6] French physicist Léon Foucault measured the speed of light with great accuracy in 1862 using a clever apparatus similar to that shown schematically in Fig. 30.2. Light emitted from a source, L, reflects from a rotating mirror, R. When R is at just the right angle, the reflected light strikes a distant mirror, D, at normal incidence and thus returns to R. Upon its arrival, R will have rotated by an angle which depends upon its rotation speed. If the rotation speed is just right, then the light reflected from R will arrive at a viewing telescope placed at location T. Suppose that Foucault arranges his apparatus such that $\angle LRT = \frac{1}{2}°$, $\angle LRD = 1°$, and $RD = 300$ m. Based on the speed of light originally determined by Römer in 1676, what must be the rotation speed in order for the reflected light to enter the telescope? If a tank of water were inserted between the mirrors, would the rotation rate need to be increased or decreased? (ANSWER: 210,000,000 rpm)

---

[6] The first successful determination of the speed of light was performed by Danish astronomer Olaf Römer; see Huygens' description of Römer's method in Chap. 13 of the present volume (especially Ex. 13.2).

Ex. 30.3 (EXISTENCE OF FIELDS).   Do you believe that electric and magnetic fields physically exist? If so, in what sense? For example, do they exist in the same sense that your hand exists? That light and gravity exist? That justice exists? That God exists? Or is the magnetic field perhaps just a useful conceptual hypotheses which we have inherited from the creative mind of Faraday?

30.5   Vocabulary

1. Electro-magnet
2. Impulse
3. Mediate
4. Vis a tergo
5. Strata
6. Contiguous
7. Ascertain
8. Effluvia
9. Essential
10. Inherent
11. Innate
12. Extension
13. Mobility
14. Impenetrability
15. Congeries
16. Manifestation
17. Emanate
18. Speculative
19. Impart
20. Lateral
21. Explication
22. Metaphysician

Chapter 31
Maxwell's Equations

Certain phenomena in electricity and magnetism lead to the same conclusion as those of optics, namely, that there is an æthereal medium pervading all bodies, and modified only in degree by their presence.

—James Clerk Maxwell

31.1 Introduction

In the previous reading selection, Maxwell argued that Faraday's lines of electric and magnetic force "must not be regarded as mere mathematical abstractions." Rather, they represent deformations or strains within an elastic medium—the electromagnetic field—by which forces are communicated between electric charges and currents. But while Faraday represented these strains pictorially using elegant diagrams, Maxwell represented them mathematically in terms of space- and time-dependent variables which obey a set of partial differential equations. Specifically, in an essay read before the Royal Society in 1864, Maxwell formulated his "Dynamical theory of the Electromagnetic Field" in the form of 20 equations involving 20 variable quantities.[1] The reading selection in the present chapter is the introductory section of Maxwell's 1864 publication. Herein, he provides an overview of his "General Equations of the Electromagnetic Field."

31.2 Reading: Maxwell, *A Dynamical Theory of the Electromagnetic Field*

Maxwell, J. C., A Dynamical Theory of the Electromagnetic Field, *Philosophical Transactions of the Royal Society of London, 155*, 459–512, 1865. Part I.

---

[1] The original article is Maxwell, J. C., A Dynamical Theory of the Electromagnetic Field, *Philosophical Transactions of the Royal Society of London, 155*, 459–512, 1865. A nice historical introduction to Maxwell's 1864 essay, along with an edited version of the essay itself, was written by Thomas F. Torrance; see Maxwell, J. C., *A Dynamical Theory of the Electromagnetic Field*, Wipf and Stock Publishers, 1996.

© Springer International Publishing Switzerland 2016
K. Kuehn, *A Student's Guide Through the Great Physics Texts*,
Undergraduate Lecture Notes in Physics, DOI 10.1007/978-3-319-21816-8_31

(1) THE most obvious mechanical phenomenon in electrical and magnetical experiments is the mutual action by which bodies in certain states set each other
in motion while still at a sensible distance from each other. The first step,
therefore, in reducing these phenomena into scientific form, is to ascertain the
magnitude and direction of the force acting between the bodies, and when it
is found that this force depends in a certain way upon the relative position of
the bodies and on their electric or magnetic condition, it seems at first sight
natural to explain the facts by assuming the existence of something either at
rest or in motion in each body, constituting its electric or magnetic state, and
capable of acting at a distance according to mathematical laws.

In this way mathematical theories of statical electricity, of magnetism, of the
mechanical action between conductors carrying currents, and of the induction of currents have been formed. In these theories the force acting between
the two bodies is treated with reference only to the condition of the bodies and their relative position, and without any express consideration of the
surrounding medium.

These theories assume, more or less explicitly, the existence of substances
the particles of which have the property of acting on one another at a distance by attraction or repulsion. The most complete development of a theory
of this kind is that of M. M. WEBER,[2] who has made the same theory include
electrostatic and electromagnetic phenomena.

In doing so, however, he has found it necessary to assume that the force
between two electric particles depends on their relative velocity, as well as
on their distance.

This theory, as developed by MM. W. WEBER and C. NEUMANN,[3] is exceedingly ingenious, and wonderfully comprehensive in its application to the
phenomena of statical electricity, electromagnetic attractions, induction of currents and diamagnetic phenomena; and it comes to us with the more authority,
as it has served to guide the speculations of one who has made so great an
advance in the practical part of electric science, both by introducing a consistent system of units in electrical measurement, and by actually determining
electrical quantities with an accuracy hitherto unknown.

(2) The mechanical difficulties, however, which are involved in the assumption
of particles acting at a distance with forces which depend on their velocities
are such as to prevent me from considering this theory as an ultimate one,
though it may have been, and may yet be useful in leading to the coordination
of phenomena.

I have therefore preferred to seek an explanation of the fact in another direction, by supposing them to be produced by actions which go on in the

---

[2] Electrodynamische Maassbestimmungen. *Leipzig Trans.* vol. I. 1849, and TAYLOR'S Scientific
Memoirs, vol. V. art. XIV.

[3] "Explicare testator quo modo fiat ut lucid planum polarization's per fires electrical gel magnetic
as declinetur."—Halis Saxonum, 1838.

surrounding medium as well as in the excited bodies, and endeavouring to explain the action between distant bodies without assuming the existence of forces capable of acting directly at sensible distances.

(3) The theory I propose may therefore be called a theory of the *Electromagnetic Field*, because it has to do with the space in the neighbourhood of the electric or magnetic bodies, and it may be called a *Dynamical* Theory, because it assumes that in that space there is matter in motion, by which the observed electromagnetic phenomena are produced.

(4) The electromagnetic field is that part of space which contains and surrounds bodies in electric or magnetic conditions. It may be filled with any kind of matter, or we may endeavour to render it empty of all gross matter, as in the case of GEISSLER'S tubes and other so-called vacua.

There is always, however, enough of matter left to receive and transmit the undulations of light and heat, and it is because the transmission of these radiations is not greatly altered when transparent bodies of measurable density are substituted for the so-called vacuum, that we are obliged to admit that the undulations are those of an æthereal substance, and not of the gross matter, the presence of which merely modifies in some way the motion of the æther.

We have therefore some reason to believe, from the phenomena of light and heat, that there is an æthereal medium filling space and permeating bodies, capable of being set in motion and of transmitting that motion from one part to another, and of communicating that motion to gross matter so as to heat it and affect it in various ways.

(5) Now the energy communicated to the body in heating it must have formerly existed in the moving medium, for the undulations had left the source of heat some time before they reached the body, and during that time the energy must have been half in the form of motion of the medium and half in the form of elastic resilience. From these considerations Professor W. THOMSON has argued,[4] that the medium must have a density capable of comparison with that of gross matter, and has even assigned an inferior limit to that density.

(6) We may therefore receive, as a datum derived from a branch of science independent of that with which we have to deal, the existence of a pervading medium, of small but real density, capable of being set in motion, and of transmitting motion from one part to another with great, but not infinite, velocity.

Hence the parts of this medium must be so connected that the motion of one part depends in some way on the motion of the rest; and at the same time these connexions must be capable of a certain kind of elastic yielding, since the communication of motion is not instantaneous, but occupies time. The medium is therefore capable of receiving and storing up two kinds of energy,

---

[4] "On the Possible Density of the Luminiferous Medium, and on the Mechanical Value of a Cubic Mile of Sunlight," *Transactions of the Royal Society of Edinburgh* (1854), p. 57.

namely, the "actual" energy depending on the motions of its parts, and "potential" energy, consisting of the work which the medium will do in recovering from displacement in virtue of its elasticity.

The propagation of undulations consists in the continual transformation of one of these forms of energy into the other alternately, and at any instant the amount of energy in the whole medium is equally divided, so that half is energy of motion, and half is elastic resilience.

(7) A medium having such a constitution may be capable of other kinds of motion and displacement than those which produce the phenomena of light and heat, and some of these may be of such a kind that they may be evidenced to our senses by the phenomena they produce.

(8) Now we know that the luminiferous medium is in certain cases acted on by magnetism; for FARADAY[5] discovered that when a plane polarized ray traverses a transparent diamagnetic medium in the direction of the lines of magnetic force produced by magnets or currents in the neighbourhood, the plane of polarization is caused to rotate.

This rotation is always in the direction in which positive electricity must be carried round the diamagnetic body in order to produce the actual magnetization of the field.

VERDET[6] has since discovered that if a paramagnetic body, such as solution of perchloride of iron in ether, be substituted for the diamagnetic body, the rotation is in the opposite direction.

Now Professor W. THOMSON[7] has pointed out that no distribution of forces acting between the parts of a medium whose only motion is that of the luminous vibrations, is sufficient to account for the phenomena, but that we must admit the existence of a motion in the medium depending on the magnetization, in addition to the vibratory motion which constitutes light.

It is true that the rotation by magnetism of the plane of polarization has been observed only in media of considerable density; but the properties of the magnetic field are not so much altered by the substitution of one medium for another, or for a vacuum, as to allow us to suppose that the dense medium does anything more than merely modify the motion of the ether. We have therefore warrantable grounds for inquiring whether there may not be a motion of the ethereal medium going on wherever magnetic effects are observed, and we have some reason to suppose that this motion is one of rotation, having the direction of the magnetic force as its axis.

(9) We may now consider another phenomenon observed in the electromagnetic field. When a body is moved across the lines of magnetic force it experiences what is called an electromotive force; the two extremities of the body tend to become oppositely electrified, and an electric current tends to flow through

---

[5] *Experimental Researches*, Series 19.

[6] *Comptes Rendus* (1856, second half year, p. 529, and 1857, first half year, p. 1209).

[7] *Proceedings of the Royal Society*, June 1856 and June 1861.

the body. When the electromotive force is sufficiently powerful, and is made to act on certain compound bodies, it decomposes them, and causes one of their components to pass towards one extremity of the body, and the other in the opposite direction.

Here we have evidence of a force causing an electric current in spite of resistance; electrifying the extremities of a body in opposite ways, a condition which is sustained only by the action of the electromotive force, and which, as soon as that force is removed, tends, with an equal and opposite force, to produce a counter current through the body and to restore the original electrical state of the body; and finally, if strong enough, tearing to pieces chemical compounds and carrying their components in opposite directions, while their natural tendency is to combine, and to combine with a force which can generate an electromotive force in the reverse direction.

This, then, is a force acting on a body caused by its motion through the electromagnetic field, or by changes occurring in that field itself; and the effect of the force is either to produce a current and heat the body, or to decompose the body, or, when it can do neither, to put the body in a state of electric polarization,—a state of constraint in which opposite extremities are oppositely electrified, and from which the body tends to relieve itself as soon as the disturbing force is removed.

(10) According to the theory which I propose to explain, this "electromotive force" is the force called into play during the communication of motion from one part of the medium to another, and it is by means of this force that the motion of one part causes motion in another part. When electromotive force acts on a conducting circuit, it produces a current, which, as it meets with resistance, occasions a continual transformation of electrical energy into heat, which is incapable of being restored again to the form of electrical energy by any reversal of the process.

(11) But when electromotive force acts on a dielectric it produces a state of polarization of its parts similar in distribution to the polarity of the parts of a mass of iron under the influence of a magnet, and like the magnetic polarization, capable of being described as a state in which every particle has its opposite poles in opposite conditions.[8]

In a dielectric under the action of electromotive force, we may conceive that the electricity in each molecule is so displaced that one side is rendered positively and the other negatively electrical, but that the electricity remains entirely connected with the molecule, and does not pass from one molecule to another. The effect of this action on the whole dielectric mass is to produce a general displacement of electricity in a certain direction. This displacement does not amount to a current, because when it has attained to a certain value it remains constant, but it is the commencement of a current, and its variations

---

[8] FARADAY, *Exp. Res.* Series XI.; MOSSOTTI, *Mem. della Soc. Italiana* (Modena), vol. XXIV. part 2. p. 49.

constitute currents in the positive or the negative direction according as the displacement is increasing or decreasing. In the interior of the dielectric there is no indication of electrification, because the electrification of the surface of any molecule is neutralized by the opposite electrification of the surface of the molecules in contact with it; but at the bounding surface of the dielectric, where the electrification is not neutralized, we find the phenomena which indicate positive or negative electrification.

The relation between the electromotive force and the amount of electric displacement it produces depends on the nature of the dielectric, the same electromotive force producing generally a greater electric displacement in solid dielectrics, such as glass or sulphur, than in air.

(12) Here, then, we perceive another effect of electromotive force, namely, electric displacement, which according to our theory is a kind of elastic yielding to the action of the force, similar to that which takes place in structures and machines owing to the want of perfect rigidity of the connexions.

(13) The practical investigation of the inductive capacity of dielectrics is rendered difficult on account of two disturbing phenomena. The first is the conductivity of the dielectric, which, though in many cases exceedingly small, is not altogether insensible. The second is the phenomenon called electric absorption,[9] in virtue of which, when the dielectric is exposed to electromotive force, the electric displacement gradually increases, and when the electromotive force is removed, the dielectric does not instantly return to its primitive state, but only discharges a portion of its electrification, and when left to itself gradually acquires electrification on its surface, as the interior gradually becomes depolarized. Almost all solid dielectrics exhibit this phenomenon, which gives rise to the residual charge in the Leyden jar, and to several phenomena of electric cables described by Mr. F. JENKIN.[10]

(14) We have here two other kinds of yielding besides the yielding of the perfect dielectric, which we have compared to a perfectly elastic body. The yielding due to conductivity may be compared to that of a viscous fluid (that is to say, a fluid having great internal friction), or a soft solid on which the smallest force produces a permanent alteration of figure increasing with the time during which the force acts. The yielding due to electric absorption may be compared to that of a cellular elastic body containing a thick fluid in its cavities. Such a body, when subjected to pressure, is compressed by degrees on account of the gradual yielding of the thick fluid; and when the pressure is removed it does not at once recover its figure, because the elasticity of the substance of the body has gradually to overcome the tenacity of the fluid before it can regain complete equilibrium.

---

[9] FARADAY, *Exp. Res.* 1233–1250.

[10] *Reports of British Association*, 1859, p. 248; and *Report of Committee of Board of Trade on Submarine Cables*, pp. 136 & 464.

Several solid bodies in which no such structure as we have supposed can be found, seem to possess a mechanical property of this kind;[11] and it seems probable that the same substances, if dielectrics, may possess the analogous electrical property, and if magnetic, may have corresponding properties relating to the acquisition, retention, and loss of magnetic polarity.

(15) It appears therefore that certain phenomena in electricity and magnetism lead to the same conclusion as those of optics, namely, that there is an æthereal medium pervading all bodies, and modified only in degree by their presence; that the parts of this medium are capable of being set in motion by electric currents and magnets; that this motion is communicated from one part of the medium to another by forces arising from the connexions of those parts; that under the action of these forces there is a certain yielding depending on the elasticity of these connexions; and that therefore energy in two different forms may exist in the medium, the one form being the actual energy of motion of its parts, and the other being the potential energy stored up in the connexions, in virtue of their elasticity.

(16) Thus, then, we are led to the conception of a complicated mechanism capable of a vast variety of motion, but at the same time so connected that the motion of one part depends, according to definite relations, on the motion of other parts, these motions being communicated by forces arising from the relative displacement of the connected parts, in virtue of their elasticity. Such a mechanism must be subject to the general laws of Dynamics, and we ought to be able to work out all the consequences of its motion, provided we know the form of the relation between the motions of the parts.

(17) We know that when an electric current is established in a conducting circuit, the neighbouring part of the field is characterized by certain magnetic properties, and that if two circuits are in the field, the magnetic properties of the field due to the two currents are combined. Thus each part of the field is in connexion with both currents, and the two currents are put in connexion with each other in virtue of their connexion with the magnetization of the field. The first result of this connexion that I propose to examine, is the induction of one current by another, and by the motion of conductors in the field.

The second result, which is deduced from this, is the mechanical action between conductors carrying currents. The phenomenon of the induction of currents has been deduced from their mechanical action by HELMHOLTZ[12] and THOMSON[13] I have followed the reverse order, and deduced the mechanical action from the laws of induction. I have then described experimental methods of determining the quantities L, M, N, on which these phenomena depend.

---

[11] As, for instance, the composition of glue, treacle, &c., of which small plastic figures are made, which after being distorted gradually recover their shape.

[12] "Conservation of Force," *Physical Society of Berlin*, 1847; and TAYLOR'S *Scientific Memoirs*, 1853, p. 11:4.

[13] *Reports of the British Association*, 1848; *Philosophical Magazine*, Dec. 1851.

(18) I then apply the phenomena of induction and attraction of currents to the exploration of the electromagnetic field, and the laying down systems of lines of magnetic force which indicate its magnetic properties. By exploring the same field with a magnet, I show the distribution of its equipotential magnetic surfaces, cutting the lines of force at right angles.

In order to bring these results within the power of symbolical calculation, I then express them in the form of the General Equations of the Electromagnetic Field. These equations express—

(A) The relation between electric displacement, true conduction, and the total current, compounded of both.

(B) The relation between the lines of magnetic force and the inductive coefficients of a circuit, as already deduced from the laws of induction.

(C) The relation between the strength of a current and its magnetic effects, according to the electromagnetic system of measurement.

(D) The value of the electromotive force in a body, as arising from the motion of the body in the field, the alteration of the field itself, and the variation of electric potential from one part of the field to another.

(E) The relation between electric displacement, and the electromotive force which produces it.

(F) The relation between an electric current, and the electromotive force which produces it.

(G) The relation between the amount of free electricity at any point, and the electric displacements in the neighbourhood.

(H) The relation between the increase or diminution of free electricity and the electric currents in the neighbourhood.

There are 20 of these equations in all, involving 20 variable quantities.

(19) I then express in terms of these quantities the intrinsic energy of the Electromagnetic Field as depending partly on its magnetic and partly on its electric polarization at every point.

From this I determine the mechanical force acting, first, on a moveable conductor carrying an electric current; secondly, on a magnetic pole; thirdly, on an electrified body.

The last result, namely, the mechanical force acting on an electrified body, gives rise to an independent method of electrical measurement founded on its electrostatic effects. The relation between the units employed in the two methods is shown to depend on what I have called the "electric elasticity" of the medium, and to be a velocity, which has been experimentally determined by M.M. WEBER and KOHLRAUSCH.

I then show how to calculate the electrostatic capacity of a condenser, and the specific inductive capacity of a dielectric.

The case of a condenser composed of parallel layers of substances of different electric resistances and inductive capacities is next examined, and it is shown that the phenomenon called electric absorption will generally occur, that is, the condenser, when suddenly discharged, will after a short time show signs of a *residual* charge.

(20) The general equations are next applied to the case of a magnetic disturbance propagated through a non-conducting field, and it is shown that the only disturbances which can be so propagated are those which are transverse to the direction of propagation, and that the velocity of propagation is the velocity v, found from experiments such as those of WEBER, which expresses the number of electrostatic units of electricity which are contained in one electromagnetic unit.

This velocity is so nearly that of light, that it seems we have strong reason to conclude that light itself (including radiant heat, and other radiations if any) is an electromagnetic disturbance in the form of waves propagated through the electromagnetic field according to electromagnetic laws. If so, the agreement between the elasticity of the medium as calculated from the rapid alternations of luminous vibrations, and as found by the slow processes of electrical experiments, shows how perfect and regular the elastic properties of the medium must be when not encumbered with any matter denser than air. If the same character of the elasticity is retained in dense transparent bodies, it appears that the square of the index of refraction is equal to the product of the specific dielectric capacity and the specific magnetic capacity. Conducting media are shown to absorb such radiations rapidly, and therefore to be generally opaque. The conception of the propagation of transverse magnetic disturbances to the exclusion of normal ones is distinctly set forth by Professor FARADAY[14] in his "Thoughts on Ray Vibrations." The electromagnetic theory of light, as proposed by him, is the same in substance as that which I have begun to develope in this paper, except that in 1846 there were no data to calculate the velocity of propagation.

(21) The general equations are then applied to the calculation of the coefficients of mutual induction of two circular currents and the coefficient of self-induction in a coil. The want of uniformity of the current in the different parts of the section of a wire at the commencement of the current is investigated, I believe for the first time, and the consequent correction of the coefficient of self-induction is found.

These results are applied to the calculation of the self-induction of the coil used in the experiments of the Committee of the British Association on Standards of Electric Resistance, and the value compared with that deduced from the experiments.

---

[14] *Philosophical Magazine*, May 1846, or Experimental Researches, III. p. 447.

31.3 Study Questions

QUES. 31.1. Why did Maxwell reject Weber's early nineteenth century theory of electric and magnetic phenomena? Specifically, how does Maxwell use the principle of conservation of energy to support his theory of the electromagnetic field?

QUES. 31.2. Does the presence of "gross matter" modify the behavior of light in an essential way?

a) What happens to a medium when light propagates through it? How does the medium move? Does the presence of a magnetic field inside the medium have any effect on the light itself?

b) Speaking of magnetic fields, what happens when a body moves through a magnetic field—or when a stationary body is subjected to a time-varying magnetic field? Does it matter whether the body is conducting or insulating? Can the work done by an electromotive force on the body always be recovered, or is there sometimes a loss of useful work?

c) In what sense do imperfect dielectrics behave like a viscous fluid, or perhaps even like a fluid-filled foam? What features of light propagation do these two analogies elucidate?

d) How is energy stored and/or communicated when light propagates through a medium?

QUES. 31.3. How is the speed of light related to the specific properties of the medium through which it travels?

a) How does Maxwell attempt to understand electric and magnetic phenomena in terms of "general laws of dynamics"? How many dynamical equations will Maxwell finally obtain? And what are some of the quantities which are connected by these equations?

b) What is the most significant feature drawn from Maxwell's general equations? What does this lead Maxwell to conclude regarding the nature of light?

c) In what sense is aether like a perfectly elastic medium? How is the index of refraction of a medium connected to its electric and magnetic properties? Why are conducting bodies typically opaque?

d) To whom does Maxwell finally attribute his conception of the electromagnetic theory of light?

31.4 Exercises

In the following exercises, we will explore Maxwell's equations *not* as Maxwell originally wrote them in 1865, but in the modern form developed some 30 years later by Oliver Heaviside, a self-taught electrical engineer and physicist.[15] Heaviside's

---

[15] Heaviside, O., *Electromagnetic Theory*, vol. 1, "The Electrician" Printing and Publishing Company, London, 1893. See especially Chap. 2, §33–§36 and Chap. 3.

Fig. 31.1 Gauss's law relates
the quantity of electrical
charge to the surface integral
of the electric field through
a closed surface surrounding
the charge

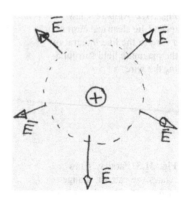

formulation of Maxwell's theory relies heavily on the techniques of vector calculus. For the student who is unacquainted with vector calculus (or for the student who needs a brief refresher course) I would highly recommend that you work through the exercises in Appendix A before proceeding.

Ex. 31.1 (GAUSS' LAW). The first of Maxwell's equations is *Gauss's law*, which states that the electric flux through a closed surface is proportional to the charge enclosed by the surface:

$$\oiint \vec{E} \cdot d\vec{A} = \frac{Q}{\epsilon}. \tag{31.1}$$

Here, the $\oiint$ symbol indicates a surface integral over a closed surface, such as a sphere, and ϵ (the proportionality constant) is the electric permittivity of the medium surrounding the charge. For the case of space free of "gross matter," the permittivity is $\epsilon_0 = 8.854 \times 10^{-12}$ C^2/Nm^2 (SI). This value is known as the permittivity of free space. As an example of the utility of Gauss's law, consider a point charge, Q, and draw a sphere of radius r around the charge (see Fig. 31.1).

Using Eq. A.8 for the equation of the electric field of a point charge and $d\vec{A} = dA\,\hat{r}$, Eq. 31.1 may be simplified to

$$\oiint \frac{1}{4\pi\epsilon_0} \frac{Q}{r^2} dA = \frac{Q}{\epsilon_0}. \tag{31.2}$$

Pulling the constants out of the integration and integrating over the surface area of a sphere, $\oiint dA = 4\pi r^2$, shows that the left and right hand sides of Eq. 31.2 are equal, just what we would expect if Gauss's law was consistent with Coulomb's law.

Now as an exercise in the application of Gauss's law, find the electric flux through (a) the surface of a spherical surface sitting next to (but not enclosing) a point charge Q, (b) the surface of a spherical surface containing both a proton and an electron, and (c) each of the faces of a cube of volume V with a point charge Q placed at one corner.

Fig. 31.2 Ampère's law
relates the electrical current in
a wire to the line integral of
the magnetic field surrounding the wire

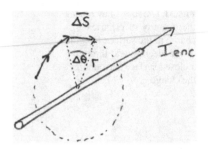

Fig. 31.3 Faraday's law
relates the rate of change
of magnetic flux through a
surface to the line integral of
the electric field around the
boundary of said surface

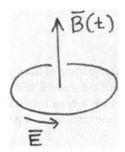

EX. 31.2 (AMPÈRE'S LAW). The second of Maxwell's equations is *Ampère's law*,
which states that the line integral of the magnetic field around a closed loop is
proportional to the electric current enclosed by the loop:

$$\oint \vec{B} \cdot d\vec{s} = \mu I. \tag{31.3}$$

Here, the symbol $\oint$ indicates a line integral around a closed loop, such as a circle,
and μ (the proportionality constant) is the magnetic permeability of the medium
surrounding the electric current. For the case of space free of "gross matter," the
permeability is $\mu_0 = 4\pi \times 10^{-7}$ Tm/A (SI). This value is known as the permeability
of free space. As an example of the utility of Ampère's law, consider a straight wire
carrying an electric current, I. Draw a circle of radius r around the wire, as depicted
in Fig. 31.2.

The differential length elements which comprise the loop around the wire may
be written as $d\vec{s} = r\, d\theta\, \hat{\theta}$, where $\hat{\theta}$ is the unit vector in a polar coordinate system
which points in the direction of increasing θ. Also, the magnetic field is directed
azimuthally around the wire and so may be written as $\vec{B} = B\hat{\theta}$. Hence, Eq. 31.3
becomes

$$\oint_0^{2\pi} (B\hat{\theta}) \cdot (r\, d\theta\, \hat{\theta}) = \mu I. \tag{31.4}$$

As an exercise, continue this procedure so as to obtain a mathematical expression
for the strength of the magnetic field, $B(r)$, at a distance r from the straight wire.
Does your result agree with Eq. 7.1?

EX. 31.3 (FARADAY'S LAW). The third of Maxwell's equations is *Faraday's law*, which states that the line integral of the electric field around a closed loop is proportional to the time rate of change of magnetic flux through the area enclosed by the loop:

$$\oint \vec{E} \cdot d\vec{s} = \frac{d}{dt} \iint \vec{B} \cdot d\vec{A}. \tag{31.5}$$

Using the definitions of the electric field (Eq. 30.1 or A.6), the electro-motive force (Eq. 25.1) and the magnetic flux (Eq. 25.5), we find that Eq. 31.5 is identical to our previous expression of Faraday's law:

$$\varepsilon = \frac{d\Phi}{dt}. \tag{25.4}$$

These equations say that whenever there exists a time-dependent magnetic field, an electric field will be generated in its vicinity, *even if there is no conducting medium which might carry an electric current*. This observation, depicted schematically in Fig. 31.3, will be very important when we come to study Maxwell's electro-magnetic theory of radiation.

As an example of the application of Faraday's law, suppose that a wire is wrapped around a glass cylinder which is 10 cm in diameter. When a current flows throughout the coil, a uniform magnetic field is produced inside the coil. As the current is reduced, the magnetic field strength decreases at the rate of 5 mT/s. Assuming that the magnetic field outside of the coil is negligible (compared to its value inside the coil), (a) determine the strength of the induced electric field, E, (in SI units) at distance $r = 2.0$ and (b) $r = 6.0$ cm from the axis of the coil. Finally, (c) make a sketch of the magnitude $E(r)$. At what value of r is $E(r)$ maximum? Explain. (d) Does this electric field act to increase or decrease the electric current in the wires? What insight does this give you into the phenomenon of an electrical circuit's *self-inductance* (see Ex. 31.7).

EX. 31.4 (THE NO-NAME LAW). The fourth and final Maxwell's equations is *The no-name law*, which states that the magnetic flux through any closed surface is always exactly zero:

$$\oiint \vec{B} \cdot d\vec{A} = 0. \tag{31.6}$$

The difference between this law and Gauss's law is due to the empirical fact that one may not isolate magnetic poles in the same way that one may isolate electric charges. Whenever a magnet is broken, each half contains both a north and a south pole. As Faraday noted, this implies that magnetic field lines do not terminate on magnetic poles, but rather form closed loops, as depicted in Fig. 31.4. (a) In this figure, what is the total magnetic flux through the closed surface shown? (b) How would this flux change if the surface surrounded the entire magnet? (c) What if it surrounded only the upper half of the magnet?

Fig. 31.4 The fourth equation
of Maxwell states that the
magnetic flux through any
closed surface is always
exactly zero

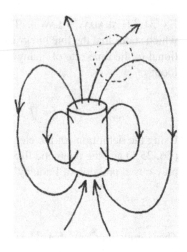

Ex. 31.5 (THE AMPÈRE-MAXWELL LAW). One of Maxwell's most significant insights was that Ampère's law, as stated as in Eq. 31.3, misses something. Consider Fig. 31.5, which depicts the charging up of a parallel plate capacitor *via* two connecting wires.

As the electric current, I, flows down the left wire, positive charge accumulates on the left capacitor plate. This charge accumulation repels positive charges from the right capacitor plate, inducing an electric current in the right wire. In accordance with Ampère's law, a magnetic field is generated around the left and right wires. But what about in the region between the capacitor plates? There is no electric current flowing between the plates, so it would seem that there would be no magnetic field in this region. As the capacitor plates are charged, however, an electric field of increasing strength is created which points rightward between the plates. Maxwell suspected that this time-varying electric field also acts as source of magnetic fields. He thus added a second term to Ampère's law to make what is now referred to as the *Ampère-Maxwell law*:

$$\oint \vec{B} \cdot d\vec{s} = \mu I + \mu \left[\epsilon \frac{d}{dt} \iint \vec{E} \cdot d\vec{A} \right]. \tag{31.7}$$

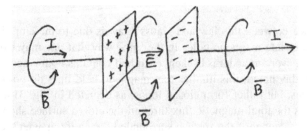

Fig. 31.5 A magnetic field surrounds not only an electrical current, but also a region in which a time-varying electric field exists, such as the region between the plates of a charging capacitor

The first term on the right side shows that electric currents give rise to magnetic fields; the second term shows that time-varying electric fields give rise to magnetic fields. In fact, the quantity inside the square brackets in the second term on the right side of Eq. 31.7 is sometimes referred to as a *displacement current*, to suggest that it behaves like a *bona fide* electrical current.

As an example of the application of the Ampère-Maxwell law, consider a parallel plate capacitor, similar to the one depicted in Fig. 31.5, but consisting of circular metal plates of radius R. The capacitor is being charged up in such a way that the electric field inside the plates is increasing linearly with time. For simplicity, suppose that the plates are close enough together that the electric field strength in the region outside the plates is essentially zero. (a) Find a mathematical expression for $B(r)$ midway between the plates at a distance $r < R$ from the axis. (b) Now find a mathematical expression for $B(r)$ midway between the plates at a distance $r > R$ from the axis. (c) Sketch $B(r)$. Does your result make sense?

EX. 31.6 (SPEED OF LIGHT CALCULATION). The goal of this laboratory exercise is to calculate the speed of electromagnetic waves using (previously) measured values of the electrical permittivity of space, ϵ_0, and the magnetic permeability of space, μ_0. Recall that these constants show up in Coulomb's law,

$$F = \frac{1}{4\pi\epsilon_0} \frac{q_1 q_2}{r^2} \tag{31.8}$$

and in the law of force acting between two current carrying wires:

$$F = \frac{\mu_0 I_1 I_2 L}{2\pi r} \tag{31.9}$$

First you must measure the force between electrically charged spheres and between current carrying wires. To this end, you should carry out the electro-static and magneto-static experiments described in Exercises 6.4 and 8.1 in the present volume. Now, fit Eqs. 31.8 and 31.9 to appropriate data plots so as to determine your values of ϵ_0 and μ_0. Finally, using Maxwell's theory, determine the speed of an electromagnetic disturbance.

EX. 31.7 (ELECTRIC INERTIA, SELF-INDUCTANCE AND CURRENT DECAY). Both Faraday and Maxwell recognized that an electrical current seems to have *inertia*. For example, when a voltaic cell is connected across the ends of a helical coil of wire, it takes a few moments for the coil to exert action upon a nearby compass needle.[16] This implies that while the electro-motive force might be established instantly, the resulting electric current takes a few moments to achieve its full magnitude. Likewise, when a current carrying helix is abruptly disconnected from the same voltaic cell, the electric current jumps across the gap, as evidenced by a spark. This is just

---

[16] Faraday refers to an "electro-tonic state" established within matter subject to an electro-motive force. See Faraday, M., *Experimental Researches in Electricity*, vol. 1, Taylor and Francis, London, 1839, §3. New Electrical State or Condition of Matter, 59–80.

Fig. 31.6 A circuit consisting
of a voltaic battery attached
to a loop of wire having resistance R and inductance L.
The battery may be removed
from the circuit by throwing
the switch S

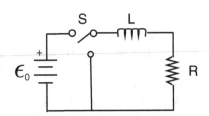

what one would expect if the electric current possessed some form of inertia. Why
might this happen? Recall that a loop of current-carrying wire creates a magnetic
field in its vicinity which is proportional to the magnitude of the electric current
itself. If, for some reason, the source of the electric current around a loop of wire is
weakened or removed altogether, the magnetic field produced by this wire must be
correspondingly reduced. The reduction of the magnetic field causes the magnetic
flux through the loop to change, so by Faraday's law an electro-motive force will
be generated so as to attempt to restore the lost magnetic flux. This implies that
the electric current in a circuit resists change. The tendency of a current-carrying
loop to resist change can be expressed in terms of the *self-inductance* of the circuit.
More specifically, the electro-motive force, ε, which attempts to restore the lost flux
through the loop is directly proportional to the rate of change of the electric current
in the loop:

$$\varepsilon = -L\frac{dI}{dt}. \tag{31.10}$$

Eq. 31.10 provides the definition of L, the self-inductance of the circuit. The self-
inductance of a circuit depends on the size and shape of the circuit itself. The
negative sign in Eq. 31.10 appears as a reminder that if the current decreases, the
induced electro-motive force acts to increase the current.

Suppose now that a voltaic cell, whose terminals produce an electro-motive force
ε_0, is attached to a loop of wire whose resistance is R and whose self-inductance
is L. This is shown in Fig. 31.6. While a steady current, I_0, is flowing through
the circuit, suddenly, a switch (S) is thrown, removing the voltaic cell from the
circuit. As a result, the current begins to immediately decay due to the resistance of
the circuit. As it does so, the self-inductance of the circuit acts to establish a new
electro-motive force, $\varepsilon(t)$, which attempts to keep the current circulating through
the wire.

a) Using Ohm's law (Eq. 25.2), find an expression for the constant electric current
   in the circuit, I_0, just before the terminals of the voltaic cell are shorted. What do
   you expect the electric current to be a long time after the terminals are shorted?
b) Using Ohm's law and Eq. 31.10, find a mathematical expression for the time-
   dependence of the electric current flowing through the circuit, $I(t)$. Make a
   sketch of this function.
c) If $R = 100$ Ohms, $L = 1$ micro-Henry, and $\varepsilon_o = 10$ volts, at what time, τ, has
   the electric current fallen by a factor of e^{-1}? More generally, how does the decay
   time, τ, of the electric current depend upon R and L?

31.5 Vocabulary

1. Mutual
2. Ascertain
3. Induction
4. Diamagnetic
5. Hitherto
6. Endeavor
7. Gross
8. Permeate
9. Undulate
10. Datum
11. Paramagnetic
12. Luminous
13. Warrantable
14. Electromotive
15. Dielectric
16. Render
17. Viscous
18. Tenacity
19. Retention
20. Capacity

Chapter 32
Propagating Electromagnetic Fields

There is more exercise for the brains in the electromagnetic than in the electrostatic problems.

—Oliver Heaviside

32.1 Introduction

In the introduction to his *Dynamical Theory of the Electromagnetic field*,[1] Maxwell explained how the region between electrified bodies acts (in certain respects) like an elastic material whose properties could be experimentally determined by measuring the forces between nearby static electric charges and currents. Motivated by this concept, Maxwell was able to establish a set of general equations which govern the relationships between charges and currents, on the one hand, and the properties and motions of the medium, on the other hand. Perhaps most remarkably, he found that these same general equations implied the possibility of self-sustaining waves within the medium which propagate at the speed of light.[2] Indeed, light itself may be understood as such a vibration.

Unfortunately, Maxwell's theory was expressed in a rather complicated form involving 20 equations linking 20 variables. Today, Maxwell's equations are written in a more compact form involving only four equations. This modern form was developed by Oliver Heaviside (1850–1925). Heaviside was born in London and underwent formal schooling until the age of 16, after which he began working as a telegraph operator. A very capable but private man, Heaviside resigned from the telegraph service after just 6 years and retired to his parents' home in London, where he lived for most of his remaining years. During this time, Heaviside studied scientific treatises, such as those of John Tyndall and James Clerk Maxwell, and began to publish sophisticated articles in *The Electrician*, a weekly trade journal for manufacturers, electricians and hobbyists. So alongside editorials on insulating copper wires and tips for avoiding electrocution, Heaviside published his complete reformulation of Maxwell's dynamical equations using the beautiful methods of vector

---

[1] See Chap. 31 of the present volume.

[2] This was mentioned by Maxwell in point (20) of the introduction to his *Dynamical Theory of the Electromagnetic Field*; see Chap. 31 of the present volume.

© Springer International Publishing Switzerland 2016
K. Kuehn, *A Student's Guide Through the Great Physics Texts,*
Undergraduate Lecture Notes in Physics, DOI 10.1007/978-3-319-21816-8_32

calculus which he himself developed for this very purpose.[3] We encountered these equations in Chap. 31 and they are reproduced here for convenience.

Gauss's law:

$$\oiint \vec{E} \cdot d\vec{A} = \frac{Q}{\epsilon}. \tag{31.1}$$

Faraday's law:

$$\oint \vec{E} \cdot d\vec{s} = \frac{d}{dt} \iint \vec{B} \cdot d\vec{A}. \tag{31.5}$$

No-name law:

$$\oiint \vec{B} \cdot d\vec{A} = 0. \tag{31.6}$$

Ampère-Maxwell law:

$$\oint \vec{B} \cdot d\vec{s} = \mu I + \mu\epsilon \frac{d}{dt} \iint \vec{E} \cdot d\vec{A}. \tag{31.7}$$

These equations—which form the basis of the classic electromagnetic theory of light—relate electric and magnetic fields to the presence of electric currents and charges. The electric permittivity, ϵ, and the magnetic permeability, μ, which appear in these equations characterize the properties of the material in which the fields exist.

To get a sense of the physical meaning of the electric permittivity, consider a slab of insulating material (such as glass) placed between two charged parallel plates. The electric field, $\vec{E}$, between the plates causes the molecules within the slab to become polarized, as depicted cartoonishly in Fig. 32.1. Thus, a small electric field is generated within each molecule which is antiparallel to the applied electric field; its direction is denoted by the vector $\vec{P}$. The effect of all these polarizations is to reduce the strength of the field within the slab. One may then speak of a *displacement field*, $\vec{D}$, which is proportional to the applied electric field but which reflects the diminution of the field due to the aforementioned polarization:

$$\vec{D} = \epsilon \vec{E}. \tag{32.1}$$

The permittivity thus characterizes the electric polarizability of the material.[4] Similarly, the permeability, μ, characterizes the magnetic polarizability of the material. When a magnetic field is applied across a slab of material, the material responds in such a way as to alter the field within the material. Unlike with electric fields, however, this may either increase (in the case of a paramagnetic material) or decrease (in the case of a diamagnetic material) the field within the material. Historically, the applied magnetic field was denoted by $\vec{H}$, while $\vec{B}$ was referred to as the magnetic induction within the material. Nowadays, $\vec{B}$ is typically called the magnetic field and $\vec{H}$ is simply called the H-field. In any case, the relationship between $\vec{H}$ and $\vec{B}$ for a particular material is given by

$$\vec{B} = \mu \vec{H}. \tag{32.2}$$

---

[3] See Heaviside, O., *Electromagnetic Theory*, vol. 1, "The Electrician" Printing and Publishing Company, London, 1893, especially Chap. 2, §33–§36 and Chap. 3.

[4] The relative permittivity of a material, ϵ_r, is defined as the ratio of its permittivity to that of free space.

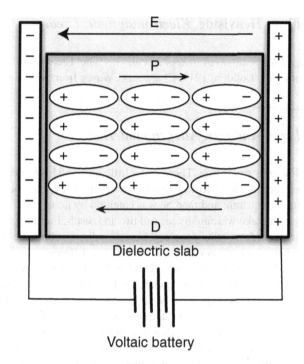

Fig. 32.1 The molecules comprising a dielectric material are polarized when inserted between charged parallel plates. Their polarization tends to reduce the field strength within the dielectric

The so-called *constitutive relations* expressed in Eqs. 32.1 and 32.2 describe the response of materials to applied electric and magnetic fields. With these preliminary comments in mind, let us now turn to Heaviside's explanation of electromagnetic waves in materials. In the reading selection below, Heaviside demonstrates in a (relatively) simple way how Maxwell's equations—particularly the two involving line integrals or *circulation*—imply the existence of traveling electromagnetic waves. After some preliminary personal and pedagogical comments, he goes on to imagine that perpendicular electric and magnetic fields, $\vec{E}$ and $\vec{H}$, are (somehow) established within a slab of material occupying a finite region of space. He then explains how the Ampère-Maxwell law and Faraday's law can be simultaneously fulfilled by requiring this initial field-filled region to propagate forward, in a direction perpendicular to both the electric and magnetic fields. The speed of this propagating electromagnetic wave may be expressed in terms of the aforementioned permittivity and permeability of the wave-supporting material alone.[5]

---

[5] A similar approach is employed in Chap. 18 of Vol. II of Feynman, R.P., Leighton, R.B., and Sands, M.L., *The Feynman Lectures on Physics*, Commemorative ed., Addison-Wesley Publishing Co., 1989.

32.2   Reading: Heaviside, *Electromagnetic Theory*

Heaviside, O., *Electromagnetic Theory*, vol. 3, "The Electrician" Printing and Publishing Company, London, 1912. Chapter IX: Waves from Moving Sources.

32.2.1   Adagio. Andante. Allegro moderato

§450. The following story is true. There was a little boy, and his father said, "Do try to be like other people. Don't frown." And he tried and tried, but could not. So his father beat him with a strap; and then he was eaten up by lions.

Reader, if young, take warning by his sad life and death. For though it may be an honour to be different from other people, if Carlyle's dictum about the 30 millions be still true, yet other people do not like it. So, if you are different, you had better hide it, and pretend to be solemn and wooden-headed. Until you make your fortune. For most wooden-headed people worship money; and, really, I do not see what else they can do. In particular, if you are going to write a book, remember the wooden-headed. So be rigorous; that will cover a multitude of sins. And do not frown.

There is a time for all things: for shouting, for gentle speaking, for silence; for the washing of pots and the writing of books. Let now the pots go black, and set to work. It is hard to make a beginning, but it must be done.

Electric and magnetic force. May they live for ever, and never be forgot, if only to remind us that the science of electromagnetics, in spite of the abstract nature of the theory, involving quantities whose nature is entirely unknown at present, is really and truly founded upon the observation of real Newtonian forces, electric and magnetic respectively. I cannot appreciate much the objection that they are not forces; because they are the forces per unit electric and magnetic pole. All the same, however, I think Dr. Fleming's recent proposal that electric force and magnetic force shall be called the voltivity and the gaussivity a very good one; not as substitutes for with abolition of the old terms, but as alternatives; and beg to recommend their use if found useful, even though I see no reason for giving up my own use of electric and magnetic force until they become too antiquated.

Having thus got to the electric and magnetic forces, it is only a short step farther to near the end of the book—namely, to the simple cases in which they occur simultaneously. It does not follow that the matter which comes towards the end of a treatise—for instance, Maxwell's great work—is harder than that in the first chapter of his Vol. I. On the contrary, some parts of it are easier out of all comparison. In the course of the next generation many treatises on electromagnetics will probably be written; and there is no reason whatever (and much good reason against it) why the old-fashioned way of beginning with electrostatics (unrelated to the general theory) should be followed. After all, should not the easier parts of a subject come first, to help the reader and widen his mind? I think it would be perfectly practical to begin the serious development of the theory with electromagnetic waves of the easy kind.

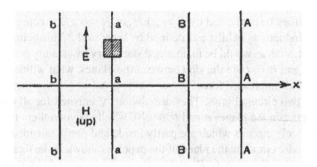

Fig. 32.2 Heaviside's construction for illustrating the propagation of electric and magnetic fields initially confined to a slab.—[*K.K.*]

First of all, of course, there should be a good experimental knowledge all round, not necessarily very deep. Then, considering the structure of a purely theoretical work to co-ordinate the previous, a general survey is good to begin with, with consideration in more detail of the properties of circuits and the circuital laws. Then, coming to developments, start with plane electromagnetic waves in a dielectric non-conductor. The algebra thereof, even when pursued into the details of reflections, &c., is perhaps more simple than in any other part of the science, save Ohm's law and similar things; and the physical interest is immense. You can then pass to waves along wires. First the distortionless theory in detail, and then make use of it to establish the general nature of the effects produced by practical departures from such perfection, leaving the difficult mathematics of the exact results for later treatment. Now, all this and much more is ever so much easier than the potential functions and spherical harmonics and conjugate transformations with which electrostatics is loaded, and there is more exercise for the brains in the electromagnetic than in the electrostatic problems. The subsequent course may be left open. There are all sorts of ways.

32.2.2   Simple Proof of Fundamental Property of a Plane Wave

§451. At present, in dealing with some elementary properties, the object is to smooth the road to the later matter. First of all, how to prove the fundamental property of a plane wave, that it travels at constant speed undistorted, if there be no conductivity, or, more generally, no molecular interference causing dispersion and other disturbances? We have merely to show that the two circuital laws are satisfied, and that can be done almost by inspection. Thus, let the region between two parallel planes *aaa* and *bbb* be an electric field and a magnetic field at the same time, the electric force E being uniform and in (and parallel to) the plane of the paper, whilst H is also uniform, perpendicular to E and directed up through the paper (Fig. 32.2). Also let their intensities be connected by $E = \mu v H$, or $H = cvE$; c being the permittivity and μ the inductivity, whilst v is defined by $\mu c v^2 = 1$.[6] This being the state at a given moment, such as would be maintained stationary by steadily acting impressed forces $e = E$ and $h = H$ in the slab between the planes, what will happen later, in the absence of all impressed force?

Apply the two circuital laws. They are obviously satisfied for all circuits which are wholly between the planes a and b, or else wholly beyond them to right or left. There are left only circuits which are partly inside and partly outside the slab. Consider a unit square circuit in the plane of the paper, as shown in the figure (Fig. 32.2). The circuitation of E is simply E, and by the second circuital law this must be the rate of decrease of induction $B = \mu H$ through the circuit downward, or its rate of increase upward. Then turn the square circuit at right angles to the paper. The circuitation of H is then simply H, and by the first circuital law this must be the rate of increase of displacement $D = Ec$ through the circuit.[7] Now let the plane aaa move to the right at speed v. The two rates of increase are made to be $v\mu H$ and vcE respectively. That is, $E = \mu v H$ and $H = cvE$ express the circuital laws. They are harmonised by the definition of v. We prove that the circuital laws are satisfied in the above way. That there is no other way of putting induction and displacement in the two circuits may be seen by considering circuits two of whose sides are infinitely near the plane a on opposite sides. The fluxes must be added on just at the plane itself, extending the region occupied by E and H. Similar reasoning applied to the plane bbb proves that it must also move to the right at speed v. Thus the whole slab moves bodily to the right at speed v, so that a moves to A and b moves to B in the time given by $vt = aA$ or bB.

The disturbance transferred in this way constitutes a pure wave. It carries all its properties with it unchanged. The density of the electric energy, or $U = \frac{1}{2}cE^2$, equals the density of the magnetic energy, or $T = \frac{1}{2}\mu H^2$. The flux of energy is $W = v(U + T)$. The simplest case of the general formula $W = V(E - e)(H - h)$. See Vol. I., §70.

32.2.3 The General Plane Wave

§452. What is proved for a discontinuity is proved for any sort of variation. For the slab may be of any depth and any strength, and there may be any number of slabs side by side behaving in the same way, all moving along independently and unchanged. So $E = \mu v H$ expresses the general solitary wave, where, at a given moment, E may be an arbitrary function of x, real and single-valued of course, but

---

[6] Shortly, Heaviside will demonstrate how these relationships may be deduced from Maxwell's equations.—[K.K.].

[7] Note that Heaviside denotes the electric permittivity by c, not ϵ (as is typical today).—[K.K.].

without any necessary continuity in itself or in any of its differential coefficients. Denoting it by $f(x)$ when $t = 0$, it becomes $f(x - vt)$ at the time t. If we change the sign, and make $E = -\mu v H$, this will represent a negative wave, going from right to left. There may be a positive and a negative wave coexistent, separate in position, or superimposed. This constitutes the complete solution for plane waves with straight lines of E and H. If E_0 and H_0 are given arbitrarily (with no connection) at the moment $t = 0$, the two waves at that moment are

(positive) $$E_1 = +\mu v H_1 = \frac{1}{2}(E_0 + \mu v H_0),\tag{32.3}$$

(negative) $$E_2 = -\mu v H_2 = \frac{1}{2}(E_0 - \mu v H_0),\tag{32.4}$$

as may be immediately verified. Move E_1 to the right, and E_2 to the left, at speed v, to produce the later states.

Since every slab is independent of the rest, there need be no connection between the directions of E in one slab and the next. The direction may vary anyhow along the wave. This makes a mathematical complication of no present importance, the behaviour of individual slabs being always the same.

The overlapping of positive and negative waves should be studied to illustrate the conversion of electric to magnetic energy, or conversely. For two equal waves moving oppositely, which fit when they coincide, there is a complete temporary disappearance and conversion of one or the other energy, according as E is doubled, leaving no H, or H is doubled, leaving no E. If E_0 exists alone initially, it makes two equal oppositely-going waves, of half strength as regards E. Similarly as regards initial H_0.

The reversal of sign of both E and H in either a positive or a negative wave does not affect the direction of motion. But if only one be reversed, it is turned from a positive to a negative wave, or conversely. Slabs of uniform strength should be studied, not simply periodic trains of waves, for simplicity of ideas. Only when there is dispersion, and the wave speed varies, is it necessary to consider a train of waves of given frequency or given wave-length; because then a slab spreads out behind as it travels, producing a diffused wave of difficult mathematical representation, by reason of the partial reflection of its different parts as it progresses.

32.3 Study Questions

QUES. 32.1. What advice does Heaviside offer to the young? How would you describe the mood of his introductory comments? And what might this reveal about his own character?

QUES. 32.2. What, according to Heaviside, is the basis of the science of electromagnetics? In what order does he believe it should be taught to students?

QUES. 32.3. What do Maxwell's equations imply about the speed of a traveling electromagnetic disturbance?

a) Consider Fig. 32.2. In what region are the electric and magnetic fields initially confined? In which direction(s) are they oriented?
b) Consider the square circuit straddling the boundary aaa. What are the lengths of its sides? What is the line integral (or circulation) of the electric field around this circuit? According to Faraday's law, what must accompany such circulation?
c) Suppose now that the square circuit is turned vertically. What is the circulation of the magnetic field around the circuit? According to the Ampère-Maxwell law, what must accompany such circulation?
d) At what rate would the electric and magnetic fluxes through the (stationary) square circuit change if the boundary aaa moved rightward at some speed, v? In particular, whence does Heaviside get the equations $E = \mu v H$ and $H = c v E$? And what do these equations imply about the speed, v?
e) Similarly, how can the Ampère-Maxwell and Faraday laws be fulfilled at the boundary bbb? More generally, what does Heaviside's analysis of boundaries aaa and bbb imply about the initial configuration of E and H? Is it stationary or is it moving? Is it retaining its shape or is it expanding?
f) According to Heaviside, do propagating electromagnetic fields carry energy? If so, how much, and at what speed?

QUES. 32.4. Is Heaviside's analysis unique to the initial configuration of E and H which he has chosen? Or does it apply to more general electric and magnetic field configurations?

QUES. 32.5. What happens when two identical strength electromagnetic waves moving in opposite directions pass through one another? For instance, do the electric and magnetic fields both grow twice as strong?

QUES. 32.6. What is the relationship between the orientations of the electric field, the magnetic field, and the direction of propagation of an electromagnetic disturbance? Do their orientations vary with time? Do their *relative* orientations vary with time?

32.4   Exercises

EX. 32.1 (PEDAGOGICAL STRATEGIES). What do you think: should teaching and learning proceed deductively from elementary principles? Or from the easiest to the most difficult material? Or perhaps starting with the most fascinating or interesting aspects of a subject?

EX. 32.2 (ELECTROMAGNETIC PLANE WAVES). Mathematically, the simplest form of a traveling wave is the plane wave. The relative strength and orientation of the electric and magnetic fields making up an electromagnetic plane wave at a particular

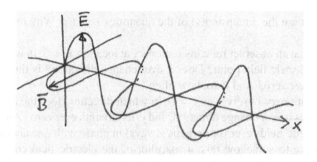

Fig. 32.3 A snapshot of the relative orientation of the electric and magnetic fields in an electromagnetic plane wave traveling along the $+x$-axis

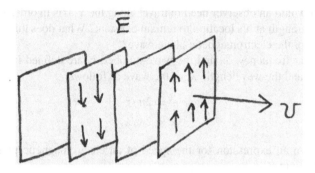

Fig. 32.4 For an electromagnetic plane wave, the electric field is uniform throughout any plane lying perpendicular its direction of propagation

instant in time are depicted schematically in Fig. 32.3. Mathematically, such an *electromagnetic plane wave* propagating along the x-axis of a cartesian coordinate system may be expressed as follows.

$$\vec{E}(x, t) = E_0 \sin\left[2\pi(t/\tau - x/\lambda)\right]\hat{y}$$
$$\vec{B}(x, t) = B_0 \sin\left[2\pi(t/\tau - x/\lambda)\right]\hat{z}$$
(32.5)

Here, E_0 and B_0 are scalar quantities which represent the amplitudes of the electric and magnetic fields, respectively. The relative orientations of the fields are indicated by the unit vectors $\hat{y}$ and $\hat{z}$. The argument of the sinusoidally varying functions indicate the phase of the field at a particular space and time coordinate. Notice that there is no y or z dependence of $\vec{E}$ or $\vec{B}$. Thus, one should understand that the the magnitude and direction of the electric field shown at any particular value of x actually represents the magnitude and direction of the electric field *at all points in a plane* which intersects the x-axis perpendicularly at that point. This is precisely why it is called an electro-magnetic *plane-wave*. A sequence of planes at equally spaced x-coordinates at a particular instant of time are schematically illustrated in Fig. 32.4. In order to better understand the nature of electromagnetic plane waves, answer the following questions in terms of quantities appearing in Eq. 32.5.

a) First, what are the dimension(s) of the quantities λ and τ. Why must this be the case?

b) Suppose that an observer remains stationary at location $x = 0$. In which direction does the electric field point? Does it ever change direction? Is the electric field strength ever zero? And if so, how often?

c) For the stationary observer at $x = 0$, in which direction does the magnetic field point? Does it ever change direction, and is its strength ever zero? Are the electric and magnetic fields ever (or perhaps always) in phase with one another?

d) If one were to somehow take a snapshot of the electric field configuration at any particular instant (such as $t = 0$), how far apart are the planes in which the electric field strength is zero? What about the zeroes of the magnetic field strength?

e) How fast would an observer need to travel along the x-axis in order for the electric field strength at his location to remain constant? What does this imply about the speed of the electromagnetic plane wave?

f) The angular frequency, ω, and the wave-number, k, are defined in terms of the period, τ, and the wavelength, λ, of the wave as follows:

$$\omega = 2\pi/\tau$$
$$k = 2\pi/\lambda. \tag{32.6}$$

Write down an expression for the speed of an electromagnetic plane wave in terms of ω and k.

g) Finally, rewrite Eq. 32.5 so as to represent an electromagnetic plane wave which is traveling along the y-axis and whose electric field points in the $+\hat{x}$ direction at $t = 0$.

Ex. 32.3 (ELECTROMAGNETIC ENERGY). Energy is required to generate electromagnetic waves; electromagnetic waves, in turn, exert forces on charged objects which they strike. Therefore if the principle of conservation of energy is correct, then energy must somehow be stored in the electric and magnetic fields which comprise a propagating electromagnetic wave. Suppose that the work done by an electric field, $\vec{E}$, in polarizing a unit volume of some medium is obtained by the integral

$$U = \int_0^D \vec{E} \cdot d\vec{D}, \tag{32.7}$$

and that, likewise, the work done by a magnetic field, $\vec{H}$, in polarizing a unit volume of some medium is given by the integral

$$T = \int_0^B \vec{H} \cdot d\vec{B}. \tag{32.8}$$

At the time of Maxwell and Heaviside, the electric energy was understood as an elastic potential energy, while the magnetic energy was understood as a type of rotational kinetic energy stored in the medium.

a) First, show that Eqs. 32.7 and 32.8 yield the following expressions for the *electromagnetic energy* stored in a unit volume:

$$U = \frac{1}{2}\epsilon E^2 \qquad\qquad T = \frac{1}{2}\mu H^2 \qquad\qquad (32.9)$$

b) Now identify the dimensions (and the SI units) of the *electromagnetic energy flux*:

$$W = v(U + T). \qquad\qquad (32.10)$$

EX. 32.4 (POYNTING VECTOR AND RADIATION INTENSITY). The *Poynting vector*, $\vec{S}$, is defined as the cross product of the electric and magnetic fields which exists at a particular point in space:

$$\vec{S} = \vec{E} \times \vec{H}. \qquad\qquad (32.11)$$

In order to understand the significance of the Poynting vector, consider the following exercises.

a) First, demonstrate that the Poynting vector has dimensions of energy per second per unit area. Thus, it may be understood as a measure of the rate at which electromagnetic energy strikes a given cross-sectional area.

b) In a traveling electromagnetic wave, the electric and magnetic fields are perpendicular to one another. Demonstrate that the magnitude of the Poynting vector may be expressed as

$$S = \frac{1}{c\mu}E^2 \qquad\qquad (32.12)$$

In which direction is the Poynting vector oriented?

c) Write down a mathematical expression which explicitly displays the space and time dependence of the Poynting vector for an electromagnetic plane wave traveling along the x-axis.

d) Many detectors of electromagnetic radiation, such as the human eye, have a slow response compared to the period of oscillation of the radiation. To such detectors, the time-average of the Poynting vector over an entire period of oscillation provides a measure of the perceived brightness of the radiation. Generally speaking, the brightness or *intensity* of electromagnetic radiation is defined as the time-average of the Poynting vector:

$$I \equiv S_{avg}$$
$$= \frac{1}{\tau}\int_0^\tau \frac{1}{c\mu}E(t)^2 dt \qquad\qquad (32.13)$$

Show that the intensity of the radiation produced by an electromagnetic plane wave of amplitude E_0 may then be simply written as

$$I = \frac{\epsilon c}{2}E_0^2. \qquad\qquad (32.14)$$

e) Finally, use the previous mathematical expressions for the electromagnetic energy to show that the electromagnetic energy flux is equal to the radiation intensity.

EX. 32.5 (POLARIZATION).   Suppose that a vertically polarized electromagnetic plane wave is normally incident on a polarizing filter. The filter is tilted at an angle θ with respect to the vertical. What is the intensity of the radiation, as a function of θ, which is transmitted through the filter? For the special case that $\theta = 30°$, what fraction of the initial light intensity is transmitted?

32.5   Vocabulary

1. Dictum
2. Antiquated
3. Electrostatic
4. Electromagnetic
5. Dielectric
6. Spherical harmonic
7. Conjugate transformation
8. Subsequent
9. Permittivity
10. Inductivity
11. Circuitation
12. Flux
13. Discontinuity
14. Coefficient
15. Superimpose
16. Dispersion

Chapter 33
Circuits, Antennae and Radiation

> *The experiments described appear to me, at any rate, eminently*
> *adapted to remove any doubt as to the identity of light, radiant*
> *heat, and electromagnetic wave-motion.*
>
> —Heinrich Hertz

33.1 Introduction

Heinrich Rudolph Hertz (1857–1894) was born in the sovereign city-state of Ham-
burg in what is now Germany. As a youth, he showed proficiency in classical and
modern languages as well as drafting and woodworking. After some initial training
for a career in engineering, he changed direction and began to study mathematics
and science, including the works of Lagrange, Laplace and Poisson. He transferred
from the university in Munich to Berlin, where he attended scientific lectures by
Gustav Kirchoff and Hermann von Helmholtz. He went on to earn his Ph.D. from
the University of Berlin in 1880 after submitting a dissertation on electromagnetic
induction in rotating spheres. For three years, he served as a research assistant to
Hermann von Helmholtz at the the Berlin Physical Institute.[1] Then in 1883 he
moved to the University of Kiel where he gave lectures on mathematical physics. In
1885 he was appointed professor at (what is now) the University of Karlsruhe. And
in 1889 he moved to the University of Bonn to fill the position previously held by
one of the principle founders of the science of thermodynamics, Rudolf Clausius,
who had recently died.[2] Hertz remained in Bonn until his own untimely death at the
age of 36.

   During Hertz's time at Karlsruhe, he carried out his most famous work: a series
of careful experiments in which he attempted to produce, detect and measure the
properties of electromagnetic waves. In addition to providing strong experimental
support for Maxwell's theory of light, Hertz's work on electromagnetic waves led to

---

[1] For Helmoltz's famous work on the conservation of energy, see Chaps. 9–11 of the present
volume.

[2] Clausius' introduction of the concept of entropy and the second law of thermodynamics is
included in Vol. IV.

© Springer International Publishing Switzerland 2016   423
K. Kuehn, *A Student's Guide Through the Great Physics Texts*,
Undergraduate Lecture Notes in Physics, DOI 10.1007/978-3-319-21816-8_33

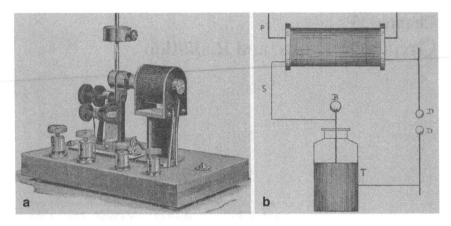

Fig. 33.1 (a) A drawing of a Ruhmkorff induction coil, used to generate sparks. (b) A circuit diagram for an induction coil consisting of a primary coil (P) which, when attached to a voltaic battery, excites oscillations in a resonant circuit containing a secondary coil (S), a leyden jar (BT), and a spark-gap (DD). (Images courtesy of Schneider, N.H., *Induction Coils*, 1901)

the eventual developments of radar and modern cell phone and satellite communication in the twentieth century. The reading selection that follows is an excerpt from Hertz's book entitled *Electric Waves*, translated by D. E. Jones.[3] Hertz begins with a careful description of the construction and operation of his experimental apparatus, which consists of transmitting and a receiving antenna. An induction coil, similar to the one depicted in Fig. 33.1, was used to drive the transmitter by generating sparks at a particular resonance frequency. In a laboratory exercise at the end of this chapter, you can explore how to design your own resonant electronic circuit.[4]

33.2   Reading: Hertz, *On Electric Radiation*

Hertz, H., *Electric Waves*, 2 ed., Macmillan and Co., London, 1900.

As soon as I had succeeded in proving that the action of an electric oscillation spreads out as a wave into space, I planned experiments with the object of concentrating this action and making it perceptible at greater distances by putting the primary conductor in the focal line of a large concave parabolic mirror. These experiments did not lead to the desired result, and I felt certain that the want of success was a necessary consequence of the disproportion between the length (4–5 m) of the waves used and the dimensions which I was able, under the most favourable

---

[3] These experiments were originally described in an article entitled "Ueber Strahlen elektrischer Kraft," which initially appeared in the *Sitzungsberichte der Berlin Adkademie der Wissenschaften* on December 13, 1888, and also in the *Wiedemann's Annalen der Physik*, **36**, p. 769.

[4] See Ex. 33.4 at the end of this chapter.

circumstances, to give to the mirror. Recently I have observed that the experiments which I have described can be carried out quite well with oscillations of more than ten times the frequency, and with waves less than one-tenth the length of those which were first discovered. I have, therefore, returned to the use of concave mirrors, and have obtained better results than I had ventured to hope for. I have succeeded in producing distinct rays of electric force, and in carrying out with them the elementary experiments which are commonly performed with light and radiant heat. The following is an account of these experiments:—

33.2.1   The Apparatus

The short waves were excited by the same method which we used for producing the longer waves. The primary conductor used may be most simply described as follows:—Imagine a cylindrical brass body,[5] 3 cm in diameter and 26 cm long, interrupted midway along its length by a spark-gap whose poles on either side are formed by spheres of 2 cm radius. The length of the conductor is approximately equal to the half wave-length of the corresponding oscillation in straight wires; from this we are at once able to estimate approximately the period of oscillation. It is essential that the pole-surfaces of the spark-gap should be frequently repolished, and also that during the experiments they should be carefully protected from illumination by simultaneous side discharges; otherwise the oscillations are not excited. Whether the spark-gap is in a satisfactory state can always be recognised by the appearance and sound of the sparks. The discharge is led to the two halves of the conductor by means of two gutta-percha- covered wires which are connected near the spark-gap on either side. I no longer made use of the large Ruhmkorff, but found it better to use a small induction-coil by Keiser and Schmidt; the longest sparks, between points, given by this were 4.5 cm long. It was supplied with current from three accumulators, and gave sparks 1–2 cm long between the spherical knobs of the primary conductor. For the purpose of the experiments the spark-gap was reduced to 3 mm.

Here, again, the small sparks induced in a secondary conductor were the means used for detecting the electric forces in space. As before, I used partly a circle which could be rotated within itself and which had about the same period of oscillation as the primary conductor. It was made of copper wire 1 mm thick, and had in the present instance a diameter of only 7.5 cm. One end of the wire carried a polished brass sphere a few millimetres in diameter; the other end was pointed and could be brought up, by means of a fine screw insulated from the wire, to within an exceedingly short distance from the brass sphere. As will be readily understood, we have here to deal only with minute sparks of a few hundredths of a millimetre in length; and after a little practice one judges more according to the brilliancy than the length of the sparks.

The circular conductor gives only a differential effect, and is not adapted for use in the focal line of a concave mirror. Most of the work was therefore done

---

[5] See Figs. 33.2 and 33.3 and the description of them at the end of this paper.

with another conductor arranged as follows:—Two straight pieces of wire, each 50 cm long and 5 mm in diameter, were adjusted in a straight line so that their near ends were 5 cm apart. From these ends two wires, 15 cm long and 1 mm in diameter, were carried parallel to one another and perpendicular to the wires first mentioned to a spark-gap arranged just as in the circular conductor. In this conductor the resonance-action was given up, and indeed it only comes slightly into play in this case. It would have been simpler to put the spark-gap directly in the middle of the straight wire; but the observer could not then have handled and observed the spark-gap in the focus of the mirror without obstructing the aperture. For this reason the arrangement above described was chosen in preference to the other which would in itself have been more advantageous.

33.2.2   The Production of the Ray

If the primary oscillator is now set up in a fairly large free space, one can, with the aid of the circular conductor, detect in its neighbourhood on a smaller scale all those phenomena which I have already observed and described as occurring in the neighbourhood of a larger oscillation. The greatest distance at which sparks could be perceived in the secondary conductor was 1.5 m, or, when the primary spark-gap was in very good order, as much as 2 m. When a plane reflecting plate is set up at a suitable distance on one side of the primary oscillator, and parallel to it, the action on the opposite side is strengthened. To be more precise:—If the distance chosen is either very small, or somewhat greater than 30 cm, the plate weakens the effect; it strengthens the effect greatly at distances of 8–15 cm, slightly at a distance of 45 cm, and exerts no influence at greater distances. We have drawn attention to this phenomenon in an earlier paper, and we conclude from it that the wave in air corresponding to the primary oscillation has a half wave-length of about 30 cm. We may expect to find a still further reinforcement if we replace the plane surface by a concave mirror having the form of a parabolic cylinder, in the focal line of which the axis of the primary oscillation lies. The focal length of the mirror should be chosen as small as possible, if it is properly to concentrate the action. But if the reflected wave is not to annul immediately the action of the reflected wave, the focal length must not be much smaller than a quarter wavelength. I therefore fixed on 12½ cm as the focal length, and constructed the mirror by bending a zinc sheet 2 m long, 2 m broad, and ½ mm thick into the desired shape over a wooden frame of the exact curvature. The height of the mirror was thus 2 m, the breadth of its aperture 1.2 m, and its depth 0.7 m. The primary oscillator was fixed in the middle of the focal line. The wires which conducted the discharge were led through the mirror; the induction-coil and the cells were accordingly placed behind the mirror so as to be out of the way. If we now investigate the neighbourhood of the oscillator with our conductors, we find that there is no action behind the mirror or at either side of it; but in the direction of the optical axis of the mirror the sparks can be perceived up to a distance of 5–6 m. When a plane conducting surface was set up so as to oppose the advancing waves at right angles, the sparks could be detected in its neighbourhood at even greater distances—up to about 9–10 m. The waves reflected from the conducting surface reinforce the advancing waves at certain points. At other points again the two sets of waves weaken one another. In front of the plane wall one can recognize with the rectilinear conductor very distinct maxima and minima, and with the circular conductor the characteristic interference-phenomena of stationary waves which I have described in an earlier paper. I was able to distinguish four nodal points, which were situated at the wall and at 33, 65 and 98 cm distance from it. We thus get 33 cm as a closer approximation to the half wavelength of the waves used, and 1.1 thousand millionth of a second as their period of oscillation, assuming that they travel with the velocity of light. In wires the oscillation gave a wave-length of 29 cm. Hence it appears that these short waves also have a somewhat lower velocity in wires than in air; but the ratio of the two velocities comes very near to the theoretical value —unity—and does not differ from it so much as appeared to be probable from our experiments on longer waves. This remarkable phenomenon still needs elucidation. Inasmuch as the phenomena are only exhibited in the neighbourhood of the optic axis of the mirror, we may speak of the result produced as an electric ray proceeding from the concave mirror.

I now constructed a second mirror, exactly similar to the first, and attached the rectilinear secondary conductor to it in such a way that the two wires of 50 cm length lay in the focal line, and the two wires connected to the spark-gap passed directly through the walls of the mirror without touching it. The spark-gap was thus situated directly behind the mirror, and the observer could adjust and examine it without obstructing the course of the waves. I expected to find that, on intercepting the ray with this apparatus, I should be able to observe it at even greater distances; and the event proved that I was not mistaken. In the rooms at my disposal I could now perceive the sparks from one end to the other. The greatest distance to which I was able, by availing myself of a doorway, to follow the ray was 16 m; but according to the results of the reflection-experiments (to be presently described), there can be no doubt that sparks could be obtained at any rate up to 20 m in open spaces. For the remaining experiments such great distances are not necessary, and it is convenient that the sparking in the secondary conductor should not be too feeble; for most of the experiments a distance of 6–10 m is most suitable. We shall now describe the simple phenomena which can be exhibited with the ray without difficulty. When the contrary is not expressly stated, it is to be assumed that the focal lines of both mirrors are vertical.

33.2.3   Rectilinear Propagation

If a screen of sheet zinc 2 m high and 1 m broad is placed on the straight line joining both mirrors, and at right angles to the direction of the ray, the secondary sparks disappear completely. An equally complete shadow is thrown by a screen of tinfoil or gold-paper. If an assistant walks across the path of the ray, the secondary spark-gap becomes dark as soon as he intercepts the ray, and again lights up when he leaves the path clear. Insulators do not stop the ray—it passes right through a wooden partition or door; and it is not without astonishment that one sees the sparks appear inside a closed room. If two conducting screens, 2 m; high and 1 m broad, are set up symmetrically on the right and left of the ray, and perpendicular to it, they do not interfere at all with the secondary spark so long as the width of the opening between them is not less than the aperture of the mirrors, *vis.* 1.2 m. If the opening is made narrower the sparks become weaker, and disappear when the width of the opening is reduced below 0.5 m. The sparks also disappear if the opening is left with a breadth of 1.2 m, but is shifted to one side of the straight line joining the mirrors. If the optical axis of the mirror containing the oscillator is rotated to the right or left about 10° out of the proper position, the secondary sparks become weak, and a rotation through 15° causes them to disappear.

There is no sharp geometrical limit to either the ray or the shadows; it is easy to produce phenomena corresponding to diffraction. As yet, however, I have not succeeded in observing maxima and minima at the edge of the shadows.

33.2.4   *Polarization*

From the mode in which our ray was produced we can have no doubt whatever that it consists of transverse vibrations and is plane-polarised in the optical sense. We can also prove by experiment that this is the case. If the receiving mirror be rotated about the ray as axis until its focal line, and therefore the secondary conductor also, lies in a horizontal plane, the secondary sparks become more and more feeble, and when the two focal lines are at right angles, no sparks whatever are obtained even if the mirrors are moved close up to one another. The two mirrors behave like the polariser and analyser of a polarisation apparatus.

I next had made an octagonal frame, 2 m high and 2 m broad; across this were stretched copper wires 1 mm thick, the wires being parallel to each other and 3 cm apart. If the two mirrors were now set up with their focal lines parallel, and the wire screen was interposed perpendicularly to the ray and so that the direction of the wires was perpendicular to the direction of the focal lines, the screen practically did not interfere at all with the secondary sparks. But if the screen was set up in such a way that its wires were parallel to the focal lines, it stopped the ray completely. With regard, then, to transmitted energy the screen behaves towards our ray just as a tourmaline plate behaves towards a plane-polarised ray of light. The receiving mirror was now placed once more so that its focal line was horizontal; under these circumstances, as already mentioned, no sparks appeared. Nor were any sparks produced when the screen was interposed in the path of the ray, so long as the wires in the screen were either horizontal or vertical. But if the frame was set up in such a position that the wires were inclined at 45° to the horizontal on either side, then the interposition of the screen immediately produced sparks in the secondary spark-gap. Clearly the screen resolves the advancing oscillation into two components and transmits only that component which is perpendicular to the direction of its wires. This component is inclined at 45° to the focal line of the second mirror, and may thus, after being again resolved by the mirror, act upon the secondary conductor.

The phenomenon is exactly analogous to the brightening of the dark field of two crossed Nicols by the interposition of a crystalline plate in a suitable position.

With regard to the polarisation it may be further observed that, with the means employed in the present investigation, we are only able to recognise the electric force. When the primary oscillator is in a vertical position the oscillations of this force undoubtedly take place in the vertical plane through the ray, and are absent in the horizontal plane. But the results of experiments with slowly alternating currents leave no room for doubt that the electric oscillations are accompanied by oscillations of magnetic force which take place in the horizontal plane through the ray and are zero in the vertical plane. Hence the polarisation of the ray does not so much consist in the occurrence of oscillations in the vertical plane, but rather in the fact that the oscillations in the vertical plane are of an electrical nature, while those in the horizontal plane are of a magnetic nature. Obviously, then, the question, in which of the two planes the oscillation in our ray occurs, cannot be answered unless one specifies whether the question relates to the electric or the magnetic oscillation. It was Herr Kolaček[6] who first pointed out clearly that this consideration is the reason why an old optical dispute has never been decided.

33.2.5 *Reflection*

We have already proved the reflection of the waves from conducting surfaces by the interference between the reflected and the advancing waves, and have also made use of the reflection in the construction of our concave mirrors. But now we are able to go further and to separate the two systems of waves from one another. I first placed both mirrors in a large room side by side, with their apertures facing in the same direction, and their axes converging to a point about 3 m off. The spark-gap of the receiving mirror naturally remained dark. I next set up a plane vertical wall made of thin sheet zinc, 2 m high and 2 m broad, at the point of intersection of the axes, and adjusted it so that it was equally inclined to both. I obtained a vigorous stream of sparks arising from the reflection of the ray by the wall. The sparking ceased as soon as the wall was rotated around a vertical axis through about 15° on either side of the correct position; from this it follows that the reflection is regular, not diffuse. When the wall was moved away from the mirrors, the axes of the latter being still kept converging towards the wall, the sparking diminished very slowly. I could still recognise sparks when the wall was 10 m away from the mirrors, *i.e.* when the waves had to traverse a distance of 20 m. This arrangement might be adopted with advantage for the purpose of comparing the rate of propagation through air with other and slower rates of propagation, *e.g.* through cables.

In order to produce reflection of the ray at angles of incidence greater than zero, I allowed the ray to pass parallel to the wall of the room in which there was a doorway.

---

[6] [F. Kolaček, *Wied. Ann.* **34**, p. 676, 1888.]

In the neighbouring room to which this door led I set up the receiving mirror so that its optic axis passed centrally through the door and intersected the direction of the ray at right angles. If the plane conducting surface was now set up vertically at the point of intersection, and adjusted so as to make angles of 45° with the ray and also with the axis of the receiving mirror, there appeared in the secondary conductor a stream of sparks which was not interrupted by closing the door. When I turned the reflecting surface about 10° out of the correct position the spark disappeared. Thus the reflection is regular, and the angles of incidence and reflection are equal. That the action proceeded from the source of disturbance to the plane mirror, and hence to the secondary conductor, could also be shown by placing shadow-giving screens at different points of this path. The secondary sparks then always ceased immediately; whereas no effect was produced when the screen was placed anywhere else in the room. With the aid of the circular secondary conductor it is possible to determine the position of the wave-front in the ray; this was found to be at right angles to the ray before and after reflection, so that in the reflection it was turned through 90°.

Hitherto the focal lines of the concave mirrors were vertical, and the plane of oscillation was therefore perpendicular to the plane of incidence. In order to produce reflection with the oscillations in the plane of incidence, I placed both mirrors with their focal lines horizontal. I observed the same phenomena as in the previous position; and, moreover, I was not able to recognize any difference in the intensity of the reflected ray in the two cases. On the other hand, if the focal line of the one mirror is vertical, and of the other horizontal, no secondary sparks can be observed. The inclination of the plane of oscillation to the plane or incidence is therefore not altered by reflection, provided this inclination has one of the two special values referred to; but in general this statement cannot hold good. It is even questionable whether the ray after reflection continues to be plane-polarised. The interferences which are produced in front of the mirror by the intersecting wave-system, and which, as I have remarked, give rise to characteristic phenomena in the circular conductor, are most likely to throw light upon all problems relating to the change of phase and amplitude produced by reflection.

One further experiment on reflection from an electrically eolotropic surface may be mentioned. The two concave mirrors were again placed side by side, as in the reflection-experiment first described; but now there was placed opposite to them, as a reflecting surface, the screen of parallel copper wires which has already been referred to. It was found that the secondary spark-gap remained dark when the wires intersected the direction of the oscillations at right angles, but that sparking began as soon as the wires coincided with the direction of the oscillations. Hence the analogy between the tourmaline plate and our surface which conducts in one direction is confined to the transmitted part of the ray. The tourmaline plate absorbs the part which is not transmitted; our surface reflects it. If in the experiment last described the two mirrors are placed with their focal lines at right angles, no sparks can be excited in the secondary conductor by reflection from an isotropic screen; but I proved to my satisfaction that sparks are produced when the reflection takes place from the eolotropic wire grating, provided this is adjusted so that the wires are inclined at 45° to the focal lines. The explanation of this follows naturally from what has been already stated.

33.2.6   *Refraction*

In order to find out whether any refraction of the ray takes place in passing from air into another insulating medium, I had a large prism made of so-called hard pitch, a material like asphalt. The base was an isosceles triangle 1.2 m in the side, and with a refracting angle of nearly 30°. The refracting edge was placed vertical, and the height of the whole prism was 1.5 m. But since the prism weighed about 12 cwt., and would have been too heavy to move as a whole, it was built up of three pieces, each 0.5 m high, placed one above the other. The material was cast in wooden boxes which were left around it, as they did not appear to interfere with its use. The prism was mounted on a support of such height that the middle of its refracting edge was at the same height as the primary and secondary spark-gaps. When I was satisfied that refraction did take place, and had obtained some idea of its amount, I arranged the experiment in the following manner:—The producing mirror was set up at a distance of 2.6 m from the prism and facing one of the refracting surfaces, so that the axis of the beam was directed as nearly as possible towards the centre of mass of the prism, and met the refracting surface at an angle of incidence of 25° (on the side of the normal towards the base). Near the refracting edge and also at the opposite side of the prism were placed two conducting screens which prevented the ray from passing by any other path than that through the prism. On the side of the emerging ray there was marked upon the floor a circle of 2.5 m radius, having as its centre the centre of mass of the lower end of the prism. Along this the receiving mirror was now moved about, its aperture being always directed towards the centre of the circle. No sparks were obtained when the mirror was placed in the direction of the incident ray produced; in this direction the prism threw a complete shadow. But sparks appeared when the mirror was moved towards the base of the prism, beginning when the angular deviation from the first position was about 11°. The sparking increased in intensity until the deviation amounted to about 22°, and then again decreased. The last sparks were observed with a deviation of about 34°. When the mirror was placed in a position of maximum effect, and then moved away from the prism along the radius of the circle, the sparks could be traced up to a distance of 5–6 m. When an assistant stood either in front of the prism or behind it the sparking invariably ceased, which shows that the action reaches the secondary conductor through the prism and not in any other way. The experiments were repeated after placing both mirrors with their focal lines horizontal, but without altering the position of the prism. This made no difference in the phenomena observed. A refracting angle of 30° and a deviation of 22° in the neighbourhood of the minimum deviation corresponds to a refractive index of 1.69. The refractive index of pitch-like materials for light is given as being between 1.5 and 1.6. We must not attribute any importance to the magnitude or even the sense of this difference, seeing that our method was not an accurate one, and that the material used was impure.

We have applied the term rays of electric force to the phenomena which we have investigated. We may perhaps further designate them as rays of light of very great wave-length. The experiments described appear to me, at any rate, eminently

Fig. 33.2 Hertz's electric
wave transmitter. A resonant
circuit drives a spark-gap
antenna which is placed vertically along the focal line of
a large parabolic mirror.—
[*K.K.*]

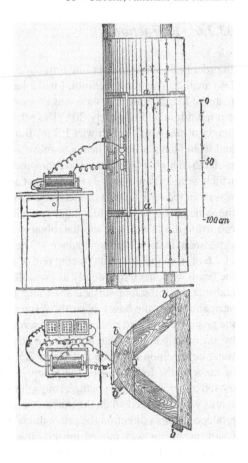

adapted to remove any doubt as to the identity of light, radiant heat, and electromagnetic wave-motion. I believe that from now on we shall have greater confidence
in making use of the advantages which this identity enables us to derive both in the
study of optics and of electricity.

33.2.7 Explanation of the Figures

In order to facilitate the repetition and extension of these experiments, I append
in the accompanying Figs. 33.2, 33.3a, and b, illustrations of the apparatus which I
used, although these were constructed simply for the purpose of experimenting at the
time and without any regard to durability. Figure 33.2 shows in plan and elevation
(section) the producing mirror. It will be seen that the framework of it consists of two
horizontal frames (*a, a*) of parabolic form, and four vertical supports (*b, b*) which
are screwed to each of the frames so as to support and connect them. The sheet metal
reflector is clamped between the frames and the supports, and fastened to both by

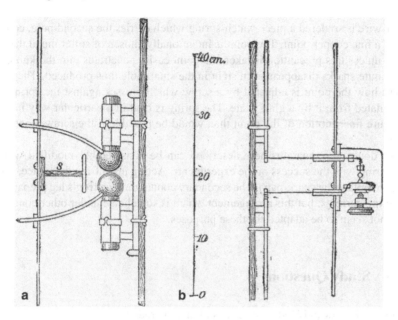

Fig. 33.3 The transmitting antenna consists of an adjustable spark gap between two conducting balls fixed to a vertical wooden rod (**a**); the receiving antenna feeds an adjustable spark gap located behind the receiving mirror (**b**).—[*K.K.*]

numerous screws. The supports project above and below beyond the sheet metal so that they can be used as handles in handling the mirror. Figure 33.3a represents the primary conductor on a somewhat larger scale.

The two metal parts slide with friction in two sleeves of strong paper which are held together by indiarubber bands. The sleeves themselves are fastened by four rods of sealing-wax to a board which again is tied by indiarubber bands to a strip of wood forming part of the frame which can be seen in Fig. 33.2. The two leading wires (covered with gutta-percha) terminate in two holes bored in the knobs of the primary conductor. This arrangement allows of all necessary motion and adjustment of the various parts of the conductor; it can be taken to pieces and put together again in a few minutes, and this is essential in order that the knobs may be frequently repolished. Just at the points where the leading wires pass through the mirror, they are surrounded during the discharge by a bluish light. The smooth wooden screen *s* is introduced for the purpose of shielding the spark-gap from this light, which otherwise would interfere seriously with the production of the oscillations. Lastly, Fig. 33.3b represents the secondary spark-gap. Both parts of the secondary conductor are again attached by sealing-wax rods and indiarubber bands to a slip forming part of the wooden framework. From the inner ends of these parts the leading wires, surrounded by glass tubes, can be seen proceeding through the mirror and bending towards one another. The upper wire carries at its pole a small brass knob. To the lower wire is soldered a piece watch-spring which carries the second pole, consisting of a fine copper point. The point is intentionally chosen of softer metal than the knob; unless this precaution is taken the point easily penetrates into the knob, and the minute sparks disappear from sight in the small hole thus produced. The figure shows how the point is adjusted by a screw which presses against the spring that is insulated from it by a glass plate. The spring is bent in a particular way in order to secure finer motion of the point than would be possible if the screw alone were used.

No doubt the apparatus here described can be considerably modified without interfering with the success of the experiments. Acting upon friendly advice, I have tried to replace the spark-gap in the secondary conductor by a frog's leg prepared for detecting currents; but this arrangement which is so delicate under other conditions does not seem to be adapted for these purposes.

33.3   Study Questions

QUES. 33.1. How did Hertz create rays of electric radiation?

a) What was the purpose, design, and size of Hertz's primary conductor? How was this primary conductor excited, and how did he know if it was indeed working?
b) Out of what material was the producing mirror constructed? What was its shape, orientation, and purpose? What was the purpose of the wooden screen, s? And why did the brass knobs need to be polished frequently?

QUES. 33.2. How did Hertz detect rays of electric radiation?

a) What was the purpose and design of the secondary conductor? How did Hertz's spark-gap work, and how could it be made more, or less, sensitive?
b) What was the purpose, shape, and orientation of the concave mirror near the secondary conductor? What other method of detection did Hertz briefly consider? And why did he reject this option?

QUES. 33.3. How was Hertz's electric radiation like, or unlike, light?

a) What was the range of the ray produced? How did Hertz succeed in increasing the range, and the sensitivity of his apparatus?
b) What was the wavelength and the frequency of the observed ray? How did Hertz measure, or calculate, these?
c) Do the rays have the same speed when they travel along wires as when they travel through air?
d) Did the rays travel in straight lines? Do they exhibit diffraction? Interference?
e) How did Hertz determine the polarization of the rays? How is the direction of polarization related to the direction of the electric and magnetic oscillations?
f) Do the rays obey the law of reflection? the law of refraction?

g) To what other phenomena does Hertz liken rays of electric force? What general conclusion does Hertz draw from his experiments? Is he justified in his conclusion?

33.4 Exercises

EX. 33.1 (TRAVELING ELECTROMAGNETIC WAVES). Write down mathematical expressions, similar to Eq. 32.5, for the electric and magnetic fields generated by Hertz's transmitting antenna. You should incorporate into your equations the particular wavelength and frequency of Hertz's waves. According to your equations, in which direction is the electric field (vector) pointing? The magnetic field? In which direction is the wave traveling?

EX. 33.2 (STANDING ELECTROMAGNETIC WAVES). At the end of the reading of Chap. 32, Heaviside briefly considered what happens when two identical electromagnetic waves pass through one another while travelling in opposite directions. In the present exercise, we will explore this situation in a bit more detail. Suppose that a small transmitting antenna is placed midway between two parallel reflecting mirrors. Generally speaking, the electromagnetic waves generated by the antenna will bounce back and forth, forming a complicated distribution of electric and magnetic fields between the mirrors. But if the wavelength and the mirror spacing are carefully chosen (so that a nodal point of the superimposed electric waves exists at the surface of each mirror) then an electromagnetic *standing wave* forms in the region between the mirrors.

a) Use appropriate trigonometric identities to mathematically prove that when two identical counter-propagating electromagnetic plane waves (having amplitude E_0 and frequency ω) superimpose, the resulting electric field strength may be written as

$$E(x, t) = 2E_0 \sin(kx) \sin(\omega t). \qquad (33.1)$$

b) Using Hertz's value of the wavelength, λ, plot Eq. 33.1 for several values of t. Then use these plots to explain why Eq. 33.1 describes a standing (as opposed to a traveling) wave.

c) What is the minimum mirror spacing possible to create a standing wave using an antenna oscillating with a period of one billionth of a second? Is there a maximum mirror spacing?

d) Do the (relative) phases of the electric and magnetic fields differ for the case of traveling and standing electromagnetic waves? If so, by how much?

EX. 33.3 (MICROWAVE LABORATORY). Measure the wavelength of electromagnetic radiation produced by a commercially available microwave optics system.[7] Do

---

[7] Basic Microwave Optics System, (Model WA-9314C), Pasco Scientific, Roseville, CA.

microwaves obey the same laws of reflection and refraction as visible light? Can you observe the interference of microwaves passing through a multiple-slit plate? Do microwaves behave in the same way as visible light?

EX. 33.4 (ELECTROMAGNETIC RESONANT CIRCUIT LABORATORY).   To produce electromagnetic radiation, Hertz used an oscillator to drive an alternating current through a transmitting antenna. To detect the electromagnetic radiation thereby created, he used a receiving antenna hooked up to some electronic circuitry. This circuitry consisted of a loop of wire interrupted by a tiny "secondary spark gap." The loop of wire acted as an inductor and the spark gap acted as a capacitor. To detect the tiny electric currents generated in the circuit by the captured radiation, Hertz had to carefully tune the circuit so that it would *resonate* at the precise frequency of the radiation. He did this by varying the spark-gap (and hence the capacitance) of his circuit. This worked because the resonance frequency of a single-loop circuit having an inductance L and capacitance C is given by

$$f_0 = \frac{1}{2\pi\sqrt{LC}.} \tag{33.2}$$

When driven by an electromotive force produced by radiation at the frequency f_0, the current in Hertz's circuit was amplified to such an extent that sizable sparks were seen to jump back and forth across the spark gap.

In this laboratory exercise, we will build a similar electronic circuit which resonates when driven by an oscillating electromotive force. The oscillating electromotive force in our circuit will be created using a function generator rather than by electromagnetic radiation, as was Hertz's.[8] The current generated in the circuit, and the voltages across various circuit elements, will be measured using an oscilloscope.[9] In order to construct our circuits, we will need variable resistance, inductance and capacitance boxes and some patch cords to connect these components.[10]

Before building our resonant circuit, let us begin by setting up a simpler circuit consisting of a resistor and a capacitor in series with the function generator. A schematic diagram for this series RC circuit is shown in Fig. 33.4. For the resistor, use a black decade resistor box set initially at 1000 Ohms (Ω); for the capacitor, use a variable capacitor box set initially at 25 nanofarads (nF). Set up the function generator to output a 2 kHz *square wave*, and use the oscilloscope to measure the voltage across the resistor, V_R. By Ohm's law, the current through the resistor, I, is proportional to V_R. You should notice that when the square-wave voltage is high, the capacitor is charging, and hence the initial electric current drops exponentially from an initial value down to zero; when the square wave voltage is low, current flows in

---

[8] Function Generator (5 MHz, Model 4011A), BK Precision, Yorba Linda, CA.

[9] Two Channel Digital Storage Oscilloscope (Model TDS 1002), Tektronix, Beaverton, OR.

[10] These components are available from scientific supply companies such as Sargent Welch, Buffalo Grove, IL.

Fig. 33.4 Schematic diagram for a driven *RC* circuit

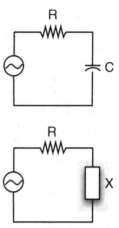

Fig. 33.5 The capacitor in Fig. 33.4 has been replaced by another element, *X*

the other direction, discharging the capacitor. The time constant, τ, is defined as the time it takes for the current to decrease by a factor of $1/e \cong 2.7$; it is conveniently measured by adjusting the function generator so that the initial voltage on the oscilloscope display is at 2.7 divisions and the final voltage is at 0. Measure τ by seeing how long it takes to fall from 2.7 to 1 division for 5 different values of R between 200 and 1000 Ω. Since the theoretical value is $\tau = RC$ for a series RC circuit, you can plot the measured values of τ versus R and thereby determine the value of C from the slope of the graph. Do your measurements match the known value of C?

Next, assemble the circuit depicted in Fig. 33.5. For the resistance, R, use a 1 kΩ resistor. Element X will be, in turn, (i) another resistor, r, (ii) a capacitor, (iii) an inductor, and (iv) all three of these components in series. Our goal will now be to determine the relationship between the driving frequency of the function generator, f, the current flowing through the circuit, I, and the voltage across circuit element X. One input channel of the oscilloscope (CH1) should be used to measure the voltage across the resistor, R; from this the current in the circuit can be inferred from Ohm's law. Since we have a one-loop circuit, the current through R must be the same as the current through any other part of the circuit, including element X.[11] The other channel of the oscilloscope (CH2) can be used to directly measure the voltage across element X.

When setting up this circuit, be sure that the common (ground) terminals of both oscilloscope probes are connected to the same point in the circuit, namely between R and X. If these two ground connections were to somehow end up in different places in the circuit, then there could be no voltage across any component located between them, and hence no current through this component. Be aware, however,

---

[11] This is one of *Kirchoff's circuit rules*; it follows from the requirement that electric charge is conserved (*i.e.* it cannot be created or destroyed at any point in the circuit).

that connecting the oscilloscope in this manner does switch the polarity of the signal on CH2 relative to the signal in CH1. Also, it is important that the function generator and the oscilloscope both be "floating"—that is, not grounded.[12]

Now with a 200 Ω resistor as element X, drive the circuit with a 3 kHz *sinusoidally* varying voltage (*not* a square wave). Can you see two sinusoidal traces on the scope display? Make a sketch the two traces, paying particular attention to the amplitudes and relative phases of I and V_X. Do the amplitudes or relative phases change when the driving frequency is changed?

Next, remove the 200 Ω resistor and insert a 25 nF capacitor as element X. Again, carefully observe the relative phases and amplitudes of I and V_X as the driving frequency is varied. You may notice that as the driving frequency changes the amplitudes of both I and V_X change. It is desirable to hold one of these constant to see how the other varies with the driving frequency. So when changing the driving frequency, try to keep V_X constant (at, say, 2 V peak-to-peak) by adjusting the amplitude setting of the function generator. Then measure the current amplitude, I, for several frequencies between about 500 and 15,000 Hz. Make a plot of I versus f.

After carrying out these measurements with a capacitor, swap it out and insert a 30 millihenry (mH) inductor for element X. Again, carefully observe the relative phases and amplitudes of I and V_X as the driving frequency is varied. Measure the current amplitude I for several frequencies between 500 Hz and 15 kHz and make a plot of I versus f on the same plot as your capacitor data.

Finally, insert an RLC series—consisting of a 200 Ω resistor, a 25 nF capacitor and a 30 mH inductor—for element X. Measure the current amplitude I for several frequencies between 500 Hz and 15 kHz and make a plot of I versus f on the same plot as your inductor (alone) and capacitor (alone) data. Are I and V_X ever in phase with one another? Do you observe resonance? If so, is your resonance frequency consistent with Eq. 33.2? You may need to take a few additional measurements near f_0 to obtain a high-precision measurement of the resonance frequency of your circuit.

Ex. 33.5 (RUHMKORFF GENERATOR LABORATORY). As an advanced laboratory project, try to design and build a Ruhmkorff generator, similar to the one shown in Fig. 33.1. Detailed instructions can be found in Schneider, N. H., *Induction Coils: How to Make, Use and Repair Them*, Spon & Chamberlain, E. & F.N. Spon, New York and London, 1901.

---

[12] Normally, the shield of the coaxial terminal is attached through the third prong of the power plug to the earth ground through a 120 VAC wall socket. You might therefore use a 3-prong to 2-prong adaptor to plug in the function generator and the oscilloscope so that they are not connected to earth ground.

33.5 Vocabulary

1. Venture
2. Coincide
3. Annul
4. Perceptible
5. Isosceles
6. Symmetric
7. Nicols

Chapter 34
The Michelson-Morley Experiment

If this experiment gave a positive result, it would determine the velocity, not merely of the earth in its orbit, but of the earth through the ether.

—Albert Michelson

34.1 Introduction

Albert Abraham Michelson (1852–1931) was born in the town of Strelno in the Kingdom of Prussia.[1] His family emigrated to California by way of Panama when he was three, and he grew up in mining towns during the gold rush era. After attending high school in San Francisco, he traveled to Washington to personally petition President Grant for an appointment to the U.S. Naval Academy in Annapolis. It was granted, and after graduating in 1873 he returned to the Academy to serve as an instructor of physics and chemistry in 1875. It was during this time that he became interested in measuring the speed of light. His first attempts at doing so were inspired by the rotating mirror technique of Léon Foucault.[2] His next set of light speed measurements were carried out in collaboration with Simon Newcomb in Washington, DC. After obtaining a leave of absence from the navy, he traveled to Europe where he studied modern experimental techniques in Paris, Heidelberg, and in the laboratory of Hermann von Helmholtz in Berlin. Upon returning to the United States in 1883, Michelson took a position as Professor of Physics at the Case School of Applied Science in Cleveland, Ohio. While in Cleveland, he and his collaborator Edward Morley carried out their famous interferometric experiments of the speed of Earth through the æther. In 1889 Michelson moved to Clark University in Massachusetts, and then in 1892 to the University of Chicago where he served as the head of the new physics department. Michelson received the Nobel Prize in Physics—the first American to do so—in the year 1907 "for his optical precision

---

[1] Michelson's biography in this introduction was based largely on information from Robert Millikan's *Biographical Memoir of Albert Abraham Michelson 1852–1931*, vol. XIX, National Academy of Sciences, 1938. Michelson's daughter tells her father's story in Livingston, D. M., *The Master of Light: A biography of Albert A. Michelson*, The University of Chicago Press, 1973.

[2] For an explanation of Foucault's experiment, see Ex. 30.2 in the present volume.

© Springer International Publishing Switzerland 2016
K. Kuehn, *A Student's Guide Through the Great Physics Texts*,
Undergraduate Lecture Notes in Physics, DOI 10.1007/978-3-319-21816-8_34

instruments and the spectroscopic and metrological investigations carried out with their aid."

In the reading selection that follows—taken from Michelson's Lowell Lectures of 1899, which were published in 1903 in the form of a book entitled *Light Waves and their Uses*—Michelson describes what is now called the Michelson-Morley experiment. This experiment is seen today as providing compelling evidence in support of Einstein's 1905 special theory of relativity.[3] Michelson relies almost exclusively upon the wave theory of light, since it is able to account for the interference of light, a phenomenon upon which his apparatus relies.[4] Michelson only mentions the particle theory of light insofar as it provides a simple explanation of the phenomenon of stellar aberration. In the course of the reading, Michelson speaks highly of Fizeau's 1851 measurement of the velocity of light in flowing water. He does so in order to motivate his own experiments on the speed of light. Interestingly, in the concluding section of this reading, Michelson endorses Kelvin's vortex model of the atom—a theory which treats atoms not as tiny particles, but rather as localized smoke-ring-like excitations. Does Michelson reject the idea of æther?

34.2   Reading: Michelson, *The Ether*

Michelson, A. A., *Light Waves and Their Uses*, University of Chicago Press, Chicago, IL, 1903. Lecture VIII.

The velocity of light is so enormously greater than anything with which we are accustomed to deal that the mind has some little difficulty in grasping it. A bullet travels at the rate of approximately half a mile a second. Sound, in a steel wire, travels at the rate of 3 miles a second. From this if we agree to except the velocities of the heavenly bodies there is no intermediate step to the velocity of light, which is about 186,000 miles a second. We can, perhaps, give a better idea of this velocity by saying that light will travel around the world seven times between two ticks of a clock.

Now, the velocity of wave propagation can be seen, without the aid of any mathematical analysis, to depend on the elasticity of the medium and its density; for we can see that if a medium is highly elastic the disturbance would be propagated at a great speed. Also, if the medium is dense the propagation would be slower than if it were rare. It can easily be shown that if the elasticity were represented by E, and the density by D, the velocity would be represented by the square root of E divided by D. So that, if the density of the medium which propagates light waves were as great as the density of steel, the elasticity, since the velocity of light is some 60,000 times

---

[3] See Part I of Einstein's *Relativity*, included in Chaps. 29–32 of Vol. II.

[4] See Chaps. 13–17 and 20–21 of the present volume for an extended discussion of the wave theory of light.

as great as that of the propagation of sound in a steel wire, must be 60,000 squared times as great as the elasticity of steel. Thus, this medium which propagates light vibrations would have to have an elasticity of the order of 3,600,000,000 times the elasticity of steel. Or, if the elasticity of the medium were the same as that of steel, the density would have to be 3,600,000,000 times as small as that of steel, that is to say, roughly speaking, about 50,000 times as small as the density of hydrogen, the lightest known gas. Evidently, then, a medium which propagates vibrations with such an enormous velocity must have an enormously high elasticity or abnormally low density. In any case, its properties would be of an entirely different order from the properties of the substances with which we are accustomed to deal, so that it belongs in a category by itself.

Another course of reasoning which leads to this same conclusion—namely, that this medium is not any ordinary form of matter, such as air or gas or steel—is the following: Sound is produced by a bell under a receiver of an air pump. When the air has the same density inside the receiver as outside, the sound reaches the ear of an observer without difficulty. But when the air is gradually pumped out of the receiver, the sound becomes fainter and fainter until it ceases entirely. If the same thing were true of light, and we exhausted a vessel in which a source of light—an incandescent lamp, for example—had been placed, then, after a certain degree of exhaustion was reached, we ought to see the light less clearly than before. We know, however, that the contrary is the case, *i.e.*, that the light is actually brighter and clearer when the exhaustion of the receiver has been carried to the highest possible degree. The probabilities are enormously against the conclusion that light is transmitted by the very small quantity of residual gas. There are other theoretical reasons, into which we will not enter.

Whatever the process of reasoning, we are led to the same result. We know that light vibrations are transverse to the direction of propagation, while sound vibrations are in the direction of propagation. We know also that in the case of a solid body transverse vibrations can be readily transmitted.—Thus, if we have a long cylindrical rod and we give one end of it a twist, the twist will travel along from one end to the other. If the medium, instead of being a solid rod, were a tube of liquid, and were twisted at one end, there would be no corresponding transmission of the twist to the other end, for a liquid cannot transmit a torsional strain. Hence this reasoning leads to the conclusion that if the medium which propagates light vibrations has the properties of ordinary matter, it must be considered to be an elastic solid rather than a fluid.

This conclusion was considered one of the most formidable objections to the undulatory theory that light consists of waves. For this medium, notwithstanding the necessity for the assumption that it has the properties of a solid, must yet be of such a nature as to offer little resistance to the motion of a body through it. Take, for example, the motion of the planets around the sun. The resistance of the medium is so small that the earth has been traveling around the sun millions of years without any appreciable increase in the length of the year. Even the vastly lighter and more attenuated comets return to the same point periodically, and the time of such periodical returns has been carefully noted from the earliest historical times,

and yet no appreciable increase in it has been detected. We are thus confronted with
the apparent inconsistency of a solid body which must at the same time possess in
such a marked degree the properties of a perfect fluid as to offer no appreciable
resistance to the motion of bodies so very light and extended as the comets. We are,
however, not without analogies, for, as was stated in the first lecture, substances such
as shoemaker's wax show the properties of an elastic solid when reacting against
rapid motions, but act like a liquid under pressures.

In the case of shoemaker's wax both of these contradictory properties are very
imperfectly realized, but we can argue from this fact that the medium which we
are considering might have the various properties which it must possess in an enormously exaggerated degree. It is, at any rate, not at all inconceivable that such a
medium should at the same time possess both properties. We know that the air itself
does not possess such properties, and that no matter which we know possesses them
in sufficient degree to account for the propagation of light. Hence the conclusion
that light vibrations are not propagated by ordinary matter, but by something else.
Cogent as these three lines of reasoning may be, it is undoubtedly true that they do
not always carry conviction. There is, so far as I am aware, no process of reasoning
upon this subject which leads to a result which is free from objection and absolutely
conclusive.

But these are not the only paradoxes connected with the medium which transmits
light. There was an observation made by Bradley a great many years ago, for quite
another purpose. He found that when we observe the position of a star by means
of the telescope, the star seems shifted from its actual position, by a certain small
angle called the angle of aberration. He attributed this effect to the motion of the
earth in its orbit, and gave an explanation of the phenomenon which is based on
the corpuscular theory and is apparently very simple. We will give this explanation,
notwithstanding the fact that we know the corpuscular theory to be erroneous.

Let us suppose a raindrop to be falling vertically and an observer to be carrying,
say, a gun, the barrel being as nearly vertical as he can hold it. If the observer is not
moving and the raindrop falls in the center of the upper end of the barrel, it will fall
centrally through the lower end. Suppose, however, that the observer is in motion in
the direction *bd* (Fig. 34.1); the raindrop will still fall exactly vertically, but if the
gun advances laterally while the raindrop is within the barrel, it strikes against the
side. In order to make the raindrop move centrally along the axis of the barrel, it
is evidently necessary to incline the gun at an angle such as *bad*. The gun barrel is
now pointing, apparently, in the wrong direction, by an angle whose tangent is the
ratio of the velocity of the observer to the velocity of the raindrop.

According to the undulatory theory, the explanation is a trifle more complex;
but it can easily be seen that, if the medium we are considering is motionless and
the gun barrel represents a telescope, and the waves from the star are moving in
the direction *ad*, they will be concentrated at a point which is in the axis of the
telescope, unless the latter is in motion. But if the earth carrying the telescope is
moving with a velocity something like 20 miles a second, and we are observing
the stars in a direction approximately at right angles to the direction of that motion,
the light from the star will not come to a focus on the axis of the telescope, but

Fig. 34.1 Stellar aberration:
a vertically aligned telescope,
carried horizontally, must
be tipped so that a light
beam travels straight down its
axis.—[*K.K.*]

will form an image in a new position, so that the telescope appears to be pointing
in the wrong direction. In order to bring the image on the axis of the instrument,
we must turn the telescope from its position through an angle whose tangent is the
ratio of the velocity of the earth in its orbit to the velocity of light. The velocity
of light is, as before stated, 186,000 miles a second—200,000 in round numbers—
and the velocity of the earth in its orbit is roughly 20 miles a second. Hence the
tangent of the angle of aberration would be measured by the ratio of 1 to 10,000.
More accurately, this angle is $20''.445$. The limit of accuracy of the telescope, as was
pointed out in several of the preceding lectures, is about one-tenth of a second; but,
by repeating these measurements under a great many variations in the conditions of
the problem, this limit may be passed, and it is practically certain that this number
is correct to the second decimal place.

   When this variation in the apparent position of the stars was discovered, it was
accounted for correctly by the assumption that light travels with a finite velocity, and
that, by measuring the angle of aberration, and knowing the speed of the earth in its
orbit, the velocity of light could be found. This velocity has since been determined
much more accurately by experimental means, so that now we use the velocity of
light to deduce the velocity of the earth and the radius of its orbit.

The objection to this explanation was, however, raised that if this angle were the ratio of the velocity of the earth in its orbit to the velocity of light, and if we filled a telescope with water, in which the velocity of light is known to be only three-fourths of what it is in air, it would take one and one-third times as long for the light to pass from the center of the objective to the cross-wires, and hence we ought to observe, not the actual angle of aberration, but one which should be one-third greater. The experiment was actually tried. A telescope was filled with water, and observations on various stars were continued throughout the greater part of the year, with the result that almost exactly the same value was found for the angle of aberration.

This result was considered a very serious objection to the undulatory theory until an explanation was found by Fresnel. He proposed that we consider that the medium which transmits the light vibrations is carried along by the motion of the water in the telescope in the direction of the motion of the earth around the sun. Now, if the light waves were carried along with the full velocity of the earth in its orbit, we should be in the same difficulty, or in a more serious difficulty, than before. Fresnel, however, made the further supposition that the velocity of the carrying along of the light waves by the motion of the medium was less than the actual velocity of the medium itself, by a quantity which depended on the index of refraction of the substance. In the case of water the value of this factor is seven-sixteenths.

This, at first sight, seems a rather forced explanation; indeed, at the time it was proposed it was treated with considerable incredulity. An experiment was made by Fizeau, however, to test the point—in my opinion one of the most ingenious experiments that have ever been attempted in the whole domain of physics. The problem is to find the increase in the velocity of light due to a motion of the medium. We have an analogous problem in the case of sound, but in this case it is a very much simpler matter. We know by actual experiment, as we should infer without experiment, that the velocity of sound is increased by the velocity of a wind which carries the air in the same direction, or diminished if the wind moves in the opposite direction. But in the case of light waves the velocity is so enormously great that it would seem, at first sight, altogether out of the question to compare it with any velocity which we might be able to obtain in a transparent medium such as water or glass. The problem consists in finding the change in the velocity of light produced by the greatest velocity we can get—about 20 ft a second—in a column of water through which light waves pass. We thus have to find a difference of the order of 20 ft in 186,000 miles, *i.e.*, of one part in 50,000,000. Besides, we can get only a relatively small column of water to pass light through and still see the light when it returns.

The difficulty is met, however, by taking advantage of the excessive minuteness of light waves themselves. This double length of the water column is something like 40 ft. In this 40 ft there are, in round numbers, 14,000,000 waves. Hence the difference due to a velocity of 20 ft/s, which is the velocity of the water current, would produce a displacement of the interference fringes (produced by two beams, one of which passes down the column and the other up the column of the moving liquid) of about one-half a fringe, which corresponds to a difference of one-half a light wave in the paths. Reversing the water current should produce a shifting of one-half a fringe in the opposite direction, so that the total shifting would actually be of

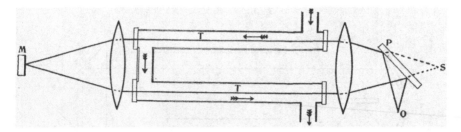

Fig. 34.2 Fizeau's experiment to measure the difference between the speed of light travelling upstream and downstream.—[*K.K.*]

the order of one interference fringe. But we can easily observe one-tenth of a fringe, or in some cases even less than that. Now, one fringe would be the displacement if water is the medium which transmits the light waves. But this other medium we have been talking about moves, according to Fresnel, with a smaller velocity than the water, and the ratio of the velocity of the medium to the velocity of the water should be a particular fraction, namely, seven-sixteenths. In other words, then, instead of the whole fringe we ought to get a displacement of seven-sixteenths of a fringe by the reversal of the water current. The experiment was actually tried by Fizeau, and the result was that the fringes were shifted by a quantity less than they should have been if water had been the medium; and hence we conclude that the water was not the medium which carried the vibrations.

The arrangement of the apparatus which was used in the experiment is shown in Fig. 34.2. The light starts from a narrow slit *S*, is rendered parallel by a lens *L*, and separated into two pencils by apertures in front of the two tubes *T T*, which carry the column of water. Both tubes are closed by pieces of the same plane-parallel plate of glass. The light passes through these two tubes and is brought to a focus by the lens in condition to produce interference fringes. The apparatus might have been arranged in this way but for the fact that there would be changes in the position of the interference fringes whenever the density or temperature of the medium changed; and, in particular, whenever the current changes direction there would be produced alterations in length and changes in density; and these exceedingly slight differences are quite sufficient to account for any motion of the fringes. In order to avoid this disturbance, Fresnel had the idea of placing at the focus of the lens the mirror *M*, so that the two rays return, the one which came through the upper tube going back through the lower, and vice versa for the other ray. In this way the two rays pass through identical paths and come together at the same point from which they started. With this arrangement, if there is any shifting of the fringes, it must be due to the reversal of the change in velocity due to the current of water. For one of the two beams, say the upper one, travels with the current in both tubes; the other, starting at the same point, travels against the current in both tubes. Upon reversing the direction of the current of water the circumstances are exactly the reverse: the beam which before traveled with the current now travels against it, *etc.* The result of the experiment, as before stated, was that there was produced a displacement of

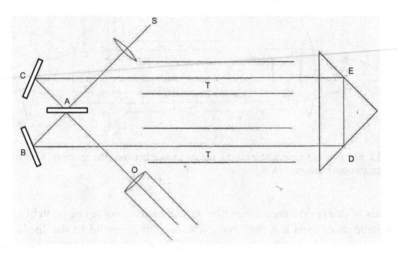

Fig. 34.3 Michelson's improvement of Fizeau's apparatus.—[*K.K.*]

less than should have been produced by the motion of the liquid. How much less was not determined. To this extent the experiment was imperfect.

On this account, and also for the reason that the experiment was regarded as one of the most important in the entire subject of optics, it seemed to me that it was desirable to repeat it in order to determine, not only the fact that the displacement was less than could be accounted for by the motion of the water, but also, if possible, how much less. For this purpose the apparatus was modified in several important points, and is shown in Fig. 34.3.

It will be noted that the principle of the interferometer has been used to produce interference fringes of considerable breadth without at the same time reducing the intensity of the light. Otherwise, the experiment is essentially the same as that made by Fizeau. The light starts from a bright flame of ordinary gas light, is rendered parallel by the lens, and then falls on the surface, which divides it into two parts, one reflected and one transmitted. The reflected portion goes down one tube, is reflected twice by the total reflection prism *P* through the other tube, and passes, after necessary reflection, into the observing telescope. The other ray pursues the contrary path, and we see interference fringes in the telescope as before, but enormously brighter and more definite. This arrangement made it possible to make measurements of the displacement of the fringes which were very accurate. The result of the experiment was that the measured displacement was almost exactly seven-sixteenths of what it would have been had the medium which transmits the light waves moved with the velocity of the water.

It was at one time proposed to test this problem by utilizing the velocity of the earth in its orbit. Since this velocity is so very much greater than anything we can produce at the earth's surface, it was supposed that such measurements could be made with considerable ease; and they were actually tried in quite a considerable number of different ways and by very eminent men. The fact is, we cannot utilize the velocity of the earth in its orbit for such experiments, for the reason that we have to determine our directions by points outside of the earth, and the only thing we have is the stars, and the stars are displaced by this very element which we want to measure; so the results would be entirely negative. It was pointed out by Lorentz that it is impossible by any measurements made on the surface of the earth to detect any effect of the earth's motion.

Maxwell considered it possible, theoretically at least, to deal with the square of the ratio of the two velocities; that is, the square of 1/10000, or 1/100000000. He further indicated that if we made two measurements of the velocity of light, one in the direction in which the earth is traveling in its orbit, and one in a direction at right angles to this, then the time it takes light to pass over the same length of path is greater in the first case than in the second.

We can easily appreciate the fact that the time is greater in this case, by considering a man rowing in a boat, first in a smooth pond and then in a river. If he rows at the rate of 4 miles an hour, for example, and the distance between the stations is 12 miles, then it would take him 3 h to pull there and three to pull back—6 h in all. This is his time when there is no current. If there is a current, suppose at the rate of 1 mile an hour, then the time it would take to go from one point to the other, would be, not 12 divided by 4, but 12 divided by 4 + 1, *i.e.*, 2.4 h. In coming back the time would be 12 divided by 4 − 1, which would be 4 h, and this added to the other time equals 6.4 instead of 6 h. It takes him longer, then, to pass back and forth when the medium is in motion than when the medium is at rest. We can understand, then, that it would take light longer to travel back and forth in the direction of the motion of the earth. The difference in the times is, however, so exceedingly small, being of the order of 1 in 100,000,000, that Maxwell considered it practically hopeless to attempt to detect it.

In spite of this apparently hopeless smallness of the quantities to be observed, it was thought that the minuteness of the light waves might again come to our rescue. As a matter of fact, an experiment was devised for detecting this small quantity. The conditions which the apparatus must fulfill are rather complex. The total distance traveled must be as great as possible, something of the order of 100 million waves, for example. Another condition requires that we be able to interchange the direction without altering the adjustment by even the one hundredth-millionth part. Further, the apparatus must be absolutely free from vibration.

The problem was practically solved by reflecting part of the light back and forth a number of times and then returning it to its starting-point. The other path was at right angles to the first, and over it the light made a similar series of excursions, and was also reflected back to the starting-point. This starting-point was a separating plane in an interferometer, and the two paths at right angles were the two arms of an interferometer. Notwithstanding the very considerable difference in path, which must involve an exceedingly high order of accuracy in the reflecting surfaces and a constancy of temperature in the air between, it was possible to see fringes and to keep them in position for several hours at a time.

These conditions having been fulfilled, the apparatus was mounted on a stone support, about 4 ft^2 and 1 ft thick, and this stone was mounted on a circular disc of

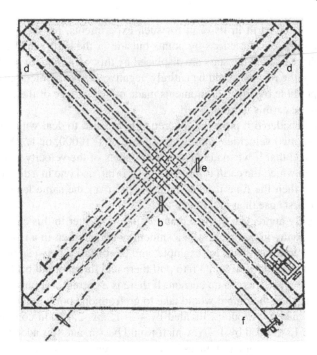

Fig. 34.4 A ray of light is split into two perpendicular beams which each traverse Michelson's interferometer several times before recombining to form an interference pattern.—[K.K.]

wood which floated in a tank of mercury. The resistance to motion is thus exceedingly small, so that by a very slight pressure on the circumference the whole could be kept in slow and continuous rotation. It would take, perhaps, 5 min to make one single turn. With this slight motion there is practically no oscillation; the observer has to follow around and at intervals to observe whether there is any displacement of the fringes.

It was found that there was no displacement of the interference fringes, so that the result of the experiment was negative and would, therefore, show that there is still a difficulty in the theory itself; and this difficulty, I may say, has not yet been satisfactorily explained. I am presenting the case, not so much for solution, but as an illustration of the applicability of light waves to new problems.

The actual arrangement of the experiment is shown in Fig. 34.4. A lens makes the rays nearly parallel. The dividing surface and the two paths are easily recognized. The telescope was furnished with a micrometer screw to determine the amount of displacement of the fringes, if there were any. The last mirror is mounted on a slide; so these two paths may be made equal to the necessary degree of accuracy—something of the order of one fifty-thousandth of an inch.

Figure 34.5 represents the actual apparatus. The stone and the circular disc of wood supporting the stone in the tank filled with mercury are readily recognized; also the dividing surface and the various mirrors.

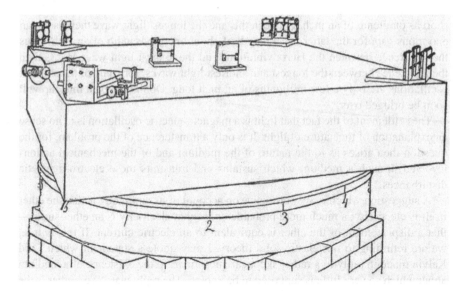

Fig. 34.5 The lenses and mirrors which comprise Michelson's interferometer are arranged atop a stone slab which is free to rotate about a central vertical axis.—[*K.K.*]

It was considered that, if this experiment gave a positive result, it would determine the velocity, not merely of the earth in its orbit, but of the earth through the ether. With good reason it is supposed that the sun and all the planets as well are moving through space at a rate of perhaps 20 miles per second in a certain particular direction. The velocity is not very well determined, and it was hoped that with this experiment we could measure this velocity of the whole solar system through space. Since the result of the experiment was negative, this problem is still demanding a solution.

The experiment is to me historically interesting, because it was for the solution of this problem that the interferometer was devised. I think it will be admitted that the problem, by leading to the invention of the interferometer, more than compensated for the fact that this particular experiment gave a negative result.

From all that precedes it appears practically certain that there must be a medium whose proper function it is to transmit light waves. Such a medium is also necessary for the transmission of electrical and magnetic effects. Indeed, it is fairly well established that light is an electro-magnetic disturbance, like that due to a discharge from an induction coil or a condenser. Such electric waves can be reflected and refracted and polarized, and be made to produce vibrations and other changes, just as the light waves can. The only difference between them and the light waves is in the wave length.

This difference may be enormous or quite moderate. For example, a telegraphic wave, which is practically an electromagnetic disturbance, may be as long as 1000 miles. The waves produced by the oscillations of a condenser, like a Leyden jar, may be as short as 100 ft; the waves produced by a Hertz oscillator may be as short as one-tenth of an inch. Between this and the longest light wave there is not an enormous gap, for the latter has a length of about one-thousandth of an inch. Thus the difference between the Hertz vibrations and the longest light wave is less than the difference between the longest and shortest light waves, for some of the shortest oscillations are only a few millionths of an inch long. Doubtless even this gap will soon be bridged over.

The settlement of the fact that light is a magneto-electric oscillation is in no sense an explanation of the nature of light. It is only a transference of the problem, for the question then arises as to the nature of the medium and of the mechanical actions involved in such a medium which sustains and transmits these electro-magnetic disturbances.

A suggestion which is very attractive on account of its simplicity is that the ether itself is electricity; a much more probable one is that electricity is an ether strain— that a displacement of the ether is equivalent to an electric current. If this is true, we are returning to our elastic-solid theory. I may quote a statement which Lord Kelvin made in reply to a rather skeptical question as to the existence of a medium about which so very little is supposed to be known. The reply was: "Yes, ether is the only form of matter about which we know anything at all." In fact, the moment we begin to inquire into the nature of the ultimate particles of ordinary matter, we are at once enveloped in a sea of conjecture and hypotheses—all of great difficulty and complexity.

One of the most promising of these hypotheses is the "ether vortex theory," which, if true, has the merit of introducing nothing new into the hypotheses already made, but only of specifying the particular form of motion required. The most natural form of such vortex motions with which to deal is that illustrated by ordinary smoke rings, such as are frequently blown from the stack of a locomotive. Such vortex rings may easily be produced by filling with smoke a box which has a circular aperture at one end and a rubber diaphragm at the other, and then tapping the rubber. The friction against the side of the opening, as the puff of smoke passes out, produces a rotary motion, and the result will be smoke rings or vortices.

Investigation shows that these smoke rings possess, to a certain degree, the properties which we are accustomed to associate with atoms, notwithstanding the fact that the medium in which these smoke rings exists is far from ideal. If the medium were ideal, it would be devoid of friction, and then the motion, when once started, would continue indefinitely, and that part of the ether which is differentiated by this motion would ever remain so.

Another peculiarity of the ring is that it cannot be cut—it simply winds around the knife. Of course, in a very short time the motion in a smoke ring ceases in consequence of the viscosity of the air, but it would continue indefinitely in such a frictionless medium as we suppose the ether to be.

There are a number of other analogies which we have not time to enter into— quite a number of details and instances of the interactions of the various atoms which have been investigated. In fact, there are so many analogies that we are tempted to think that the vortex ring is in reality an enlarged image of the atom. The mathematics of the subject is unfortunately very difficult, and this seems to be one of the principal reasons for the slow progress made in the theory.

Suppose that an ether strain corresponds to an electric charge, an ether displacement to the electric current, these ether vortices to the atoms—if we continue these suppositions, we arrive at what may be one of the grandest generalizations of modern science—of which we are tempted to say that it ought to be true even if it is not—namely, that all the phenomena of the physical universe are only different manifestations of the various modes of motions of one all-pervading substance—the ether.

All modern investigation tends toward the elucidation of this problem, and the day seems not far distant when the converging lines from many apparently remote regions of thought will meet on this common ground. Then the nature of the atoms, and the forces called into play in their chemical union; the interactions between these atoms and the non-differentiated ether as manifested in the phenomena of light and electricity; the structures of the molecules and molecular systems of which the atoms are the units; the explanation of cohesion, elasticity, and gravitation all these will be marshaled into a single compact and consistent body of scientific knowledge.

Summary

1) A number of independent courses of reasoning lead to the conclusion that the medium which propagates light waves is not an ordinary form of matter. Little as we know about it, we may say that our ignorance of ordinary matter is still greater.
2) In all probability, it not only exists where ordinary matter does not, but it also permeates all forms of matter. The motion of a medium such as water is found not to add its full value to the velocity of light moving through it, but only such a fraction of it as is perhaps accounted for on the hypothesis that the ether itself does not partake of this motion.
3) The phenomenon of the aberration of the fixed stars can be accounted for on the hypothesis that the ether does not partake of the earth's motion in its revolution about the sun. All experiments for testing this hypothesis have, however, given negative results, so that the theory may still be said to be in an unsatisfactory condition.

34.3 Study Questions

QUES. 34.1. What paradoxes are associated with the wave theory of light?

a) How, exactly, does the speed of a wave depend upon the density of the medium through which it propagates? What does this imply about the medium of light transmission?

b) How does the intensity of sound depend upon the density of the medium? Do light and sound behave in the same manner?

c) Can transverse waves, or perhaps torsional waves, be propagated through a fluid? How does this issue raise one of the most "formidable objections" to the wave theory of light?

d) What does Michelson conclude regarding the wave theory of light?

QUES. 34.2. What paradox, regarding the nature of light, arises from the observation of stellar aberration?

a) What is meant by the angle of aberration of starlight? How is it measured using a telescope?

b) How can the corpuscular theory of light account for stellar aberration? And how can stellar aberration, in turn, be used to measure the speed of light?

c) What puzzle was raised by measuring stellar aberration with a water-filled telescope? What solution to this puzzle was offered by Fresnel? Was this a good solution?

QUES. 34.3. What is the speed of light in a moving medium (such as water)?

a) How is the speed of sound affected by motion of the air through which it propagates? And what makes the measurement of a similar effect in light difficult?

b) Describe Fizeau's experimental apparatus (Fig. 34.2) and procedure. Why was it necessary to reverse the direction of water flow during the experiment, and what did Fresnel observe when doing so?

c) What conclusion did Fresnel draw from his results? Is there any other (or perhaps even a better) conclusion which he could have drawn?

d) Did Michelson's later experiment verify or disprove Fresnel's results?

QUES. 34.4. Can one measure the absolute speed of the earth?

a) What is meant by *absolute speed*?

b) Does it take longer to row a boat (i) back and forth across a river which is 1 mile across, or (ii) 1 mile up river then 1 mile back down the river? What does this rowing example have to do with measurement of the speed of light?

c) What experimental difficulties must be overcome in order to measure the absolute speed of the earth?

d) Describe Michelson's interferometer (Figs. 34.4 and 34.5), and his experimental procedure. Specifically, why must the interferometer be rotated, and what did Michelson observe when doing so?

e) To what conclusion(s) was Michelson led by his results? Are his conclusions justified?

QUES. 34.5. What, according to Michelson, is the nature of the ether?

a) What is the difference between Hertz's electromagnetic waves and light waves? Does the identification of light with electromagnetic waves solve the problem of the nature of light?

b) What is the ether vortex theory? What support does Michelson offer for this theory? How might this theory account for electrical charge, electrical current, and even the existence of atoms?
c) What is Michelson's ultimate goal for scientific knowledge? Do you agree with this?

34.4 Exercises

EX. 34.1 (SPEED OF SOUND). Suppose a hammer strikes the end of a 10 ft long steel beam. How long does it take the sound to travel the length of the beam (i) through the beam and (ii) through the air. Would the difference be greater or smaller on a humid day?

EX. 34.2 (ANGULAR ABERRATION). Suppose a raindrop falls straight down at a constant speed of 10 m/s.

a) At what angle must a gun barrel be aimed so that the raindrop passes directly down the axis of the barrel if the gun is traveling horizontally at a speed of 1 m/s?
b) If the rain were falling near the speed of light, and the gun were traveling at the orbital speed of the earth, by how much would your answer change?
c) What does this imply about the apparent position of a star when viewed with a moving telescope?

EX. 34.3 (RELATIVE MOTION). A 100 m wide river flows due south at a speed of 3 m/s. A motorboat is able to travel at 10 m/s with respect to stationary water.

a) How long would it take this boat to travel (i) upstream 100 m (with respect to the bank)? (ii) downstream 100 m (with respect to the bank)? (iii) due east across the river? (iv) due west across the river?
b) Based on your calculations, would it take this boat more time to take a round trip (i) up and down the river, or (ii) back and forth across the river? What is the time difference between these round trips?

EX. 34.4 (INTERFEROMETRY AND REFRACTIVE INDEX). An airtight chamber 5 cm long with glass windows on its ends is placed in one arm of a Michelson interferometer (see Fig. 34.6). Slowly evacuating the air from the chamber causes a shift of 60 fringes at the detection screen when 500 nm light is used. From this data, find the refractive index of air at atmospheric pressure.

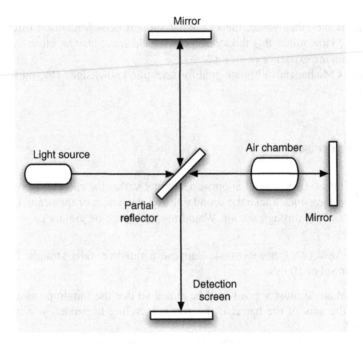

Fig. 34.6 A michelson interferometer for measuring the refractive index of gas

34.5   Vocabulary

1. Cogent
2. Aberration
3. Incredulity
4. Interferometer
5. Conjecture
6. Eminent

Appendix A
Vector Algebra and Vector Calculus

A.1 Exercises

EX. A.1 (VECTOR AND SCALAR QUANTITIES). Many of the quantities with which we often deal, such as temperature and age, are called *scalar quantities*. Other quantities, such as velocity and force, are called *vector quantities*. What distinguishes these two classes of quantities is that the latter possess an orientation in space, the former do not. One speaks of a car moving "ten miles per hour eastward," but not of a person being "ten years old eastward." A vector quantity, $\vec{v}$, may be represented graphically by an arrow, as shown in Fig. A.1. The length of the vector provides a measure of its magnitude, such as 10 miles per hour. The orientation of the vector represents its direction in space, such as eastward. A vector can also be represented mathematically in the form of an equation:

$$\vec{v} = v\hat{x}. \tag{A.1}$$

On the right side of this equation, the scalar quantity, v, describes the magnitude of the vector, and the *unit vector*, $\hat{x}$, describes the direction in which the vector is oriented. In this case, vector $\vec{v}$ points along the x-axis of a particular cartesian system of coordinates.

As an exercise describe each of the following using either a scalar or a vector quantity; if it is a vector quantity, represent it both graphically and mathematically. (a) Water has a density of 1 g per cubic centimeter. (b) A compressed gas exerts a pressure of 100 p.s.i. on the bottom of a cylindrical flask. (c) A ball is whirled around in a circle by a string which exerts a constant force of 5 N inward.

EX. A.2 (VECTOR ADDITION). The rules of vector algebra are a bit different than those of scalar algebra. For example, when adding vector quantities, one must consider not only magnitudes, but also their relative orientations. Only in the case in which they are in the same orientation can their magnitudes be simply added like scalar quantities. Generally speaking, the addition of two vectors, $\vec{a}$ and $\vec{b}$, may be represented graphically by placing the "tail" of one vector against the "tip" of the other, and forming a new vector, $\vec{c} = \vec{a} + \vec{b}$, as shown in Fig. A.2. Vector addition can be expressed mathematically by selecting a coordinate system and writing

© Springer International Publishing Switzerland 2016   457
K. Kuehn, *A Student's Guide Through the Great Physics Texts,*
Undergraduate Lecture Notes in Physics, DOI 10.1007/978-3-319-21816-8

Fig. A.1 A vector

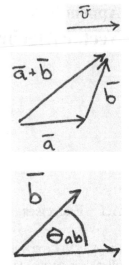

Fig. A.2 Vector addition

Fig. A.3 The dot product
formed from two vectors
depends on the angle between
them

each vector in terms of this coordinate system. For example, the sum of two vectors,
$\vec{m} = m\hat{x}$ and $\vec{n} = n\hat{y}$, produces a vector quantity, $\vec{g}$ which may be written as

$$\vec{g} = m\hat{x} + n\hat{y}. \tag{A.2}$$

Conversely, any vector may be *decomposed* into a (non-unique) sum of two or more
vectors which, when summed, would form the vector.

As an exercise in vector decomposition, suppose that a 100 g block rests atop
a wedge whose acute angle is 30°. Choose a convenient coordinate system, and
express the force of gravity which is acting upon the block as a sum of two vectors,
one which is directed parallel to the surface of the ramp, and one which is directed
perpendicular to the surface of the ramp. What are the magnitudes and directions of
these two vectors?

Ex. A.3 (VECTOR MULTIPLICATION, PART 1: THE DOT PRODUCT).   There are two
methods of multiplying vector quantities: the *dot product* and the *cross product*. In
this exercise we will consider the first of these. The *dot product* (or *scalar product*)
of two vector quantities, $\vec{a} \cdot \vec{b}$, produces a scalar quantity, c, whose magnitude is
given by

$$c = ab \cos\theta_{ab} \tag{A.3}$$

Here, a and b are the magnitudes of $\vec{a}$ and $\vec{b}$ and θ_{ab} is the angle between $\vec{a}$ and $\vec{b}$,
as shown in Fig. A.3.

As an example of the dot product, the *work* done on an object by a force $\vec{F}$ in
moving the object a distance $\vec{d}$ is given by the dot product $W = \vec{F} \cdot \vec{d}$. Consider

Fig. A.4 The cross product formed from two vectors is perpendicular to both of them

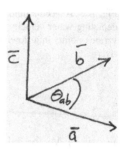

once again a 100 g block resting at the top of a 30° wedge. Suppose that the surface of the ramp is 1 m long. Calculate the work done on the block as it slides from the top to the bottom of the ramp (a) by the force which the earth exerts on the block, and (b) by the force which the ramp exerts on the block.

EX. A.4 (VECTOR MULTIPLICATION, PART 2: THE CROSS PRODUCT). The *cross product* (or *vector product*) of two vector quantities, $\vec{a} \times \vec{b}$, produces a vector quantity, $\vec{c}$, whose magnitude is given by

$$c = ab \sin \theta_{ab} \tag{A.4}$$

and whose direction is perpendicular to the plane defined by vectors $\vec{a}$ and $\vec{b}$, as shown in Fig. A.4. The relative orientations of $\vec{a}$, $\vec{b}$ and $\vec{c}$ may be expressed using the (second) right hand rule:[1] if you point the index finger of your right hand in the direction of $\vec{a}$ and extend the remaining fingers of your right hand in the direction of $\vec{b}$, then your thumb, when extended, points in the direction of $\vec{c}$.

As an example of a cross product, the *Lorentz force* acting upon a charge, q, which is traveling at velocity $\vec{v}$ through a magnetic field $\vec{B}$ is given by the cross product

$$\vec{F} = q\vec{v} \times \vec{B}. \tag{A.5}$$

Suppose that a singly ionized cosmic particle travels horizontally and northwards above Ellesmere Island in Canada (the location of Earth's magnetic pole) at a speed of approximately 400 km/s. The magnetic field strength here is approximately 5 n-T. (a) Calculate the magnitude of the Lorentz force acting on this ion. (b) Describe the trajectory of the ion; does it continue in a straight line, or does it curve? If it curves, in which direction? (c) How much work is done on the ion by the earth's magnetic field? Does the magnitude of its velocity change? Does its kinetic energy change?

EX. A.5 (VECTOR FIELDS). A *vector field* is a function which assigns a vector quantity to each location of space. For example, the fluid velocity at each spatial point (x, y, z) within a stream of flowing water can be described by a vector field

---

[1] Recall that we used the first right hand rule to determine the orientation of the magnetic field in the vicinity of a current-carrying wire; see Ex. 7.4 of the present volume.

Fig. A.5 A vector field depicting water velocity at various points in a flowing river

$\vec{v}(x, y, z)$. Vector fields are often represented graphically by selecting a discrete set of points in the region under consideration and drawing a vector at each of these points so as to give a sense of the overall flow within the region (see Fig. A.5). The electric field, $\vec{E}$, is another example of a vector field. It may be defined in terms of the force that would be exerted on a test charge, q, if it were placed in the vicinity of one or more source charges.

$$\vec{E} \equiv \vec{F}/q \qquad (A.6)$$

Generally speaking, the force field of a particular configuration of source charges is quite complicated, but the electric field surrounding a single point charge, Q, may be obtained readily by using Coulomb's law,

$$\vec{F}(r) = k\frac{qQ}{r^2}\hat{r}. \qquad (A.7)$$

Here k is Coulomb's constant, r is the magnitude of the distance from the source charge, Q, to the test charge, q, and $\hat{r}$ is a radially directed unit vector in a spherical polar coordinate system centered on Q. Combining Eqs. 30.1 and A.7, we find that the electric field in the vicinity of a charge Q is given by

$$\vec{E}(r) = k\frac{Q}{r^2}\hat{r} \qquad (A.8)$$

Notice that the electric field does not depend on the test charge q. In this sense, the electric field caused by the source charge is said to have an existence which is independent of any other charges which might respond to the electric field.

As an exercise, write down an expression for the gravitational field, $\vec{g}(r)$, surrounding the earth. The gravitational field of the earth is analogous to the electric field surrounding a point charge. Can you make a sketch of this gravitational vector field? In which direction are the vectors pointing? Is the magnitude of this vector field uniform throughout space?

EX. A.6 (VECTOR CALCULUS, PART 1: LINE INTEGRATION). One may integrate a vector field along a particular path between two points in space. This is referred to as a *line integral* of the vector field. Consider a vector field, $\vec{F}$, defined over a particular region of space. Two points, a and b, lie in this region and are connected by a curve C (see Fig. A.6). Curve C may be divided into a large number, N, of consecutive vectors of length Δs which connect points a and b. Now at each point along the curve, one may calculate the dot product of the vector field with the vector

Fig. A.6 A curve may be described by a set of infinitesimal tangent vectors at each point along the curve

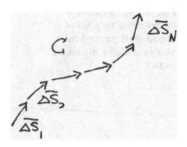

$\overrightarrow{\Delta s}$. In the limit that $N \to \infty$ and $\Delta s \to 0$, the sum of these dot products becomes the line integral of the vector field along C:

$$\lim_{\substack{N \to \infty \\ \Delta s \to 0}} \sum_{i=0}^{N} \vec{F}_i \cdot \overrightarrow{\Delta s_i} = \int_a^b \vec{F} \cdot d\vec{s} \tag{A.9}$$

Generally speaking, this integration is difficult to perform. But as a simple example, suppose that a spatially uniform magnetic field, $\vec{B}$, is directed along the x-axis of a particular cartesian system of coordinates. We wish to calculate the line integral of the vector field along a straight line, C, from $x = 0$ to $x = x_o$. The line integral may be written as:

$$\int_C \vec{B} \cdot d\vec{s} = \int_0^{x_o} (B_o \hat{x}) \cdot (dx\,\hat{x})$$
$$= \int_0^{x_o} B_o\,dx \tag{A.10}$$
$$= B_o x_o$$

In the second line of Eq. A.10, we have written the magnetic field as $\vec{B} = B_o\hat{x}$ and the displacement vector along the x-axis as $d\vec{s} = dx\,\hat{x}$. We then performed the dot product and pulled B_o out of the integral, since it is spatially uniform. As an exercise, consider the line integral of the same magnetic field ($B_o\hat{x}$) around a complete square loop of side length l lying in the $x - y$ plane. What is the line integral along each side of the square? What is the value of the line integral around the entire loop?

Ex. A.7 (ELECTRIC FIELDS AND ELECTRIC POTENTIAL). Recall from the electric field mapping laboratory (Ex. 28.2 in the present volume) that electric field lines are always perpendicular to equipotential lines. More specifically, the electric field at any location is equal to the gradient (or more simply put, the slope) of the electric potential. Thus, the electric field points in the direction in which the potential decreases most rapidly. Conversely, the electrical potential difference, ΔV, between two locations a and b may be calculated from the line integral of the electric field along a path between these two points:

$$\Delta V = -\int_a^b \vec{E} \cdot d\vec{s}. \tag{A.11}$$

Fig. A.7 A surface may
be described by a set of
infinitesimal perpendicular
vectors at each point on the
surface

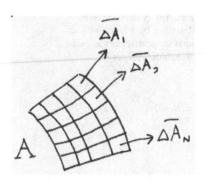

As an exercise, suppose that the electric field in the region between two narrowly separated and oppositely charged parallel plates is given by $\vec{E} = \frac{\sigma}{2\epsilon}\hat{x}$. Here, σ is the surface charge density on each plate (in Coulombs per square meter), ϵ is the electrical permittivity of the medium between the plates, and $\hat{x}$ is a unit vector which is normal to the plates and which points from the positively to the negatively charged plate. (a) What is the potential difference between the plates in terms of their separation, d, and the surface charge density of each plate, σ? (b) For a fixed electric potential difference, which can store more electric charge, a capacitor having *nothing*, or having *mylar*, filling the space between the plates?

EX. A.8 (VECTOR CALCULUS, PART 2: SURFACE INTEGRATION). One may also integrate a vector field over a particular area. This is referred to as a *surface integral* of the vector field. Consider a vector field, $\vec{F}$, defined over a particular region of space. A surface, A, lies in this region (see Fig. A.7). This surface may be divided into a number, N, of smaller areas which are described by vectors which are perpendicular to the surface and which have length ΔA. At each location on the surface one may calculate the dot product of the vector field with the area vector $\overrightarrow{\Delta A}$. In the limit that $N \to \infty$ and $\Delta A \to 0$, the sum of these dot products becomes the surface integral of the vector field over A:

$$\lim_{\substack{N\to\infty \\ \Delta A\to 0}} \sum_{i=0}^{N} \vec{F}_i \cdot \overrightarrow{\Delta A_i} = \int_a^b \int \vec{F} \cdot d\vec{A} \tag{A.12}$$

Generally speaking, this involves a complicated two-dimensional integration. But as a simple example, consider a spatially uniform magnetic field directed along the x-axis and a rectangular area which is described by the vector $d\vec{A} = A_o\hat{x}$. The surface integral is then

$$\iint_A \vec{B} \cdot d\vec{A} = \iint (B_o\hat{x}) \cdot (dA\,\hat{x})$$

$$= \iint B_o\, dA \tag{A.13}$$

$$= B_o A_o$$

In Eq. A.13, we have again performed the dot product and pulled B_o out of the integral. As an exercise, calculate the surface integral of the same magnetic field $(B_0\hat{x})$ through a square area of side length l whose normal is in the direction $\vec{n} = a\hat{x} + b\hat{y}$. For what values of a and b is this surface integral maximum? minimum?

Bibliography

Biographical Memoir of Albert Abraham Michelson 1852–1931, vol. XIX, National Academy of Sciences, 1938.

Ampère, A. M., New Names and Actions Between Currents, in *A Source Book in Physics*, edited by W. F. Magie, Source books in the history of science, pp. 447–460, Harvard University Press, Cambridge, Massachusetts, 1963.

Bence, J., *The life and letters of Faraday*, Longmans, Green and Co., London, 1870.

Brewster, D., On the laws which regulate the polarisation of light by reflexion from transparent bodies., *Philosophical Transactions of the Royal Society of London (1776–1886)*, *105*, 125–159, 1815.

Campbell, L., and W. Garnett, *The Life of James Clerk Maxwell*, Macmillan and Co., London, 1882.

Coulomb, C., Law of Electric Force and Fundamental Law of Electricity, in *Source Book in Physics*, edited by W. F. Magie, Source books in the history of science, pp. 408–413, Harvard University Press, Cambridge, Massachusetts, 1963.

Descartes, R., *The World or Treatise on Light*, Abaris Books, 1979.

Drosdoff, D., and A. Widom, Snell's law from an elementary particle viewpoint, *American Journal of Physics*, *73*(10), 973–975, 2005.

Faraday, M., *Experimental Researches in Electricity*, vol. 1, Taylor and Francis, London, 1839.

Faraday, M., *Experimental Researches in Electricity*, vol. 3, Taylor and Francis, London, 1855.

Faraday, M., *A Course of Six Lectures on the Forces of Matter and Their Relations to Each Other*, Richard Griffin and Company, London and Glasgow, 1860.

Faraday, M., *The Chemical History of a Candle: to which is added a Lecture on Platinum*, Harper & Brothers, New York, 1861.

Feynman, R.P., Leighton, R.B., and Sands, M.L., *The Feynman Lectures on Physics*, Commemorative ed., Addison-Wesley Publishing Co., 1989.

Franklin, B., *Experiments and Observations on Electricity*, fourth ed., Henry, David, London, 1769.

Franklin, B., *Fart Proudly*, Enthea Press, Columbus, Ohio, 1990.

Fresnel, A., Diffraction of Light, in *Source Book in Physics*, edited by W. F. Magie, Source books in the history of science, pp. 318–324, Harvard University Press, Cambridge, Massachusetts, 1963.

Gilbert, W., *On the Loadstone and Magnetic Bodies and on the Great Magnet the Earth: a New Physiology, Demonstrated with Many Arguments and Experiments*, Bernard Quaritch, London, 1893.

Heaviside, O., *Electromagnetic Theory*, vol. 1, "The Electrician" Printing and Publishing Company, London, 1893.

Heaviside, O., *Electromagnetic Theory*, vol. 3, "The Electrician" Printing and Publishing Company, London, 1912.

© Springer International Publishing Switzerland 2016

K. Kuehn, *A Student's Guide Through the Great Physics Texts*,
Undergraduate Lecture Notes in Physics, DOI 10.1007/978-3-319-21816-8

Helmholtz, H., On the Conservation of Force, in *Popular Lectures on Scientific Subjects*, edited by E. Atkinson, D. Appleton and Company, New York, 1885.

Helmholtz, H., *On the Sensation of Tone as a Physiological Basis for the Theory of Music*, fourth ed., Longmans, Green, and Co., New York, Bombay and Calcutta, 1912.

Helmholtz, H., *Science and Culture*, The University of Chicago Press, 1995.

Hertz, H., *Electric Waves*, 2 ed., Macmillan and Co., London, 1900.

Hofmann, J. R., *André-Marie Ampère: Enlightenment and Electrodynamics*, Cambridge Science Biographies, Cambridge University Press, 1996.

Huygens, C., *Treatise on Light*, Macmillan, London, 1912.

Jackson, J. D., *Classical Electrodynamics*, second ed., John Wiley & Sons, New York, 1975.

Lemay, J. A. L., *The Life of Benjamin Franklin*, vol. 3, University of Pennsylvania Press, 2009.

Livingston, D. M., *The Master of Light: A biography of Albert A. Michelson*, The University of Chicago Press, 1973.

Magie, W. F. (Ed.), *A Source Book in Physics*, Harvard University Press, Cambridge, Massachusetts, 1963.

Maxwell, J. C., A Dynamical Theory of the Electromagnetic Field, *Philosophical Transactions of the Royal Society of London*, *155*, 459–512, 1865.

Maxwell, J. C., On action at a distance, *Proceedings of the Royal Institution of Great Britain*, *VII*, 48–49, 1873–1875.

Maxwell, J. C., *A Dynamical Theory of the Electromagnetic Field*, Wipf and Stock Publishers, 1996.

McVittie, G., Laplace's alleged "black hole", *The Observatory*, *98*, 272–274, 1978.

Meleshko, V. V., Coaxial axisymmetric vortex rings: 150 years after Helmholtz, *Theor. Comput. Fluid Dyn.*, *24*, 403–431, 2010.

Michelson, A. A., *Light Waves and Their Uses*, University of Chicago Press, Chicago, IL, 1903.

Newton, I., *Opticks: or A Treatise of the Reflections, Refractions, Inflections & Colours of Light*, 4th ed., William Innes at the West-End of St. Pauls, London, 1730.

Oersted, H. C., *Selected Scientific Works of Hans Christian Oersted*, Princeton University Press, 1998.

Ørsted, H. C., Experiments on the effect of a current of electricity on the magnetic needle, *Annals of Philosophy*, *16*(4), 273–276, 1820.

Schneider, N. H., *Induction Coils: How to Make, Use and Repair Them*, Spon & Chamberlain, E. & F.N. Spon, New York and London, 1901.

Smith, A. M., Descartes's Theory of Light and Refraction: A Discourse on Method, *Transactions of the American Philosophical Society*, *77*(3), 1–92, 1987.

Tait, P. G., James Clerk Maxwell, *Proceedings of the Royal Society of Edinburgh*, *10*, 331–339, 1878–1880.

Tricker, R. A. R., *Early Electrodynamics*, Pergamon, New York, 1965.

Tyndall, J., *Address Delivered Before The British Association Assembled At Belfast with Additions*, Longmans, Green, and Co., 1874.

Tyndall, J., *Six lectures on light delivered in America in 1872–1878*, D. Appleton and Company, New York, 1886.

Wallace, W. A., The Problem of Causality in Galileo's Science, *The Review of Metaphysics*, *36*(3), 607–632, 1983.

Young, T., *Course of lectures on natural philosophy and the mechanical arts*, vol. 1, Printed for Joseph Johnson, London, 1807.

Index

© Springer International Publishing Switzerland 2016
K. Kuehn, *A Student's Guide Through the Great Physics Texts,*
Undergraduate Lecture Notes in Physics, DOI 10.1007/978-3-319-21816-8

Printed in the United States
By Bookmasters